Computational Fluid Dynamics

A von Karman Institute Book

John F. Wendt (Ed.)

Computational Fluid Dynamics

An Introduction

With Contributions by
John D. Anderson Jr., Joris Degroote, Gérard Degrez, Erik Dick,
Roger Grundmann and Jan Vierendeels

Third Edition

 Springer

Editor
Prof. Dr. John F. Wendt
Director von Karman Institute for Fluid Dynamics (ret.)
72 Chaussée de Waterloo
1640 Rhode-Saint-Genèse
Belgium

ISBN: 978-3-642-09873-4   e-ISBN: 978-3-540-85056-4

Cover design: xxxxx

Printed on acid-free paper

9 8 7 6 5 4 3 2 1

springer.com

Preface

Computational Fluid Dynamics: An Introduction grew out of a von Karman Institute (VKI) Lecture Series by the same title first presented in 1985 and repeated with modifications every year since that time.

The objective, then and now, was to present the subject of computational fluid dynamics (CFD) to an audience unfamiliar with all but the most basic numerical techniques and to do so in such a way that the practical application of CFD would become clear to everyone.

A second edition appeared in 1995 with updates to all the chapters and when that printing came to an end, the publisher requested that the editor and authors consider the preparation of a third edition. Happily, the authors received the request with enthusiasm.

The third edition has the goal of presenting additional updates and clarifications while preserving the introductory nature of the material.

The book is divided into three parts. John Anderson lays out the subject in Part I by first describing the governing equations of fluid dynamics, concentrating on their mathematical properties which contain the keys to the choice of the numerical approach. Methods of discretizing the equations are discussed and transformation techniques and grids are presented. Two examples of numerical methods close out this part of the book: source and vortex panel methods and the explicit method.

Part II is devoted to four self-contained chapters on more advanced material. Roger Grundmann treats the boundary layer equations and methods of solution. Gerard Degrez treats implicit time-marching methods for inviscid and viscous compressible flows; relative to the second edition, figures in the section on stability properties have been added and the section on numerical dissipation has been expanded with examples. Eric Dick, in two separate articles, treats both finite volume and finite element methods; the sections on current developments have been updated and references to a number of essential recent publications have been added.

Part III brings a new contribution by Jan Vierendeels and Joris Degroote which provides insight into the steps that are needed to obtain a CFD solution of a flow field using commercial CFD software packages. The wide availability of such codes provides advantages for the non-specialist in numerical techniques, but requires an appreciation of their limitations and knowledge of an application methodology.

The editor and authors will consider this book to have been successful if the readers conclude they have been well prepared to examine the literature in the field and to begin the application of CFD methods to the resolution of problems in their area of interest.

The editor takes this opportunity to thank the authors for their contributions to this book and for their enthusiasm to continue the tradition of continually improving the VKI Lecture Series on which it is based.

Eagle River, WI, USA   *John F. Wendt*

Biographical Sketches of the Authors

Professor John D. Anderson, Jr.
National Air and Space Museum, Smithsonian Institution, Washington, DC.

John D. Anderson, Jr. is the Curator for Aerodynamics at the National Air and Space Museum, Smithsonian Institution. He graduated from the University of Florida with a B. Eng. degree, and from The Ohio State University with a PhD in Aeronautical and Astronautical Engineering. He served as a Lieutenant and Task Scientist at Wright Field in Dayton, as Chief of the Hypersonics Group at the Naval Ordnance Laboratory in White Oak, Maryland and became Chairman of the Department of Aerospace Engineering at the University of Maryland in 1973. He was designated a Distinguished Scholar/Teacher in 1982. In 1993 he was made a full faculty member of the Committee for the History and Philosophy of Science, and in 1996 an affiliate member of the History Department at the University of Maryland. In 1996 he became the Glenn L. Martin Distinguished Professor in Aerospace Engineering, retired from the University in 1999, and is now Professor Emeritus. Dr. Anderson has published ten books and over 120 professional papers in the areas of high temperature gas dynamics, computational fluid dynamics, applied aerodynamics, and the history of aeronautics. He is an Honorary Fellow of the American Institute of Aeronautics and Astronautics and a Fellow of the Royal Aeronautical Society. His e-mail ID is AndersonJA@si.edu

Professor Gérard Degrez
Université Libre de Bruxelles, Brussels, Belgium

Gérard Degrez, Full Professor at the Faculty of Engineering at Université Libre de Bruxelles (ULB), received his initial engineering degree (Ingénieur civil mécanicien & électricien) from ULB, a Master of Science degree in engineering from Princeton University and a PhD degree from ULB. He held academic positions successively at the University of Sherbrooke (Canada), at the von Karman Institute for Fluid Dynamics (Belgium) and at Université Libre de Bruxelles (Belgium) where he is now Head of the Aero-Thermo-Mechanics Laboratory, while having a part-time appointment as Adjunct Professor at the von Karman Institute. Author of more than 25 archival journal publications on shock wave/boundary layer interactions, computational methods for incompressible and compressible flows and numerical simulation of high enthalpy flows, his current research interests concern numerical methods and physical models for the simulation of high enthalpy reacting flows and of turbulent flows, including magnetofluiddynamics. His e-mail ID is gdegrez@ulb.ac.be

Mr. Joris Degroote
Ghent University, Ghent, Belgium

Joris Degroote received the M.Sc. degree in electromechanical engineering from Ghent University, Ghent, Belgium, in 2006. Currently, he is a PhD Fellow of the Research Foundation of Flanders (FWO) in the Department of Flow, Heat, and Combustion Mechanics at Ghent University, working in the field of reduced-order models in computational fluid dynamics and fluid–structure interaction. His e-mail ID is joris.degroote@ugent.be

Professor Erik Dick
Ghent University, Ghent, Belgium

Erik Dick obtained the M.Sc. Degree in Mechanical Engineering from Ghent University in 1973 and the Ph.D. in Computational Fluid Dynamics from the same university in 1980. From 1974 to 1991, he worked at the Department of Mechanical Engineering, Division of Turbomachinery, of Ghent University as researcher, senior researcher, and head of research. He was associate professor at the University of Liège, from July 1991 to September 1992. He returned to Ghent University as associate professor and became full professor in 1995. Professor Dick teaches turbomachines and computational fluid mechanics. His area of research is computational methods and models for turbulence and transition for flow problems in mechanical engineering. He is author or co-author of about 80 articles in international scientific journals and about 160 papers in international scientific conferences and was the recipient of the 1990 Iwan Akerman prize for fluid machinery awarded by the Belgian national science foundation. His e-mail ID is Erik.Dick@ugent.be

Professor Roger Grundmann
Technische Universität Dresden, Dresden, Germany

Roger Grundmann received the Dipl.-Ing. and Dr.-Ing. degrees from the Technische Universität of Berlin. Since 1972 he has been a member of the Deutsches Zentrum für Luft- und Raumfahrt (DLR) at the Institute for Theoretical Fluid Dynamics. From 1985 to 1987 he was Associate Professor at the von Karman Institute for Fluid Dynamics (VKI) in Rhode-Saint-Genèse, Belgium and later spent another three years at the VKI as a Visiting Professor. In 1993 he received the Chair in Thermofluid Dynamics at the Institute for Fluid Dynamics of the Technische Universität Dresden and in 1994 became the Institute's director. In 1996 he founded the Institute for Aerospace Engineering at the T.U. Dresden and was its director for

10 years. From 1996 until 2007, Professor Grundmann was the head of an Innovation College and its successor, the Collaborative Research Centre "Electromagnetic Flow Control in Metallurgy, Crystal-Growth and Electro-Chemistry" of the German Research Foundation (DFG). He is a member of the Board of Directors and General Assembly of the VKI, the Scientific Advisory Board of the Forschungszentrum Dresden-Rossendorf (FZD), and a Review Board of the DFG. His fields of research are viscous hypersonic flows by means of numerical methods, the modelling and prediction of transition, and volume-force driven flows such as magnetofluid dynamics and acoustical fluid dynamics. His e-mail ID is grundman@tfd.mw.tu-dresden.de

Associate Professor Jan Vierendeels
Ghent University, Ghent, Belgium

Jan Vierendeels obtained the degree of MSc in electromechanical engineering in 1991 at Ghent University, Belgium. In 1993, he obtained the degree of MSc in aeronautical and astronautical engineering and in 1996, he obtained the degree of MSc in biomedical engineering, both at Ghent University. In 1998, he obtained his PhD in electromechanical engineering at Ghent University. Currently, he is an associate professor at the Department of Flow, Heat and Combustion Mechanics at Ghent University, working in the field of computational fluid dynamics and fluid-structure interaction. His e-mail ID is jan.vierendeels@ugent.be

Contents

Part I

Chapter 1
Basic Philosophy of CFD

J.D. Anderson, Jr.

1.1 Motivation: An Example

Imagine that you are an aeronautical engineer in the later 1950s. You have been given the task of designing an atmospheric entry vehicle—in those days it would have been an intercontinental ballistic missile. (Later, in the early 1960s, interest also focused on manned atmospheric entry vehicles for orbital and lunar return missions.) You are well aware of the fact that such vehicles will enter the earth's atmosphere at very high velocities, about 7.9 km/s for entry from earth orbit and about 11.2 km/s for entry after returning from a lunar mission. At these extreme hypersonic speeds, aerodynamic heating of the entry vehicle becomes very severe, and is the dominant concern in the design of such vehicles. Moreover, you are cognizant of the recent work performed at the NACA Ames Aeronautical Laboratory by H. Julian Allen and colleagues wherein a blunt-nosed hypersonic body was shown to experience considerably less aerodynamic heating than a sharp, slender body—contrary to some popular intuition at that time. (This work was finally unclassified and released to the general public in 1958 in NACA Report 1381 entitled *A Study of the Motion and Aerodynamic Heating of Ballistic Missiles Entering the Earth's Atmosphere at High Supersonic Speeds*.) Therefore, you know that your task involves the design of a *blunt body* for hypersonic speed. Moreover, you know from supersonic wind tunnel experiments that the flowfield over the blunt body is qualitatively like that sketched in Fig. 1.1. You know that a strong curved bow shock wave sits in front of the blunt nose, detached from the nose by the distance δ, called the shock detachment distance. You know that the gas temperatures between the shock and the body can be as high as 7000 K for an ICBM, and 11000 K for entry from a lunar mission. And you know that you must understand some of the details of this flowfield in order to intelligently design the entry vehicle. So, your first logical step is to perform an analysis of the aerodynamic flow over a blunt body in order to provide detailed information on the pressure and heat transfer distributions over the body surface, and to examine the properties of the high-temperature shock layer between the bow

J.D. Anderson, Jr.
National Air and Space Museum, Smithsonian Institution, Washington, DC
e-mail: AndersonJA@si.edu

Fig. 1.1 Qualitative aspects of flow over a supersonic blunt body

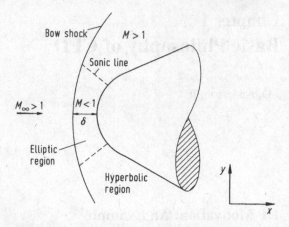

shock wave and the body. You ask such questions as: what is the shape of the bow shock wave; what is the detachment distance δ; what are the velocity, temperature and pressure distributions throughout the shock layer, etc.? However, much to your dismay, you find that no reliable, accurate aerodynamic theory exists to answer your questions. You quickly discover that an accurate and practical analysis of supersonic blunt body flows is beyond your current state-of-the-art. As a result, you ultimately resort to empirical information along with some simplified but approximate theories (such as Newtonian theory) in order to carry out your designated task of designing the entry vehicle.

The above paragraph illustrates one of the most important, yet perplexing, aerodynamic problems of the 1950s and early 1960s. The application of blunt bodies had become extremely important due to the advent of ICBMs, and later the manned space programme. Yet, no aerodynamic theory existed to properly calculate the flow over such bodies. Indeed, entire sessions of technical meetings (such as meetings of the Institute for Aeronautical Sciences in the USA, later to become the American Institute for Aeronautics and Astronautics) were devoted exclusively to research on the supersonic blunt body problem. Moreover, some of the best aerodynamicists of that day spent their time on this problem, funded and strongly encouraged by the NACA (later NASA), the US Air Force and others.

What was causing the difficulty? Why was the flowfield over a body moving at supersonic and hypersonic speeds so hard to calculate? The answer rests basically in the sketch shown in Fig. 1.1, which illustrates the steady flow over a supersonic blunt body. The region of steady flow near the nose region behind the shock is locally subsonic, and hence is governed by elliptic partial differential equations. In contrast, the flow further downstream of the nose becomes supersonic, and this locally steady supersonic flow is governed by hyperbolic partial differential equations. (What is meant by 'elliptic' and 'hyperbolic' equations, and the mathematical distinction between them, will be discussed in Chap. 4.) The dividing line between the subsonic and supersonic regions is called the sonic line, as sketched in Fig. 1.1. The change in the mathematical behaviour of the governing equations from elliptic in the subsonic region to hyperbolic in the supersonic region made a consistent mathematical analysis, which included both regions, virtually impossible to obtain. Techniques were developed for just the subsonic portion, and other techniques (such as the standard 'method of characteristics') were developed for the supersonic region. Unfortunately, the proper patching of these different techniques through the transonic region around the sonic line was extremely difficult. Hence, as late as the mid-1960s, no uniformly valid aerodynamic technique existed to treat the entire flowfield over the blunt body. This situation was clearly noted in the classic textbook by Liepmann and Roshko [1] published in 1957, where in a discussion of blunt body flows on page 105, they state:

> *The shock shape and detachment distance cannot, at present, be theoretically predicted.*

The purpose of this lengthy discussion on the status of the blunt body problem in the late 1950s is to set the background for the following important point. In 1966, a breakthrough occurred in the blunt body problem. Using the developing power of computational fluid dynamics at that time, and employing the concept of a 'time-dependent' approach to the steady state, Moretti and Abbett [2] developed a numerical, finite-difference solution to the supersonic blunt body problem which constituted the first practical, straightforward engineering solution for this flow. (This solution will be discussed in Chap. 7.) After 1966, the blunt body problem was no longer a real 'problem'. Industry and government laboratories quickly adopted this computational technique for their blunt body analyses. Perhaps the most striking aspect of this comparison is that the supersonic blunt body problem, which was one of the most serious, most difficult, and most researched theoretical aerodynamic problems of the 1950s and 1960s, is today assigned as a *homework problem* in a computational fluid dynamics graduate course at the University of Maryland.

Therein lies an example of the power of *computational fluid dynamics*. The purpose of these notes is to provide an introduction to computational fluid dynamics. The above example concerning blunt body flows serves to illustrate the importance of computational fluid dynamics to modern aerodynamic applications. Here is an important problem which was impossible to solve in a practical fashion before the advent of computational fluid dynamics (CFD), but which is now tractable and straightforward using the modern techniques of CFD. Indeed, this is but one example out of many where CFD is revolutionizing the world of aerodynamics. The purpose of the present author writing these notes, and your reading these notes and attending the VKI short course, is to introduce you to this revolution.

(As an aside, for those of you interested in more historical details concerning the blunt body problem, see Sect. 1.1 of Ref. [3]).

1.2 Computational Fluid Dynamics: What is it?

The physical aspects of any fluid flow are governed by the following three fundamental principles: (1) mass is conserved; (2) $F = ma$ (Newton's second law); and (3) energy is conserved. These fundamental principles can be expressed in terms of mathematical equations, which in their most general form are usually partial differential equations. Computational fluid dynamics is, in part, the art of replacing the governing partial differential equations of fluid flow with *numbers*, and advancing these numbers in space and/or time to obtain a final numerical description of the complete flow field of interest. This is not an all-inclusive definition of CFD; there are some problems which allow the immediate solution of the flow field without advancing in time or space, and there are some applications which involve integral equations rather than partial differential equations. In any event, all such problems involve the manipulation of, and the solution for, *numbers*. The end product of CFD is indeed a collection of numbers, in contrast to a closed-form analytical solution. However, in the long run the objective of most engineering analyses, closed form or otherwise, is a quantitative description of the problem, i.e. numbers. (See, for example, Ref. [4]).

Of course, the instrument which has allowed the practical growth of CFD is the high-speed digital computer. CFD solutions generally require the repetitive manipulation of thousands, or even millions, of numbers—a task that is humanly impossible without the aid of a computer. Therefore, advances in CFD, and its application to problems of more and more detail and sophistication, are intimately related to advances in computer hardware, particularly in regard to storage and execution speed. This is why the strongest force driving the development of new supercomputers is coming from the CFD community (see, for example, the survey article by Graves [5]).

1.3 The Role of Computational Fluid Dynamics in Modern Fluid Dynamics

First, let us make a few historical comments. Perhaps the first major example of computational fluid dynamics was the work of Kopal [6], who in 1947 compiled massive tables of the supersonic flow over sharp cones by numerically solving the governing differential equations (the Taylor–Maccoll equation [7]). These solutions were carried out on a primitive digital computer at the Massachusetts Institute of Technology. However, the first generation of computational fluid-dynamic solutions appeared during the 1950s and early 1960s, spurred by the simultaneous advent of efficient, high-speed computers and the need to solve the high velocity, high-temperature re-entry body problem. High temperatures necessitated the inclusion of vibrational energies and chemical reactions in flow problems, sometimes equilibrium and other times non-equilibrium. Such physical phenomena generally cannot be solved analytically, even for the simplest flow geometry. Therefore, numerical solutions of the governing equations on a high-speed digital computer were an absolute necessity. Examples of these first generation computations are the pioneering work of Fay and Riddell [8] and Blottner [9, 10] for boundary layers, and Hall et al. [11] for inviscid flows. Even though it was not fashionable at the time to describe such high temperature gas-dynamic calculations as 'computational fluid dynamics,' they nevertheless represented the first generation of the discipline.

The second generation of computational fluid-dynamic solutions, those which today are generally descriptive of the discipline, involve the application of the governing equations to applied fluid-dynamic problems which are in themselves so complicated (without the presence of chemical reactions, etc.) that a computer must be utilized. Examples of such inherently difficult problems are mixed subsonic–supersonic flows (such as the supersonic blunt body problem discussed in Sect. 1.1), and viscous flows which are not amenable to the boundary layer approximation, such as separated and recirculating flows. For the latter case, the full Navier–Stokes equations are required for an exact solution. In these cases, the *time-dependent technique*, introduced in a practical fashion in the mid-1960s, has created a revolution in flowfield calculations. This technique will be discussed in Chap. 7.

The role of CFD in engineering predictions has become so strong that today it can be viewed as a new 'third dimension' in fluid dynamics, the other two dimensions being the classical cases of pure experiment and pure theory. This relationship is sketched in Fig. 1.2. From 1687, with the publication of Isaac Newton's *Principia*, to the mid-1960s, advancements in fluid mechanics were made with the synergistic combination of pioneering experiments and basic theoretical analyses—analyses which almost always required the use of simplified models of the flow to obtain closed-form solutions of the governing equations. These closed-form solutions have the distinct advantage of immediately identifying some of the fundamental parameters of a given problem, and explicitly demonstrating how the answers to the problems are influenced by variations in the parameters. They frequently have the disadvantage of not including all the requisite physics of the flow. Into this picture stepped CFD in the mid-1960s. With its ability to handle the governing equations in 'exact' form, along with the inclusion of detailed physical phenomena such as finite-rate chemical reactions, CFD rapidly became a popular tool in engineering analyses. Today, CFD supports and complements both pure experiment and pure theory, and it is this author's opinion that, from now on, it always will. CFD is *not* a passing fad; rather, with the advent of the high-speed digital computer, CFD will remain a third dimension in fluid dynamics, of equal stature and importance to experiment and theory. It has taken a permanent place in all aspects of fluid dynamics, from basic research to engineering design.

One of the most important aspects of modern CFD is the impact it is having on wind tunnel testing. This is related to the rapid decrease in the cost of computations compared to the rapid increase in the cost of wind tunnel tests. In his pioneering survey of CFD in 1979, Chapman [12] shows a plot of relative computation cost

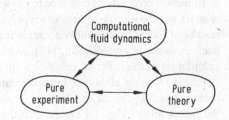

Fig. 1.2 Relationship between pure experiment and pure theory

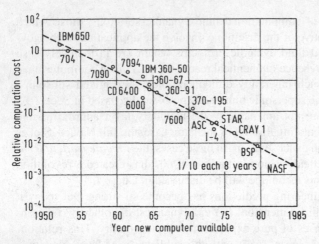

Fig. 1.3 Relative computation cost as a function of years

as a function of years since 1953. This is reproduced as Fig. 1.3, where it will be noted that the relative costs of computations has decreased by an order of magnitude every eight years since 1953—and it is still dropping today. This is due to the continued development of new computers with faster run times, leading to a class of computers that are called 'super-computers' (such as the CRAY machines, and the CYBER 205). As a result, the calculation of the aerodynamic characteristics of new aeroplane designs via application of CFD is becoming economically cheaper than measuring the same characteristics in the wind tunnel. Indeed, in much of the aircraft industry, the testing of preliminary designs for new aircraft, which used to be carried out via numerous wind tunnel tests, is today performed almost entirely on the computer; the wind tunnel is used to 'fine-tune' the final design. This is particularly true in the design of new airfoil shapes [13]. In addition to economics, CFD offers the opportunity to obtain detailed flow-field information, some of which is either difficult to measure in a wind tunnel, or is compromised by wall effects.

Of course, inherent in the above discussion is the assumption that CFD results are *accurate* as well as cost effective; otherwise, any assumption of part of the role of wind tunnels by CFD would be foolish. The results of CFD are only as valid as the physical models incorporated in the governing equations and boundary conditions, and therefore are subject to error, especially for turbulent flows. Truncation errors associated with the particular algorithm used to obtain a numerical solution, as well as round-off errors, both combine to compromise the accuracy of CFD results. (Such matters will be discussed in later sections.) In spite of these inherent drawbacks, the results of CFD are amazingly accurate for a very large number of applications. One such example is given in Ref. [12], and is reproduced in Fig. 1.4. Here we see the calculation of the lift coefficient for a space shuttle orbiter/Boeing 747 combination obtained from an elaborate implementation of the subsonic panel method (panel methods are discussed in Chap. 3). Comparison with wind tunnel data shown in the lower left of Fig. 1.4 clearly illustrates the high degree of accuracy obtained.

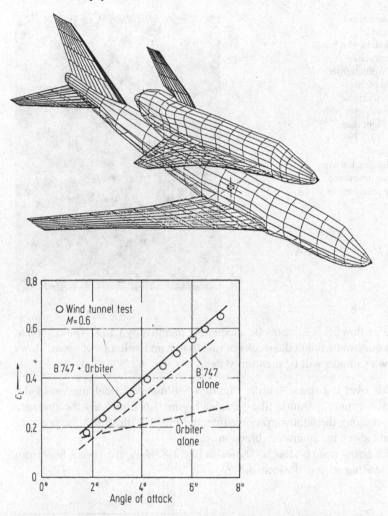

Fig. 1.4 A complex application of computational aerodynamics (from Ref. [12])

Faced with this type of comparison, and keeping in mind that the computations are frequently cheaper than the wind tunnel measurements, aeronautical engineers are more and more transferring the role of preliminary design testing from the wind tunnel to the computer.

The role of CFD in preliminary design has a corollary in basic research. Assuming that a given CFD solution to a basic flow (say, for example, the separated flow over a rearward-facing step) contains all the important physics, then this CFD solution (the computer program itself) is a *numerical tool*. In turn, this numerical tool can be used to carry out *numerical experiments* to help study the fundamental characteristics of the flow. These numerical experiments are directly analogous to actual laboratory experiments.

Fig. 1.5 Calculated shock
wave shape around a
shuttle-like vehicle at Mach 6
and an angle of attack of 26.6
degrees. (From Weilmuenser,
K.J., 'High angle of attack
inviscid flow calculations
over shuttle-like vehicles with
comparisons to flight data,'
AIAA Paper No. 83–1798,
1983.) *Note*: Fluted
appearance of the shock wave
is due to the finite-difference
grid used for the calculations

What types of flowfields can now be adequately handled by CFD? The complete
answer to this question would take weeks of discussion and volumes of notes. However, just a few examples will be mentioned here.

(1) Flow fields over the space shuttle. Figure 1.5 illustrates a calculation of the
    shock wave around a shuttle-like vehicle. Figure 1.6 illustrates the pressure
    distribution along the windward centerline and Fig. 1.7 illustrates the pressure
    distribution along the spanwise direction.
(2) Flows over arrow wing bodies, as shown in Fig. 1.8. Here, the vortex flow from
    the wing leading edge is illustrated.

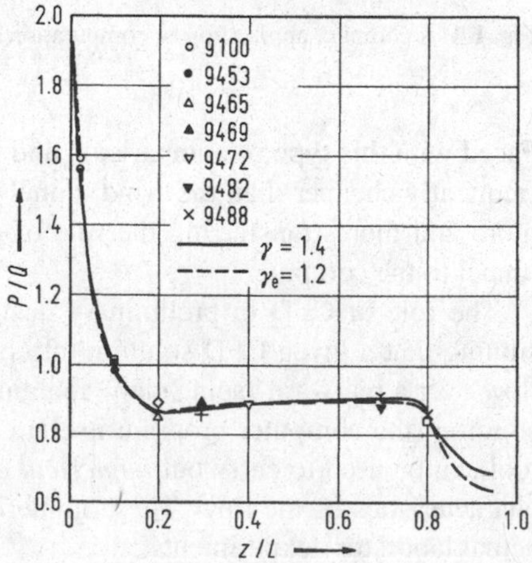

Fig. 1.6 Calculated pressure
distribution along the
windward centreline of the
space shuttle, and comparison
with flight test data (from
Weilmuenser, as referenced in
Fig. 1.5)

Fig. 1.7 Calculated spanwise pressure distribution on the windward surface of the space shuttle (from Maus, J.R. et al. 'Hypersonic Mach number and real gas effects on space shuttle orbiter aerodynamics,' *Journal* of *Spacecraft and Rockets*, Vol. 21, No. 2, March–April 1984, pp. 136–141)

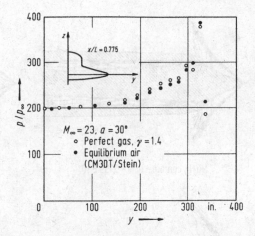

(3) Unsteady, oscillating flows through supersonic engine inlets, as shown in Fig. 1.9. Here, the contours of constant Mach number are shown for four different times.

(4) Flow field over an automobile towing a trailer, as shown by the streamlines given in Fig. 1.10.

(5) Flows through supersonic combustion ramjet engines, as shown in Fig. 1.11.

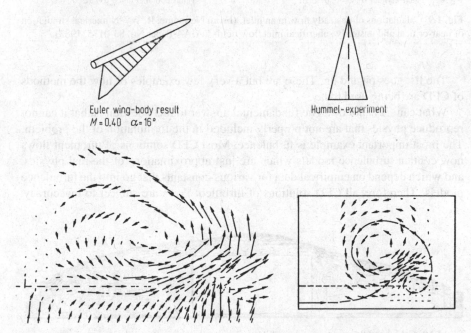

Fig. 1.8 The calculation of the leading edge vortex from a delta wing (from AIAA Short Course entitled 'Using Euler Solvers', July 1984, with material presented by Wolfgang Schmidt)

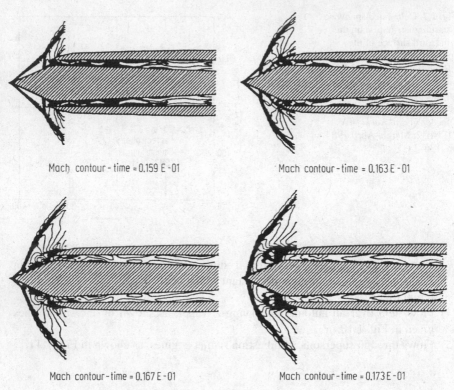

Mach contour - time = 0.159 E - 01   Mach contour - time = 0.163 E - 01

Mach contour - time = 0.167 E - 01   Mach contour - time = 0.173 E - 01

Fig. 1.9 Calculations of unsteady flow in an inlet. (From Newsome, R. W., 'Numerical simulation of near-critical and unsteady subcritical inlet flow fields,' AIAA Paper No. 83-0175, 1983)

The list goes on and on. These are but a very few examples of how the methods of CFD are being used today.

What can CFD *not* do? The fundamental answer to this question is that it cannot reproduce physics that are not properly included in the formulation of the problem. The most important example is turbulence. Most CFD solutions of turbulent flows now contain turbulence models which are just approximations of the real physics, and which depend on empirical data for various constants that go into the turbulence models. Therefore, all CFD solutions of turbulent flows are subject to inaccuracy,

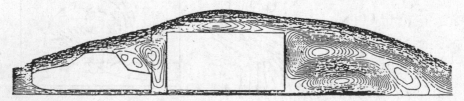

Fig. 1.10 Calculated flow over an automobile-trailer configuration (from the same reference given in Fig. 1.8)

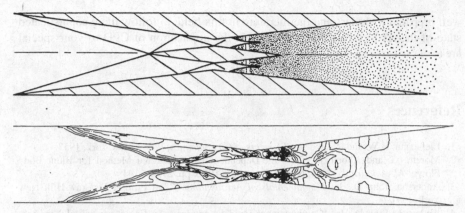

Fig. 1.11 Calculations of the flow field in a scramjet engine (from Drummond, J.P., and Weidner, E.H., 'Numerical study of a scramjet engine flowfield," *AIAA Journal*, Vol. 20, No. 9, Sept. 1982, pp. 1182–1187)

even though some calculations for some situations are reasonable. It is interesting to note that the CFD community is directly attacking this problem in the most basic sense. There is work today on the direct computation of turbulence (Ref. [12]). This is based on the assumption that, on a fine enough scale, all turbulent flows obey the Navier–Stokes equations (to be derived in Chap. 2); and if a fine enough grid can be used, with a requisite large number of grid points, maybe both the fine scale and large scale aspects of turbulence can be calculated. This is currently a wide-open area of CFD research.

Again, emphasis is made that CFD solutions are slaves to the degree of physics that goes into their formulation. Another example is the computation of chemically reacting flows. Here, the chemical kinetic rate mechanisms as well as the magnitudes of the rate constants are frequently very uncertain, and any CFD solution will be compromised by these uncertainties.

1.4 The Role of This Course

The objective of this course is somewhat different from the conventional short course in computational fluid dynamics. Our purpose here is to provide a very basic, elementary and tutorial presentation of CFD, emphasizing the fundamentals, and surveying a number of solution techniques ranging from low-speed incompressible flow to hypersonic flow. This course is aimed at the completely unititiated student—a student who has little or no experience in computational fluid dynamics. The purpose of this course is to provide such students with (a) some insight into the philosophy and power of CFD; (b) an understanding of the governing equations; (c) a familiarity with various popular solution techniques; and (d) a working vocabulary in the discipline. It is hoped that at the conclusion of this course, you will be well prepared to understand the literature in this field, to follow more sophisticated state-of-the-art lecture series and to begin the application of CFD to your special areas of concern.

References

1. Liepmann, H.W. and Roshko, A., *Elements of Gasdynamics*, Wiley, New York, 1957.
2. Moretti, G. and Abbett, M., 'A Time-Dependent Computational Method for Blunt Body Flows,' *AIAA Journal*, Vol. 4, No. 12, December 1966, pp. 2136–2141.
3. Anderson, John D., Jr., *Fundamentals of Aerodynamics*, 2nd Edition McGraw-Hill, New York, 1991.
4. Anderson, John D., Jr., 'Computational Fluid Dynamics—An Engineering Tool?' in A.A. Pouring (ed.), *Numerical Laboratory Computer Methods in Fluid Dynamics*, ASME, New York, 1976, pp. 1–12.
5. Graves, R.A., 'Computational Fluid Dynamics: The Coming Revolution,' *Astronautics and Aeronautics*, Vol. 20, No. 3, March 1982, pp. 20–28.
6. Kopal, Z., *Tables of Supersonic Flow Around Cones*, Depart of Electrical Engineering, Center of Analysis, Massachusetts Institute of Technology, Cambridge, 1947.
7. Taylor, G.I. and Maccoll, J.W., 'The Air Pressure on a Cone Moving at High Speed,' *Proceedings of the Royal Society* (A), Vol. 139, 1933, p. 278.
8. Fay, J.A. and Riddell, F.R., 'Theory of Stagnation Point Heat Transfer in Dissociated Air,' *Journal of the Aeronautical Sciences*, Vol. 25, No. 2, Feb. 1958, pp. 73–85.
9. Blottner, F.G., 'Chemical Nonequilibrium Boundary Layer,' *AIAA Journal*, Vol. 2, No. 2, Feb. 1964, pp. 232–239.
10. Blottner, F.G., 'Nonequilibrium Laminar Boundary-Layer Flow of Ionized Air,' *AIAA Journal*, Vol. 2, No. 11, Nov. 1964, pp. 1921–1927.
11. Hall, H.G., Eschenroeder, A.Q. and Marrone, P.V., 'Blunt-Nose Inviscid Airflows with Coupled Nonequilibrium Processes,' *Journal of the Aerospace Sciences*, Vol. 29, No. 9, Sept. 1962, pp. 1038–1051.
12. Chapman, D.R., 'Computational Aerodynamics Development and Outlook,' *AIAA Journal*, Vol. 17, No. 12, Dec. 1979, pp. 1293–1313.
13. *Advanced Technology Airfoil Research*, NASA Conference Publications 2045, March 1978.

Chapter 2
Governing Equations of Fluid Dynamics

J.D. Anderson, Jr.

2.1 Introduction

The cornerstone of computational fluid dynamics is the fundamental governing equations of fluid dynamics—the continuity, momentum and energy equations. *These equations speak physics*. They are the mathematical statements of three fundamental physical principles upon which all of fluid dynamics is based:

(1) mass is conserved;
(2) $F = ma$ (Newton's second law);
(3) energy is conserved.

The purpose of this chapter is to derive and discuss these equations.

The purpose of taking the time and space to derive the governing equations of fluid dynamics in this course are three-fold:

(1) Because all of CFD is based on these equations, it is important for each student to feel very comfortable with these equations before continuing further with his or her studies, and certainly before embarking on any application of CFD to a particular problem.
(2) This author assumes that the attendees of the present VKI short course come from varied background and experience. Some of you may not be totally familiar with these equations, whereas others may use them every day. For the former, this chapter will hopefully be some enlightenment; for the latter, hopefully this chapter will be an interesting review.
(3) The governing equations can be obtained in various different forms. For most aerodynamic theory, the particular form of the equations makes little difference. However, for CFD, the use of the equations in one form may lead to success, whereas the use of an alternate form may result in oscillations (wiggles) in the numerical results, or even instability. Therefore, in the world of CFD, the various forms of the equations are of vital interest. In turn, it is important to *derive* these equations in order to point out their differences and similarities, and to reflect on possible implications in their application to CFD.

J.D. Anderson, Jr.
National Air and Space Museum, Smithsonian Institution, Washington, DC
e-mail: AndersonJA@si.edu

J.F. Wendt (ed.), *Computational Fluid Dynamics*, 3rd ed.,   15
© Springer-Verlag Berlin Heidelberg 2009

2.2 Modelling of the Flow

In obtaining the basic equations of fluid motion, the following philosophy is always followed:

(1) Choose the appropriate fundamental physical principles from the laws of physics, such as

    (a) Mass is conserved.
    (b) $F = ma$ (Newton's 2nd Law).
    (c) Energy is conserved.

(2) Apply these physical principles to a suitable model of the flow.
(3) From this application, extract the mathematical equations which embody such physical principles.

This section deals with item (2) above, namely the definition of a suitable model of the flow. This is not a trivial consideration. A solid body is rather easy to see and define; on the other hand, a fluid is a 'squishy' substance that is hard to grab hold of. If a solid body is in translational motion, the velocity of each part of the body is the same; on the other hand, if a fluid is in motion the velocity may be different at each location in the fluid. How then do we visualize a moving fluid so as to apply to it the fundamental physical principles?

For a continuum fluid, the answer is to construct one of the two following models.

2.2.1 Finite Control Volume

Consider a general flow field as represented by the streamlines in Fig. 2.1(a). Let us imagine a closed volume drawn within a *finite* region of the flow. This volume defines a *control volume, V*, and a *control surface, S*, is defined as the closed surface which bounds the volume. The control volume may be *fixed* in space with the fluid moving through it, as shown at the left of Fig. 2.1(a). Alternatively, the control volume may be moving with the fluid such that the same fluid particles are always inside it, as shown at the right of Fig. 2.1(a). In either case, the control volume is a reasonably large, finite region of the flow. The fundamental physical principles are applied to the fluid inside the control volume, and to the fluid crossing the control surface (if the control volume is fixed in space). Therefore, instead of looking at the whole flow field at once, with the control volume model we limit our attention to just the fluid in the finite region of the volume itself. The fluid flow equations that we *directly* obtain by applying the fundamental physical principles to a finite control volume are in *integral form*. These integral forms of the governing equations can be manipulated to *indirectly* obtain partial differential equations. The equations so obtained from the finite control volume fixed in space (left side of Fig. 2.1a), in either integral or partial differential form, are called the *conservation* form of the governing equations. The equations obtained from the finite control volume moving

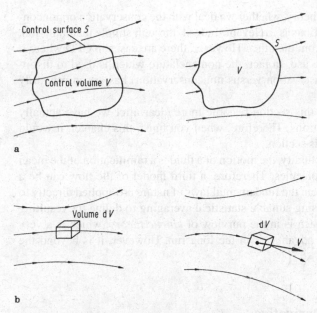

Fig. 2.1 (a) Finite control volume approach. (b) Infinitesimal fluid element approach

with the fluid (right side of Fig. 2.1a), in either integral or partial differential form, are called the *non-conservation* form of the governing equations.

2.2.2 Infinitesimal Fluid Element

Consider a general flow field as represented by the streamlines in Fig. 2.1b. Let us imagine an infinitesimally small fluid element in the flow, with a differential volume, dV. The fluid element is infinitesimal in the same sense as differential calculus; however, it is large enough to contain a huge number of molecules so that it can be viewed as a continuous medium. The fluid element may be fixed in space with the fluid moving through it, as shown at the left of Fig. 2.1(b). Alternatively, it may be moving along a streamline with a vector velocity V equal to the flow velocity at each point. Again, instead of looking at the whole flow field at once, the fundamental physical principles are applied to just the fluid element itself. This application leads *directly* to the fundamental equations in *partial differential equation form*. Moreover, the particular partial differential equations obtained directly from the fluid element fixed in space (left side of Fig. 2.1b) are again the *conservation form* of the equations. The partial differential equations obtained *directly* from the moving fluid element (right side of Fig. 2.1b) are again called the *non-conservation form* of the equations.

In general aerodynamic theory, whether we deal with the conservation or nonconservation forms of the equations is irrelevant. Indeed, through simple manipulation, one form can be obtained from the other. However, there are cases in CFD where it is important which form we use. In fact, the nomenclature which is used to distinguish these two forms (conservation versus nonconservation) has arisen primarily in the CFD literature.

The comments made in this section become more clear after we have actually derived the governing equations. Therefore, when you finish this chapter, it would be worthwhile to re-read this section.

As a final comment, in actuality, the motion of a fluid is a ramification of the mean motion of its atoms and molecules. Therefore, a third model of the flow can be a microscopic approach wherein the fundamental laws of nature are applied directly to the atoms and molecules, using suitable statistical averaging to define the resulting fluid properties. This approach is in the purview of *kinetic theory*, which is a very elegant method with many advantages in the long run. However, it is beyond the scope of the present notes.

2.3 The Substantial Derivative

Before deriving the governing equations, we need to establish a notation which is common in aerodynamics—that of the substantial derivative. In addition, the substantial derivative has an important physical meaning which is sometimes not fully appreciated by students of aerodynamics. A major purpose of this section is to emphasize this physical meaning.

As the model of the flow, we will adopt the picture shown at the right of Fig. 2.1(b), namely that of an infinitesimally small fluid element moving with the flow. The motion of this fluid element is shown in more detail in Fig. 2.2. Here, the fluid element is moving through cartesian space. The unit vectors along the x, y, and z axes are $\vec{i}, \vec{j}$, and $\vec{k}$ respectively. The vector velocity field in this cartesian space is given by

$$\vec{V} = u\vec{i} + v\vec{j} + w\vec{k}$$

where the x, y, and z components of velocity are given respectively by

$$u = u(x, y, z, t)$$
$$v = v(x, y, z, t)$$
$$w = w(x, y, z, t)$$

Note that we are considering in general an *unsteady flow*, where u, v, and w are functions of both space and time, t. In addition, the scalar density field is given by

$$\rho = \rho(x, y, z, t)$$

Fig. 2.2 Fluid element moving in the flow field—illustration for the substantial derivative

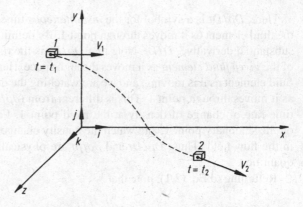

At time t_1, the fluid element is located at point 1 in Fig. 2.2. At this point and time, the density of the fluid element is

$$\rho_1 = \rho(x_1, y_1, z_1, t_1)$$

At a later time, t_2, the same fluid element has moved to point 2 in Fig. 2.2. Hence, at time t_2, the density of this *same* fluid element is

$$\rho_2 = \rho(x_2, y_2, z_2, t_2)$$

Since $\rho = \rho(x, y, z, t)$, we can expand this function in a Taylor's series about point 1 as follows:

$$\rho_2 = \rho_1 + \left(\frac{\partial \rho}{\partial x}\right)_1 (x_2 - x_1) + \left(\frac{\partial \rho}{\partial y}\right)_1 (y_2 - y_1) + \left(\frac{\partial \rho}{\partial z}\right)_1 (z_2 - z_1)$$
$$+ \left(\frac{\partial \rho}{\partial t}\right)_1 (t_2 - t_1) + (\text{higher order terms})$$

Dividing by $(t_2 - t_1)$, and ignoring higher order terms, we obtain

$$\frac{\rho_2 - \rho_1}{t_2 - t_1} = \left(\frac{\partial \rho}{\partial x}\right)_1 \left(\frac{x_2 - x_1}{t_2 - t_1}\right) + \left(\frac{\partial \rho}{\partial y}\right)_1 \left(\frac{y_2 - y_1}{t_2 - t_1}\right)$$
$$+ \left(\frac{\partial \rho}{\partial z}\right)_1 \left(\frac{z_2 - z_1}{t_2 - t_1}\right) + \left(\frac{\partial \rho}{\partial t}\right)_1 \qquad (2.1)$$

Examine the left side of Eq. (2.1). This is physically the *average* time-rate-of-change in density of the fluid element as it moves from point 1 to point 2. In the limit, as t_2 approaches t_1, this term becomes

$$\lim_{t_2 \to t_1} \left(\frac{\rho_2 - \rho_1}{t_2 - t_1}\right) \equiv \frac{D\rho}{Dt}$$

Here, $D\rho/Dt$ is a symbol for the *instantaneous* time rate of change of density of the fluid element as it moves through point 1. By definition, this symbol is called the substantial derivative, D/Dt. Note that $D\rho/Dt$ is the time rate of change of density of the *given fluid element* as it moves through space. Here, our eyes are locked on the fluid element as it is moving, and we are watching the density of the element change as it moves through point 1. This is different from $(\partial\rho/\partial t)_1$, which is physically the time rate of change of density at the fixed point 1. For $(\partial\rho/\partial t)_1$, we fix our eyes on the stationary point 1, and watch the density change due to transient fluctuations in the flow field. Thus, $D\rho/Dt$ and $\partial\rho/\rho t$ are physically and numerically different quantities.

Returning to Eq. (2.1), note that

$$\lim_{t_2 \to t_1}\left(\frac{x_2 - x_1}{t_2 - t_1}\right) \equiv u$$

$$\lim_{t_2 \to t_1}\left(\frac{y_2 - y_1}{t_2 - t_1}\right) \equiv v$$

$$\lim_{t_2 \to t_1}\left(\frac{z_2 - z_1}{t_2 - t_1}\right) \equiv w$$

Thus, taking the limit of Eq. (2.1) as $t_2 \to t_1$, we obtain

$$\frac{D\rho}{Dt} = u\frac{\partial\rho}{\partial x} + v\frac{\partial\rho}{\partial y} + w\frac{\partial\rho}{\partial z} + \frac{\partial\rho}{\partial t} \tag{2.2}$$

Examine Eq. (2.2) closely. From it, we can obtain an expression for the substantial derivative in cartesian coordinates:

$$\frac{D}{Dt} \equiv \frac{\partial}{\partial t} + u\frac{\partial}{\partial x} + v\frac{\partial}{\partial y} + w\frac{\partial}{\partial z} \tag{2.3}$$

Furthermore, in cartesian coordinates, the vector operator ∇ is defined as

$$\nabla \equiv \vec{i}\frac{\partial}{\partial x} + \vec{j}\frac{\partial}{\partial y} + \vec{k}\frac{\partial}{\partial z} \tag{2.4}$$

Hence, Eq. (2.3) can be written as

$$\frac{D}{Dt} \equiv \frac{\partial}{\partial t} + \left(\vec{V}\cdot\nabla\right) \tag{2.5}$$

Equation (2.5) represents a definition of the substantial derivative operator in vector notation; thus, it is valid for any coordinate system.

Focusing on Eq. (2.5), we once again emphasize that D/Dt is the substantial derivative, which is physically the time rate of change following a moving fluid element; $\partial/\partial t$ is called the *local derivative*, which is physically the time rate of change at a fixed point; $\vec{V}\cdot\nabla$ is called the *convective derivative*, which is physically the time rate of change due to the movement of the fluid element from one location to another in the flow field where the flow properties are spatially different. The substantial derivative applies to any flow-field variable, for example, Dp/Dt, DT/Dt, Du/Dt, etc., where p and T are the static pressure and temperature respectively. For example:

$$\underbrace{\frac{DT}{Dt} \equiv \underbrace{\frac{\partial T}{\partial t}}_{\substack{\text{local} \\ \text{derivative}}} + \underbrace{(\vec{V} \cdot \nabla)}_{\substack{\text{convective} \\ \text{derivative}}} T} \equiv \frac{\partial T}{\partial t} + u\frac{\partial T}{\partial x} + v\frac{\partial T}{\partial y} + w\frac{\partial T}{\partial z} \qquad (2.6)$$

Again, Eq. (2.6) states physically that the temperature of the fluid element is changing as the element sweeps past a point in the flow because at that point the flow field temperature itself may be fluctuating with time (the local derivative) and because the fluid element is simply on its way to another point in the flow field where the temperature is different (the convective derivative).

Consider an example which will help to reinforce the physical meaning of the substantial derivative. Imagine that you are hiking in the mountains, and you are about to enter a cave. The temperature inside the cave is cooler than outside. Thus, as you walk through the mouth of the cave, you feel a temperature decrease—this is analogous to the convective derivative in Eq. (2.6). However, imagine that, at the same time, a friend throws a snowball at you such that the snowball hits you just at the same instant you pass through the mouth of the cave. You will feel an additional, but momentary, temperature drop when the snowball hits you—this is analagous to the local derivative in Eq. (2.6). The net temperature drop you feel as you walk through the mouth of the cave is therefore a combination of both the act of moving into the cave, where it is cooler, and being struck by the snowball at the same instant—this net temperature drop is analogous to the substantial derivative in Eq. (2.6).

The above derivation of the substantial derivative is essentially taken from this author's basic aerodynamics text book given as Ref. [1]. It is used there to introduce new aerodynamics students to the full physical meaning of the substantial derivative. The description is repeated here for the same reason—to give you a physical feel for the substantial derivative. We could have circumvented most of the above discussion by recognizing that the substantial derivative is essentially the same as the total differential from calculus. That is, if

$$\rho = \rho(x, y, z, t)$$

then the chain rule from differential calculus gives

$$d\rho = \frac{\partial \rho}{\partial x}dx + \frac{\partial \rho}{\partial y}dy + \frac{\partial \rho}{\partial z}dz + \frac{\partial \rho}{\partial t}dt \qquad (2.7)$$

From Eq. (2.7), we have

$$\frac{d\rho}{dt} = \frac{\partial \rho}{\partial t} + \frac{\partial \rho}{\partial x}\frac{dx}{dt} + \frac{\partial \rho}{\partial y}\frac{dy}{dt} + \frac{\partial \rho}{\partial z}\frac{dz}{dt} \qquad (2.8)$$

Since $\dfrac{dx}{dt} = u, \dfrac{dy}{dt} = v$, and $\dfrac{dz}{dt} = w$, Eq. (2.8) becomes

$$\frac{d\rho}{dt} = \frac{\partial\rho}{\partial t} + u\frac{\partial\rho}{\partial x} + v\frac{\partial\rho}{\partial y} + w\frac{\partial\rho}{\partial z} \tag{2.9}$$

Comparing Eqs. (2.2) and (2.9), we see that $d\rho/dt$ and $D\rho/Dt$ are one-in-the-same.

Therefore, the substantial derivative is nothing more than a total derivative with respect to time. However, the derivation of Eq. (2.2) highlights more of the physical significance of the substantial derivative, whereas the derivation of Eq. (2.9) is more formal mathematically.

2.4 Physical Meaning of $\nabla \cdot \vec{V}$

As one last item before deriving the governing equations, let us consider the divergence of the velocity, $\nabla \cdot \vec{V}$. This term appears frequently in the equations of fluid dynamics, and it is well to consider its physical meaning.

Consider a control volume moving with the fluid as sketched on the right of Fig. 2.1(a). This control volume is always made up of the same fluid particles as it moves with the flow; hence, its mass is fixed, invariant with time. However, its volume $\mathscr{V}$ and control surface S are changing with time as it moves to different regions of the flow where different values of ρ exist. That is, this moving control volume of fixed mass is constantly increasing or decreasing its volume and is changing its shape, depending on the characteristics of the flow. This control volume is shown in Fig. 2.3 at some instant in time. Consider an infinitesimal element of the surface dS moving at the local velocity $\vec{V}$, as shown in Fig. 2.3. The change in the volume of the control volume $\Delta\mathscr{V}$, due to just the movement of dS over a time increment Δt, is, from Fig. 2.3, equal to the volume of the long, thin cylinder with base area dS and altitude $(\vec{V}\Delta t) \cdot \vec{n}$, where $\vec{n}$ is a unit vector perpendicular to the surface at dS. That is,

$$\Delta\mathscr{V} = \left[(\vec{V}\Delta t) \cdot \vec{n}\right] dS = (\vec{V}\Delta t) \cdot \vec{dS} \tag{2.10}$$

where the vector $\vec{dS}$ is defined simply as $\vec{dS} \equiv \vec{n}\, dS$. Over the time increment Δt, the total change in volume of the whole control volume is equal to the summation of Eq. (2.10) over the total control surface. In the limit as $dS \to 0$, the sum becomes the surface integral

Fig. 2.3 Moving control volume used for the physical interpretation of the divergence of velocity

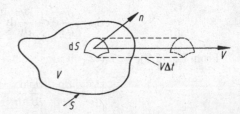

$$\oiint_S (\vec{V}\Delta t)\cdot d\vec{S}$$

If this integral is divided by Δt, the result is physically the time rate of change of the control volume, denoted by DV/Dt, i.e.

$$\frac{D\mathcal{V}}{Dt} = \frac{1}{\Delta t}\oiint_S (\vec{V}\cdot\Delta t)\cdot d\vec{S} = \oiint_S \vec{V}\cdot d\vec{S} \qquad (2.11)$$

Note that we have written the left side of Eq. (2.11) as the substantial derivative of $\mathcal{V}$, because we are dealing with the time rate of change of the control volume *as the volume moves with the flow* (we are using the picture shown at the right of Fig. 2.1a), and this is physically what is meant by the substantial derivative. Applying the divergence theorem from vector calculus to the right side of Eq. (2.11), we obtain

$$\frac{D\mathcal{V}}{Dt} = \oiiint_V (\nabla\cdot\vec{V})d\mathcal{V} \qquad (2.12)$$

Now, let us image that the moving control volume in Fig. 2.3 is shrunk to a very small volume, $\delta\mathcal{V}$, essentially becoming an infinitesimal moving fluid element as sketched on the right of Fig. 2.1(a). Then Eq. (2.12) can be written as

$$\frac{D(\delta\mathcal{V})}{Dt} = \oiiint_{\delta\mathcal{V}} (\nabla\cdot\vec{V})d\mathcal{V} \qquad (2.13)$$

Assume that $\delta\mathcal{V}$ is small enough such that $\nabla\cdot\vec{V}$ is essentially the same value throughout $\delta\mathcal{V}$. Then the integral in Eq. (2.13) can be approximated as $(\nabla\cdot\vec{V})\delta\mathcal{V}$. From Eq. (2.13), we have

$$\frac{D(\delta\mathcal{V})}{Dt} = (\nabla\cdot\vec{V})\delta\mathcal{V}$$

or

$$\boxed{\nabla\cdot\vec{V} = \frac{1}{\delta\mathcal{V}}\frac{D(\delta\mathcal{V})}{Dt}} \qquad (2.14)$$

Examine Eq. (2.14) closely. On the left side we have the divergence of the velocity; on the right side we have its physical meaning. That is,

$\nabla\cdot\vec{V}$ *is physically the time rate of change of the volume of a moving fluid element, per unit volume.*

2.5 The Continuity Equation

Let us now apply the philosophy discussed in Sect. 2.2; that is, (a) write down a fundamental physical principle, (b) apply it to a suitable model of the flow, and (c) obtain an equation which represents the fundamental physical principle. In this section we will treat the following case:

2.5.1 Physical Principle: Mass is Conserved

We will carry out the application of this principle to *both* the finite control volume and infinitesimal fluid element models of the flow. This is done here specifically to illustrate the physical nature of both models. Moreover, we will choose the finite control volume to be *fixed in space* (left side of Fig. 2.1a), whereas the infinitesimal fluid element will be *moving with the flow* (right side of Fig. 2.1b). In this way we will be able to contrast the differences between the conservation and non-conservation forms of the equations, as described in Sect. 2.2.

First, consider the model of a moving fluid element. The mass of this element is fixed, and is given by δm. Denote the volume of this element by $\delta \mathcal{V}$, as in Sect. 2.4. Then

$$\delta m = \rho \delta \mathcal{V} \tag{2.15}$$

Since mass is conserved, we can state that the time-rate-of-change of the mass of the fluid element is zero as the element moves along with the flow. Invoking the physical meaning of the substantial derivative discussed in Sect. 2.3, we have

$$\frac{D(\delta m)}{Dt} = 0 \tag{2.16}$$

Combining Eqs. (2.15) and (2.16), we have

$$\frac{D(\rho \delta \mathcal{V})}{Dt} = \delta \mathcal{V} \frac{D\rho}{Dt} + \rho \frac{D(\delta \mathcal{V})}{Dt} = 0$$

or,

$$\frac{D\rho}{Dt} + \rho \left[\frac{1}{\delta \mathcal{V}} \frac{D(\delta \mathcal{V})}{Dt} \right] = 0 \tag{2.17}$$

We recognize the term in brackets in Eq. (2.17) as the physical meaning of $\nabla \cdot \vec{V}$, discussed in Sect. 2.4. Hence, combining Eqs. (2.14) and (2.17), we obtain

$$\boxed{\frac{D\rho}{Dt} + \rho \nabla . \vec{V} = 0} \tag{2.18}$$

Equation (2.18) is the *continuity equation* in *non-conservation* form. In light of our philosophical discussion in Sect. 2.2, note that:

(1) By applying the model of an *infinitesimal fluid element*, we have obtained Eq. (2.18) *directly* in partial differential form.
(2) By choosing the model to be *moving with the flow*, we have obtained the *non-conservation form* of the continuity equation, namely Eq. (2.18).

Now, consider the model of a finite control volume fixed in space, as sketched in Fig. 2.4. At a point on the control surface, the flow velocity is $\vec{V}$ and the vector elemental surface area (as defined in Sect. 2.4) is $d\vec{S}$. Also let $d\mathcal{V}$ be an elemental volume inside the finite control volume. Applied to this control volume, our fundamental physical principle that mass is conserved means

Fig. 2.4 Finite control
volume fixed in space

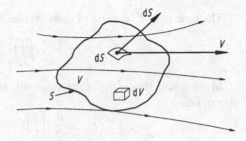

$$\begin{Bmatrix} \text{Net mass flow } out \\ \text{of control volume} \\ \text{through surface } S \end{Bmatrix} = \begin{Bmatrix} \text{time rate of decrease} \\ \text{of mass inside control} \\ \text{volume} \end{Bmatrix} \qquad (2.19a)$$

or,

$$B = C \qquad (2.19b)$$

where B and C are just convenient symbols for the left and right sides, respectively, of Eq. (2.19a). First, let us obtain an expression for B in terms of the quantities shown in Fig. 2.4. The mass flow of a moving fluid across any fixed surface (say, in kg/s, or slug/s) is equal to the product of (density) × (area of surface) × (component of velocity perpendicular to the surface). Hence the elemental mass flow across the area dS is

$$\rho V_n dS = \rho \vec{V} \cdot \vec{dS} \qquad (2.20)$$

Examining Fig. 2.4, note that by convention, $\vec{dS}$ always points in a direction *out* of the control volume. Hence, when $\vec{V}$ also points out of the control volume (as shown in Fig. 2.4), the product $\rho \vec{V} \cdot \vec{dS}$ is *positive*. Moreover, when $\vec{V}$ points out of the control volume, the mass flow is physically leaving the control volume, i.e. it is an *outflow*. Hence, a positive $\rho \vec{V} \cdot \vec{dS}$ denotes an outflow. In turn, when $\vec{V}$ points into the control volume, $\rho \vec{V} \cdot \vec{dS}$ is *negative*. Moreover, when $\vec{V}$ points inward, the mass flow is physically entering the control volume, i.e. it is an *inflow*. Hence, a negative $\rho \vec{V} \cdot \vec{dS}$ denotes an inflow. The net mass flow *out* of the entire control volume through the control surface S is the summation over S of the elemental mass flows shown in Eq. (2.20). In the limit, this becomes a surface integral, which is physically the left side of Eqs. (2.19a and b), i.e.

$$B = \oiint_S \rho \vec{V} \cdot \vec{dS} \qquad (2.21)$$

Now consider the right side of Eqs. (2.19a and b). The mass contained within the elemental volume $d\mathcal{V}$ is $\rho\, d\mathcal{V}$. The total mass inside the control volume is therefore

$$\iiint_\mathcal{V} \rho\, d\mathcal{V}$$

The time rate of *increase* of mass inside $\mathscr{V}$ is then

$$-\frac{\partial}{\partial t} \iiint_{\mathscr{V}} \rho \, d\mathscr{V}$$

In turn, the time rate of decrease of mass inside $\mathscr{V}$ is the negative of the above, i.e.

$$-\frac{\partial}{\partial t} \iiint_{\mathscr{V}} \rho \, d\mathscr{V} = C \qquad (2.22)$$

Thus, substituting Eqs. (2.21) and (2.22) into (2.19b), we have

$$\oiint_{S} \rho \vec{V} \cdot d\vec{S} = -\frac{\partial}{\partial t} \iiint_{\mathscr{V}} \rho \, d\mathscr{V}$$

or,

$$\boxed{\frac{\partial}{\partial t} \iiint_{\mathscr{V}} \rho \, d\mathscr{V} + \oiint_{S} \rho \vec{V} \cdot d\vec{S} = 0} \qquad (2.23)$$

Equation (2.23) is the *integral form of the continuity equation*; it is also in *conservation* form.

Let us cast Eq. (2.23) in the form of a differential equation. Since the control volume in Fig. 2.4 is fixed in space, the limits of integration for the integrals in Eq. (2.23) are constant, and hence the time derivative $\partial/\partial t$ can be placed inside the integral.

$$\iiint_{\mathscr{V}} \frac{\partial \rho}{\partial t} d\mathscr{V} + \oiint_{S} \rho \vec{V} \cdot d\vec{S} = 0 \qquad (2.24)$$

Applying the divergence theorem from vector calculus, the surface integral in Eq. (2.24) can be expressed as a volume integral

$$\oiint_{S} (\rho \vec{V}) \cdot d\vec{S} = \iiint_{\mathscr{V}} \nabla \cdot (\rho \vec{V}) d\mathscr{V} \qquad (2.25)$$

Substituting Eq. (2.25) into Eq. (2.24), we have

$$\iiint_{\mathscr{V}} \frac{\partial \rho}{\partial t} d\mathscr{V} + \iiint_{\mathscr{V}} \nabla \cdot (\rho \vec{V}) d\mathscr{V} = 0$$

or

$$\iiint_{\mathscr{V}} \left[\frac{\partial \rho}{\partial t} + \nabla \cdot (\rho \vec{V}) \right] d\mathscr{V} = 0 \qquad (2.26)$$

Since the finite control volume is *arbitrarily* drawn in space, the only way for the integral in Eq. (2.26) to equal zero is for the integrand to be zero at every point within the control volume. Hence, from Eq. (2.26)

$$\boxed{\frac{\partial \rho}{\partial t} + \nabla \cdot (\rho \vec{V}) = 0} \qquad (2.27)$$

Equation (2.27) is the *continuity equation* in *conservation* form.

Examining the above derivation in light of our discussion in Sect. 2.2, we note that:

(1) By applying the model of a finite control volume, we have obtained Eq. (2.23) *directly* in integral form.
(2) Only after some manipulation of the integral form did we *indirectly* obtain a partial differential equation, Eq. (2.27).
(3) By choosing the model to be *fixed in space*, we have obtained the *conservation* form of the continuity equation, Eqs. (2.23) and (2.27).

Emphasis is made that Eqs. (2.18) and (2.27) are both statements of the conservation of mass expressed in the form of partial differential equations. Eq. (2.18) is in non-conservation form, and Eq. (2.27) is in conservation form; both forms are equally valid. Indeed, one can easily be obtained from the other, as follows. Consider the vector identity involving the divergence of the product of a scalar times a vector, such as

$$\nabla \cdot (\rho \vec{V}) \equiv \rho \nabla \cdot \vec{V} + \vec{V} \cdot \nabla \rho \tag{2.28}$$

Substitute Eq. (2.28) in the conservation form, Eq. (2.27):

$$\frac{\partial \rho}{\partial t} + \vec{V} \cdot \nabla \rho + \rho \nabla \cdot \vec{V} = 0 \tag{2.29}$$

The first two terms on the left side of Eq. (2.29) are simply the substantial derivative of density. Hence, Eq. (2.29) becomes

$$\frac{D\rho}{Dt} + \rho \nabla \cdot \vec{V} = 0$$

which is the non-conservation form given by Eq. (2.18).

Once again we note that the use of conservation or non-conservation forms of the governing equations makes little difference in most of theoretical aerodynamics. In contrast, which form is used can make a difference in some CFD applications, and this is why we are making a distinction between these two different forms in the present notes.

2.6 The Momentum Equation

In this section, we apply another fundamental physical principle to a model of the flow, namely:

Physical Principle : $\vec{F} = m\vec{a}$ (Newton's 2nd law)

We choose for our flow model the moving fluid element as shown at the right of Fig. 2.1(b). This model is sketched in more detail in Fig. 2.5.

Newton's 2nd law, expressed above, when applied to the moving fluid element in Fig. 2.5, says that the net force on the fluid element equals its mass times the

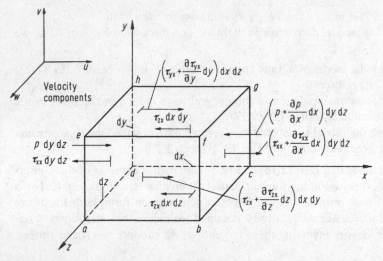

Fig. 2.5 Infinitesimally small, moving fluid element. Only the forces in the x direction are shown

acceleration of the element. This is a vector relation, and hence can be split into three scalar relations along the x, y, and z-axes. Let us consider only the x-component of Newton's 2nd law,

$$F_x = ma_x \tag{2.30}$$

where F_x and a_x are the scalar x-components of the force and acceleration respectively.

First, consider the left side of Eq. (2.30). We say that the moving fluid element experiences a force in the x-direction. What is the source of this force? There are two sources:

(1) *Body forces*, which act directly on the volumetric mass of the fluid element. These forces 'act at a distance'; examples are gravitational, electric and magnetic forces.

(2) *Surface forces*, which act directly on the surface of the fluid element. They are due to only two sources: (a) the pressure distribution acting on the surface, imposed by the outside fluid surrounding the fluid element, and (b) the shear and normal stress distributions acting on the surface, also imposed by the outside fluid 'tugging' or 'pushing' on the surface by means of friction.

Let us denote the body force per unit mass acting on the fluid element by $\vec{f}$, with f_x as its x-component. The volume of the fluid element is $(dx\, dy\, dz)$; hence,

$$\left\{ \begin{array}{l} \text{Body force on the} \\ \text{fluid element acting} \\ \text{in the } x\text{-direction} \end{array} \right\} = \rho f_x (dx\, dy\, dz) \tag{2.31}$$

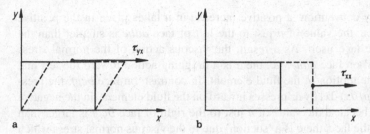

Fig. 2.6 Illustration of shear and normal stresses

The shear and normal stresses in a fluid are related to the time-rate-of-change of the deformation of the fluid element, as sketched in Fig. 2.6 for just the xy plane. The shear stress, denoted by τ_{xy} in this figure, is related to the time rate-of-change of the shearing deformation of the fluid element, whereas the normal stress, denoted by τ_{xx} in Fig. 2.6, is related to the time-rate-of-change of volume of the fluid element. As a result, both shear and normal stresses depend on velocity gradients in the flow, to be designated later. In most viscous flows, normal stresses (such as τ_{xx}) are much smaller than shear stresses, and many times are neglected. Normal stresses (say τ_{xx} in the x-direction) become important when the normal velocity gradients (say $\partial u/\partial x$) are very large, such as *inside* a shock wave.

The surface forces in the x-direction exerted on the fluid element are sketched in Fig. 2.5. The convention will be used here that τ_{ij} denotes a stress in the j-direction exerted on a plane perpendicular to the i-axis. On face *abcd*, the only force in the x-direction is that due to shear stress, $\tau_{yx}\,\mathrm{d}x\,\mathrm{d}z$. Face *efgh* is a distance $\mathrm{d}y$ above face *abcd*; hence the shear force in the x-direction on face *efgh* is $[\tau_{yx} + (\partial\tau_{yx}/\partial y)\,\mathrm{d}y]\,\mathrm{d}x\,\mathrm{d}z$. Note the directions of the shear force on faces *abcd* and *efgh*; on the bottom face, τ_{yx} is to the left (the negative x-direction), whereas on the top face, $[\tau_{yx} + (\partial\tau_{yx}/\partial y)\,\mathrm{d}y]$ is to the right (the positive x-direction). These directions are consistent with the convention that positive increases in all three components of velocity. u, v and w, occur in the positive directions of the axes. For example, in Fig. 2.5, u increases in the positive y-direction. Therefore, concentrating on face *efgh*, u is higher just above the face than on the face; this causes a 'tugging' action which tries to pull the fluid element in the positive x-direction (to the right) as shown in Fig. 2.5. In turn, concentrating on face *abcd*, u is lower just beneath the face than on the face; this causes a retarding or dragging action on the fluid element, which acts in the negative x-direction (to the left) as shown in Fig. 2.5. The directions of all the other viscous stresses shown in Fig. 2.5, including τ_{xx}, can be justified in a like fashion. Specifically on face *dcgh*, τ_{zx} acts in the negative x-direction, whereas on face *abfe*, $[\tau_{zx} + (\partial\tau_{zx}/\partial z)\,\mathrm{d}z]$ acts in the positive x-direction. On face *adhe*, which is perpendicular to the x-axis, the only forces in the x-direction are the pressure force $p\,\mathrm{d}x\,\mathrm{d}z$, which always acts in the direction *into* the fluid element, and $\tau_{xx}\,\mathrm{d}y\,\mathrm{d}z$, which is in the negative x-direction. In Fig. 2.5, the reason why τ_{xx} on face *adhe* is to the left hinges on the convention mentioned earlier for the direction of increasing velocity. Here, by convention, a positive increase in u takes place in the positive x-direction. Hence, the value of u just to the left of face *adhe* is smaller than the value of u on the face itself. As a result, the viscous action of the normal stress acts as a 'suction' on face *adhe*, i.e. there is a dragging action toward the left that wants to retard the motion of the fluid element. In contrast, on face *bcgf*, the pressure force $[p+(\partial p/\partial x)\,\mathrm{d}x]\,\mathrm{d}y\,\mathrm{d}z$ presses inward on the fluid element (in the negative x-direction), and because the value of u just to the right of face *bcgf* is larger than the value of u on the face, there is a 'suction' due to the viscous normal stress which tries to pull the element to the right (in the positive x-direction) with a force equal to $[\tau_{xx}+(\partial\tau_{xx}/\partial x)]\,\mathrm{d}y\,\mathrm{d}z$.

With the above in mind, for the moving fluid element we can write

$$
\begin{Bmatrix} \text{Net surface force} \\ \text{in the } x\text{-direction} \end{Bmatrix} = \left[p - \left(p + \frac{\partial p}{\partial x}\mathrm{d}x \right) \right] \mathrm{d}y\,\mathrm{d}z
$$

$$
+ \left[\left(\tau_{xx} + \frac{\partial \tau_{xx}}{\partial x}\mathrm{d}x \right) - \tau_{xx} \right] \mathrm{d}y\,\mathrm{d}z
$$

$$
+ \left[\left(\tau_{yx} + \frac{\partial \tau_{yx}}{\partial y}\mathrm{d}y \right) - \tau_{yx} \right] \mathrm{d}x\,\mathrm{d}z
$$

$$
+ \left[\left(\tau_{zx} + \frac{\partial \tau_{zx}}{\partial z}\mathrm{d}z \right) - \tau_{zx} \right] \mathrm{d}x\,\mathrm{d}y \tag{2.32}
$$

The total force in the x-direction, F_x, is given by the sum of Eqs. (2.31) and (2.32). Adding, and cancelling terms, we obtain

$$
F_x = \left(-\frac{\partial p}{\partial x} + \frac{\partial \tau_{xx}}{\partial x} + \frac{\partial \tau_{yx}}{\partial y} + \frac{\partial \tau_{zx}}{\partial z} \right) \mathrm{d}x\,\mathrm{d}y\,\mathrm{d}z + \rho f_x\,\mathrm{d}x\,\mathrm{d}y\,\mathrm{d}z \tag{2.33}
$$

Equation (2.33) represents the left-hand side of Eq. (2.30).

Considering the right-hand side of Eq. (2.30), recall that the mass of the fluid element is fixed and is equal to

$$
m = \rho\,\mathrm{d}x\,\mathrm{d}y\,\mathrm{d}z \tag{2.34}
$$

Also, recall that the acceleration of the fluid element is the time-rate-of-change of its velocity. Hence, the component of acceleration in the x-direction, denoted by a_x, is simply the time-rate-of-change of u; since we are following a moving fluid element, this time-rate-of-change is given by the substantial derivative. Thus,

$$
a_x = \frac{Du}{Dt} \tag{2.35}
$$

Combining Eqs. (2.30), (2.33), (2.34) and (2.35), we obtain

$$
\boxed{\rho\frac{Du}{Dt} = -\frac{\partial p}{\partial x} + \frac{\partial \tau_{xx}}{\partial x} + \frac{\partial \tau_{yx}}{\partial y} + \frac{\partial \tau_{zx}}{\partial z} + \rho f_x} \tag{2.36a}
$$

which is the x-component of the momentum equation for a viscous flow. In a similar fashion, the y and z components can be obtained as

$$\rho \frac{Dv}{Dt} = -\frac{\partial p}{\partial y} + \frac{\partial \tau_{xy}}{\partial x} + \frac{\partial \tau_{yy}}{\partial y} + \frac{\partial \tau_{zy}}{\partial z} + \rho f_y \qquad (2.36b)$$

and

$$\rho \frac{Dw}{Dt} = -\frac{\partial p}{\partial z} + \frac{\partial \tau_{xz}}{\partial x} + \frac{\partial \tau_{yz}}{\partial y} + \frac{\partial \tau_{zz}}{\partial z} + \rho f_z \qquad (2.36c)$$

Equations (2.36a, b and c) are the x-, y- and z-components respectively of the momentum equation. Note that they are partial differential equations obtained *directly* from an application of the fundamental physical principle to an infinitesimal fluid element. Moreover, since this fluid element is moving with the flow, Eqs. (2.36a, b and c) are in *non-conservation* form. They are scalar equations, and are called the *Navier–Stokes equations* in honour of two men—the Frenchman M. Navier and the Englishmen G. Stokes—who independently obtained the equations in the first half of the nineteenth century.

The Navier–Stokes equations can be obtained in conservation form as follows. Writing the left-hand side of Eq. (2.36a) in terms of the definition of the substantial derivative,

$$\rho \frac{Du}{Dt} = \rho \frac{\partial u}{\partial t} + \rho \vec{V} \cdot \nabla u \qquad (2.37)$$

Also, expanding the following derivative,

$$\frac{\partial(\rho u)}{\partial t} = \rho \frac{\partial u}{\partial t} + u \frac{\partial \rho}{\partial t}$$

or,

$$\rho \frac{\partial u}{\partial t} = \frac{\partial(\rho u)}{\partial t} - u \frac{\partial \rho}{\partial t} \qquad (2.38)$$

Recalling the vector identity for the divergence of the product of a scalar times a vector, we have

$$\nabla \cdot (\rho u \vec{V}) = u \nabla \cdot (\rho \vec{V}) + (\rho \vec{V}) \cdot \nabla u$$

or

$$\rho \vec{V} \cdot \nabla u = \nabla \cdot (\rho u \vec{V}) - u \nabla \cdot (\rho \vec{V}) \qquad (2.39)$$

Substitute Eqs. (2.38) and (2.39) into Eq. (2.37).

$$\rho \frac{Du}{Dt} = \frac{\partial(\rho u)}{\partial t} - u \frac{\partial \rho}{\partial t} - u \nabla \cdot (\rho \vec{V}) + \nabla \cdot (\rho u \vec{V})$$

$$\rho \frac{Du}{Dt} = \frac{\partial(\rho u)}{\partial t} - u \left[\frac{\partial \rho}{\partial t} + \nabla \cdot (\rho \vec{V}) \right] + \nabla \cdot (\rho u \vec{V}) \qquad (2.40)$$

The term in brackets in Eq. (2.40) is simply the left-hand side of the continuity equation given as Eq. (2.27); hence the term in brackets is zero. Thus Eq. (2.40) reduces to

$$\rho \frac{Du}{Dt} = \frac{\partial(\rho u)}{\partial t} + V \cdot (\rho u \vec{V})$$ (2.41)

Substitute Eq. (2.41) into Eq. (2.36a).

$$\frac{\partial(\rho u)}{\partial t} + V \cdot (\rho u \vec{V}) = -\frac{\partial p}{\partial x} + \frac{\partial \tau_{xx}}{\partial x} + \frac{\partial \tau_{yx}}{\partial y} + \frac{\partial \tau_{zx}}{\partial z} + \rho f_x$$ (2.42a)

Similarly, Eqs. (2.36b and c) can be expressed as

$$\frac{\partial(\rho v)}{\partial t} + V \cdot (\rho v \vec{V}) = -\frac{\partial p}{\partial y} + \frac{\partial \tau_{xy}}{\partial x} + \frac{\partial \tau_{yy}}{\partial y} + \frac{\partial \tau_{zy}}{\partial z} + \rho f_y$$ (2.42b)

and

$$\frac{\partial(\rho w)}{\partial t} + V \cdot (\rho w \vec{V}) = -\frac{\partial p}{\partial z} + \frac{\partial \tau_{xz}}{\partial x} + \frac{\partial \tau_{yz}}{\partial y} + \frac{\partial \tau_{zz}}{\partial z} + \rho f_z$$ (2.42c)

Equations (2.42a–c) are the Navier-Stokes equations in *conservation* form.

In the late seventeenth century Isaac Newton stated that shear stress in a fluid is proportional to the time-rate-of-strain, i.e. velocity gradients. Such fluids are called *Newtonian* fluids. (Fluids in which τ is *not* proportional to the velocity gradients are non-Newtonian fluids; blood flow is one example.) In virtually all practical aerodynamic problems, the fluid can be assumed to be Newtonian. For such fluids, Stokes, in 1845, obtained:

$$\tau_{xx} = \lambda V \cdot \vec{V} + 2\mu \frac{\partial u}{\partial x}$$ (2.43a)

$$\tau_{yy} = \lambda V \cdot \vec{V} + 2\mu \frac{\partial v}{\partial y}$$ (2.43b)

$$\tau_{zz} = \lambda V \cdot \vec{V} + 2\mu \frac{\partial w}{\partial z}$$ (2.43c)

$$\tau_{xy} = \tau_{yx} = \mu \left(\frac{\partial v}{\partial x} + \frac{\partial u}{\partial y} \right)$$ (2.43d)

$$\tau_{xz} = \tau_{zx} = \mu \left(\frac{\partial u}{\partial z} + \frac{\partial w}{\partial x} \right)$$ (2.43e)

$$\tau_{yz} = \tau_{zy} = \mu \left(\frac{\partial w}{\partial y} + \frac{\partial v}{\partial z} \right)$$ (2.43f)

where μ is the molecular viscosity coefficient and λ is the bulk viscosity coefficient. Stokes made the hypothesis that

$$\lambda = -\frac{2}{3}\mu$$

which is frequently used but which has still not been definitely confirmed to the present day.

Substituting Eq. (2.43) into Eq. (2.42), we obtain the complete Navier–Stokes equations in conservation form:

$$
\frac{\partial(\rho u)}{\partial t} + \frac{\partial(\rho u^2)}{\partial x} + \frac{\partial(\rho u v)}{\partial y} + \frac{\partial(\rho u w)}{\partial z}
$$

$$
= -\frac{\partial p}{\partial x} + \frac{\partial}{\partial x}\left(\lambda \nabla \cdot \vec{V} + 2\mu \frac{\partial u}{\partial x}\right) + \frac{\partial}{\partial y}\left[\mu\left(\frac{\partial v}{\partial x} + \frac{\partial u}{\partial y}\right)\right]
$$

$$
+ \frac{\partial}{\partial z}\left[\mu\left(\frac{\partial u}{\partial z} + \frac{\partial w}{\partial x}\right)\right] + \rho f_x \tag{2.44a}
$$

$$
\frac{\partial(\rho v)}{\partial t} + \frac{\partial(\rho u v)}{\partial x} + \frac{\partial(\rho v^2)}{\partial y} + \frac{\partial(\rho v w)}{\partial z}
$$

$$
= -\frac{\partial p}{\partial y} + \frac{\partial}{\partial x}\left[\mu\left(\frac{\partial v}{\partial x} + \frac{\partial u}{\partial y}\right)\right] + \frac{\partial}{\partial y}\left(\lambda \nabla \cdot \vec{V} + 2\mu \frac{\partial v}{\partial y}\right)
$$

$$
+ \frac{\partial}{\partial z}\left[\mu\left(\frac{\partial w}{\partial y} + \frac{\partial v}{\partial z}\right)\right] + \rho f_y \tag{2.44b}
$$

$$
\frac{\partial(\rho w)}{\partial t} + \frac{\partial(\rho u w)}{\partial x} + \frac{\partial(\rho v w)}{\partial y} + \frac{\partial(\rho w^2)}{\partial z}
$$

$$
= -\frac{\partial p}{\partial z} + \frac{\partial}{\partial x}\left[\mu\left(\frac{\partial u}{\partial z} + \frac{\partial w}{\partial x}\right)\right] + \frac{\partial}{\partial y}\left[\mu\left(\frac{\partial w}{\partial y} + \frac{\partial v}{\partial z}\right)\right]
$$

$$
+ \frac{\partial}{\partial z}\left(\lambda \nabla \cdot \vec{V} + 2\mu \frac{\partial w}{\partial z}\right) + \rho f_z \tag{2.44c}
$$

2.7 The Energy Equation

In the present section, we derive the energy equation using as our model an infinitesimal moving fluid element. This will be in keeping with our derivation of the Navier–Stokes equations in Sect. 2.6, where the infinitesimal element was shown in Fig. 2.5.

We now invoke the following fundamental physical principle:

2.7.1 Physical Principle: Energy is Conserved

A statement of this principle is the first law of thermodynamics, which, when applied to the moving fluid element in Fig. 2.5, becomes

$$\left\{\begin{array}{l}\text{Rate of change of} \\ \text{energy inside the} \\ \text{fluid element}\end{array}\right\} = \left\{\begin{array}{l}\text{Net flux of} \\ \text{heat into} \\ \text{the element}\end{array}\right\} + \left\{\begin{array}{l}\text{Rate of working done on} \\ \text{the element due to body} \\ \text{and surface forces}\end{array}\right\}$$

or,

$$A \quad = \quad B \quad + \quad C \qquad (2.45)$$

where A, B and C denote the respective terms above.

Let us first evaluate C, i.e. obtain an expression for the rate of work done on the moving fluid element due to body and surface forces. It can be shown that the rate of doing work by a force exerted on a moving body is equal to the product of the force and the component of velocity in the direction of the force (see References 3 and 14 for such a derivation). Hence the rate of work done by the body force acting on the fluid element moving at a velocity $\vec{V}$ is

$$\rho \vec{f} \cdot \vec{V}(\mathrm{d}x\,\mathrm{d}y\,\mathrm{d}z)$$

With regard to the surface forces (pressure plus shear and normal stresses), consider just the forces in the x-direction, shown in Fig. 2.5. The rate of work done on the moving fluid element by the pressure and shear forces in the x-direction shown in Fig. 2.5 is simply the x-component of velocity, u, multiplied by the forces, e.g. on face *abcd* the rate of work done by $\tau_{yx}\mathrm{d}x\,\mathrm{d}z$ is $u\tau_{yx}\mathrm{d}x\,\mathrm{d}z$, with similar expressions for the other faces. To emphasize these energy considerations, the moving fluid element is redrawn in Fig. 2.7, where the rate of work done on each face by surface forces in the x-direction is shown explicitly. To obtain the *net* rate of work done on the fluid element by the surface forces, note that forces in the positive x-direction do positive work and that forces in the negative x-direction do negative work. Hence,

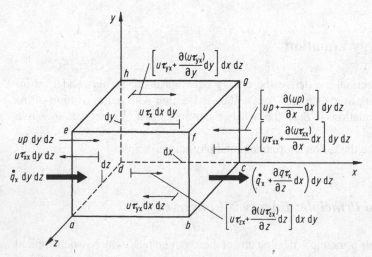

Fig. 2.7 Energy fluxes associated with an infinitesimally small, moving fluid element. For simplicity, only the fluxes in the x direction are shown

comparing the pressure forces on face *adhe* and *bcgf* in Fig. 2.7, the net rate of work done by pressure in the *x*-direction is

$$\left[up - \left(up + \frac{\partial(up)}{\partial x}dx\right)\right]dy\,dz = -\frac{\partial(up)}{\partial x}dx\,dy\,dz$$

Similarly, the net rate of work done by the shear stresses in the *x*-direction on faces *abcd* and *efgh* is

$$\left[\left(u\tau_{yx} + \frac{\partial(u\tau_{yx})}{\partial y}dy\right) - u\tau_{yx}\right]dx\,dz = \frac{\partial(u\tau_{yx})}{\partial y}dx\,dy\,dz$$

Considering all the surface forces shown in Fig. 2.7, the net rate of work done on the moving fluid element due to these forces is simply

$$\left[-\frac{\partial(up)}{\partial x} + \frac{\partial(u\tau_{xx})}{\partial x} + \frac{\partial(u\tau_{yx})}{\partial y} + \frac{\partial(u\tau_{zx})}{\partial z}\right]dx\,dy\,dz$$

The above expression considers only surface forces in the *x*-direction. When the surface forces in the *y*- and *z*-directions are also included, similar expressions are obtained. In total, the net rate of work done on the moving fluid element is the sum of the surface force contributions in the *x*-, *y*- and *z*-directions, as well as the body force contribution. This is denoted by C in Eq. (2.45), and is given by

$$\begin{aligned}
C = &\left[-\left(\frac{\partial(up)}{\partial x} + \frac{\partial(vp)}{\partial y} + \frac{\partial(wp)}{\partial z}\right) + \frac{\partial(u\tau_{xx})}{\partial x} + \frac{\partial(u\tau_{yx})}{\partial y}\right.\\
&+ \frac{\partial(u\tau_{zx})}{\partial z} + \frac{\partial(v\tau_{xy})}{\partial x} + \frac{\partial(v\tau_{yy})}{\partial y} + \frac{\partial(v\tau_{zy})}{\partial z} + \frac{\partial(w\tau_{xz})}{\partial x}\\
&\left.+ \frac{\partial(w\tau_{yz})}{\partial y} + \frac{\partial(w\tau_{zz})}{\partial z}\right]dx\,dy\,dz + \rho\vec{f}\cdot\vec{V}\,dx\,dy\,dz
\end{aligned} \qquad (2.46)$$

Note in Eq. (2.46) that the first three terms on the right-hand side are simply $\nabla\cdot(p\vec{V})$.

Let us turn our attention to B in Eq. (2.45), i.e. the net flux of heat into the element. This heat flux is due to: (1) volumetric heating such as absorption or emission of radiation, and (2) heat transfer across the surface due to temperature gradients, i.e. thermal conduction. Define $\dot{q}$ as the rate of volumetric heat addition per unit mass. Noting that the mass of the moving fluid element in Fig. 2.7 is $\rho\,dx\,dy\,dz$, we obtain

$$\left\{\begin{array}{l}\text{Volumetric heating}\\\text{of the element}\end{array}\right\} = \rho\dot{q}\,dx\,dy\,dz \qquad (2.47)$$

In Fig. 2.7, the heat transferred by thermal conduction into the moving fluid element across face *adhe* is $\dot{q}_x\,dy\,dz$ where $\dot{q}_x$ is the heat transferred in the *x*-direction per unit time per unit area by thermal conduction. The heat transferred out of the element across face *bcgf* is $[\dot{q}_x + (\partial\dot{q}_x/\partial x)\,dx]\,dy\,dz$. Thus, the net heat transferred in the *x*-direction into the fluid element by thermal conduction is

$$\left[\dot{q}_x - \left(\dot{q}_x + \frac{\partial \dot{q}_x}{\partial x} dx \right) \right] dy\, dz = -\frac{\partial \dot{q}_x}{\partial x} dx\, dy\, dz$$

Taking into account heat transfer in the y- and z-directions across the other faces in Fig. 2.7, we obtain

$$\left\{ \begin{array}{l} \text{Heating of the} \\ \text{fluid element by} \\ \text{thermal conduction} \end{array} \right\} = -\left(\frac{\partial \dot{q}_x}{\partial x} + \frac{\partial \dot{q}_y}{\partial y} + \frac{\partial \dot{q}_z}{\partial z} \right) dx\, dy\, dz \qquad (2.48)$$

The term B in Eq. (2.45) is the sum of Eqs. (2.47) and (2.48).

$$B = \left[\rho \dot{q} - \left(\frac{\partial \dot{q}_x}{\partial x} + \frac{\partial \dot{q}_y}{\partial y} + \frac{\partial \dot{q}_z}{\partial z} \right) \right] dx\, dy\, dz \qquad (2.49)$$

Heat transfer by thermal conduction is proportional to the local temperature gradient:

$$\dot{q}_x = -k\frac{\partial T}{\partial x}; \qquad \dot{q}_y = -k\frac{\partial T}{\partial y}; \qquad \dot{q}_z = -k\frac{\partial T}{\partial z}$$

where k is the thermal conductivity. Hence, Eq. (2.49) can be written

$$B = \left[\rho \dot{q} + \frac{\partial}{\partial x}\left(k\frac{\partial T}{\partial x} \right) + \frac{\partial}{\partial y}\left(k\frac{\partial T}{\partial y} \right) + \frac{\partial}{\partial z}\left(k\frac{\partial T}{\partial z} \right) \right] dx\, dy\, dz \qquad (2.50)$$

Finally, the term A in Eq. (2.45) denotes the time-rate-of-change of energy of the fluid element. The total energy of a moving fluid per unit mass is the sum of its internal energy per unit mass, e, and its kinetic energy per unit mass, $V^2/2$. Hence, the total energy is $(e + V^2/2)$. Since we are following a moving fluid element, the time-rate-of-change of energy per unit mass is given by the substantial derivative. Since the mass of the fluid element is $\rho\, dx\, dy\, dz$, we have

$$A = \rho \frac{D}{Dt}\left(e + \frac{V^2}{2} \right) dx\, dy\, dz \qquad (2.51)$$

The final form of the energy equation is obtained by substituting Eqs. (2.46), (2.50) and (2.51) into Eq. (2.45), obtaining:

$$\begin{aligned}
\rho \frac{D}{Dt}\left(e + \frac{V^2}{2} \right) = {} & \rho \dot{q} + \frac{\partial}{\partial x}\left(k\frac{\partial T}{\partial x} \right) + \frac{\partial}{\partial y}\left(k\frac{\partial T}{\partial y} \right) + \frac{\partial}{\partial z}\left(k\frac{\partial T}{\partial z} \right) \\
& - \frac{\partial(up)}{\partial x} - \frac{\partial(vp)}{\partial y} - \frac{\partial(wp)}{\partial z} + \frac{\partial(u\tau_{xx})}{\partial x} + \frac{\partial(u\tau_{yx})}{\partial y} \\
& + \frac{\partial(u\tau_{zx})}{\partial z} + \frac{\partial(v\tau_{xy})}{\partial x} + \frac{\partial(v\tau_{yy})}{\partial y} + \frac{\partial(v\tau_{zy})}{\partial z} \\
& + \frac{\partial(w\tau_{xz})}{\partial x} + \frac{\partial(w\tau_{yz})}{\partial y} + \frac{\partial(w\tau_{zz})}{\partial z} + \rho \vec{f} \cdot \vec{V}
\end{aligned} \qquad (2.52)$$

This is the *non-conservation* form of the energy equation; also note that it is in terms of the *total* energy, $(e + V^2/2)$. Once again, the non-conservation form results from the application of the fundamental physical principle to a *moving* fluid element.

The left-hand side of Eq. (2.52) involves the total energy, $(e + V^2/2)$. Frequently, the energy equation is written in a form that involves just the internal energy, e. The derivation is as follows. Multiply Eqs. (2.36a, b, and c) by u, v, and w respectively.

$$\rho \frac{D\left(\frac{u^2}{2}\right)}{Dt} = -u\frac{\partial p}{\partial x} + u\frac{\partial \tau_{xx}}{\partial x} + u\frac{\partial \tau_{yx}}{\partial y} + u\frac{\partial \tau_{zx}}{\partial z} + \rho u f_x \tag{2.53a}$$

$$\rho \frac{D\left(\frac{v^2}{2}\right)}{Dt} = -v\frac{\partial p}{\partial y} + v\frac{\partial \tau_{xy}}{\partial x} + v\frac{\partial \tau_{yy}}{\partial y} + v\frac{\partial \tau_{zy}}{\partial z} + \partial v f_y \tag{2.53b}$$

$$\rho \frac{D\left(\frac{w^2}{2}\right)}{Dt} = -w\frac{\partial p}{\partial z} + w\frac{\partial \tau_{xz}}{\partial x} + w\frac{\partial \tau_{yz}}{\partial y} + w\frac{\partial \tau_{zz}}{\partial z} + \rho w f_z \tag{2.53c}$$

Add Eqs. (2.53a, b and c), and note that $u^2 + v^2 + w^2 = V^2$. We obtain

$$\rho \frac{DV^2/2}{Dt} = -u\frac{\partial p}{\partial x} - v\frac{\partial p}{\partial y} - w\frac{\partial p}{\partial z} + u\left(\frac{\partial \tau_{xx}}{\partial x} + \frac{\partial \tau_{yx}}{\partial y} + \frac{\partial \tau_{zx}}{\partial z}\right)$$

$$+ v\left(\frac{\partial \tau_{xy}}{\partial x} + \frac{\partial \tau_{yy}}{\partial y} + \frac{\partial \tau_{zy}}{\partial z}\right) + w\left(\frac{\partial \tau_{xz}}{\partial x} + \frac{\partial \tau_{yz}}{\partial y} + \frac{\partial \tau_{zz}}{\partial z}\right)$$

$$+ \rho(u f_x + v f_y + w f_r) \tag{2.54}$$

Subtracting Eq. (2.54) from Eq. (2.52), noting that $\rho \vec{f} \cdot \vec{V} = \rho(u f_x + v f_y + w f_z)$, we have

$$\rho \frac{De}{Dt} = \rho \dot{q} + \frac{\partial}{\partial x}\left(k\frac{\partial T}{\partial x}\right) + \frac{\partial}{\partial y}\left(k\frac{\partial T}{\partial y}\right) + \frac{\partial}{\partial z}\left(k\frac{\partial T}{\partial z}\right)$$

$$- p\left(\frac{\partial u}{\partial x} + \frac{\partial v}{\partial y} + \frac{\partial w}{\partial z}\right) + \tau_{xx}\frac{\partial u}{\partial x} + \tau_{yx}\frac{\partial u}{\partial y} + \tau_{zx}\frac{\partial u}{\partial z}$$

$$+ \tau_{xy}\frac{\partial v}{\partial x} + \tau_{yy}\frac{\partial v}{\partial y} + \tau_{zy}\frac{\partial v}{\partial z} + \tau_{xz}\frac{\partial w}{\partial x}$$

$$+ \tau_{yz}\frac{\partial w}{\partial y} + \tau_{zz}\frac{\partial w}{\partial z} \tag{2.55}$$

Equation (2.55) is the energy equation in terms of internal energy, e. Note that the body force terms have cancelled; the energy equation when written in terms of e does not explicitly contain the body force. Eq. (2.55) is still in *non-conservation* form.

Equations (2.52) and (2.55) can be expressed totally in terms of flow field variables by replacing the viscous stress terms τ_{xy}, τ_{xz}, etc. with their equivalent expressions from Eqs (2.43a, b, c, d, e and f). For example, from Eq. (2.55), noting that $\tau_{xy} = \tau_{yx}$, $\tau_{xz} = \tau_{zx}$, $\tau_{yz} = \tau_{zy}$,

$$\rho\frac{De}{Dt} = \rho\dot{q} + \frac{\partial}{\partial x}\left(k\frac{\partial T}{\partial x}\right) + \frac{\partial}{\partial y}\left(k\frac{\partial T}{\partial y}\right) + \frac{\partial}{\partial z}\left(k\frac{\partial T}{\partial z}\right)$$

$$- p\left(\frac{\partial u}{\partial x} + \frac{\partial v}{\partial y} + \frac{\partial w}{\partial z}\right) + \tau_{xx}\frac{\partial u}{\partial x} + \tau_{yy}\frac{\partial v}{\partial y} + \tau_{zz}\frac{\partial w}{\partial z}$$

$$+ \tau_{yx}\left(\frac{\partial u}{\partial y} + \frac{\partial v}{\partial x}\right) + \tau_{zx}\left(\frac{\partial u}{\partial z} + \frac{\partial w}{\partial x}\right) + \tau_{zy}\left(\frac{\partial v}{\partial z} + \frac{\partial w}{\partial y}\right)$$

Substituting Eqs. (2.43a, b, c, d, e and f) into the above equation, we have

$$\rho\frac{De}{Dt} = \rho\dot{q} + \frac{\partial}{\partial x}\left(k\frac{\partial T}{\partial x}\right) + \frac{\partial}{\partial y}\left(k\frac{\partial T}{\partial y}\right) + \frac{\partial}{\partial z}\left(k\frac{\partial T}{\partial z}\right)$$

$$- p\left(\frac{\partial u}{\partial x} + \frac{\partial v}{\partial y} + \frac{\partial w}{\partial z}\right) + \lambda\left(\frac{\partial u}{\partial x} + \frac{\partial v}{\partial y} + \frac{\partial w}{\partial z}\right)^2$$

$$+ \mu\left[2\left(\frac{\partial u}{\partial x}\right)^2 + 2\left(\frac{\partial v}{\partial y}\right)^2 + 2\left(\frac{\partial w}{\partial z}\right)^2 + \left(\frac{\partial u}{\partial y} + \frac{\partial v}{\partial x}\right)^2\right.$$

$$\left.+ \left(\frac{\partial u}{\partial z} + \frac{\partial w}{\partial x}\right)^2 + \left(\frac{\partial v}{\partial z} + \frac{\partial w}{\partial y}\right)^2\right] \tag{2.56}$$

Equation (2.56) is a form of the energy equation completely in terms of the flow-field variables. A similar substitution of Eqs. (2.43a, b, c, d, e and f) can be made into Eq. (2.52); the resulting form of the energy equation in terms of the flow-field variables is lengthy, and to save time and space it will not be given here.

The energy equation in *conservation* form can be obtained as follows. Consider the left-hand side of Eq. (2.56). From the definition of the substantial derivative:

$$\rho\frac{De}{Dt} = \rho\frac{\partial e}{\partial t} + \rho\vec{V}\cdot\nabla e \tag{2.57}$$

However,

$$\frac{\partial(\rho e)}{\partial t} = \rho\frac{\partial e}{\partial t} + e\frac{\partial\rho}{\partial t}$$

or,

$$\rho\frac{\partial e}{\partial t} = \frac{\partial(\rho e)}{\partial t} - e\frac{\partial\rho}{\partial t} \tag{2.58}$$

From the vector identity concerning the divergence of the product of a scalar times a vector,

$$\nabla\cdot(\rho e\vec{V}) = e\nabla\cdot(\rho\vec{V}) + \rho\vec{V}\cdot\nabla e$$

or

$$\rho\vec{V}\cdot\nabla e = \nabla\cdot(\rho e\vec{V}) - e\nabla\cdot(\rho\vec{V}) \tag{2.59}$$

Substitute Eqs. (2.58) and (2.59) into Eq. (2.57)

$$\rho\frac{De}{Dt} = \frac{\partial(\rho e)}{\partial t} - e\left[\frac{\partial\rho}{\partial t} + \nabla\cdot(\rho\vec{V})\right] + \nabla\cdot(\rho e\vec{V}) \tag{2.60}$$

The term in square brackets in Eq. (2.60) is zero, from the continuity equation, Eq. (2.27). Thus, Eq. (2.60) becomes

$$\rho \frac{De}{Dt} = \frac{\partial(\rho e)}{\partial t} + \nabla \cdot (\rho e \vec{V})$$
(2.61)

Substitute Eq. (2.61) into Eq. (2.56):

$$
\frac{\partial(\rho e)}{\partial t} + \nabla \cdot (\rho e \vec{V}) = \rho \dot{q} + \frac{\partial}{\partial x}\left(k\frac{\partial T}{\partial x}\right) + \frac{\partial}{\partial y}\left(k\frac{\partial T}{\partial y}\right)
$$
$$
+ \frac{\partial}{\partial z}\left(k\frac{\partial T}{\partial z}\right) - p\left(\frac{\partial u}{\partial x} + \frac{\partial v}{\partial y} + \frac{\partial w}{\partial z}\right)
$$
$$
+ \lambda\left(\frac{\partial u}{\partial x} + \frac{\partial v}{\partial y} + \frac{\partial w}{\partial z}\right)^2 + \mu\left[2\left(\frac{\partial u}{\partial x}\right)^2\right.
$$
$$
+ 2\left(\frac{\partial v}{\partial y}\right)^2 + 2\left(\frac{\partial w}{\partial z}\right)^2 + \left(\frac{\partial u}{\partial y} + \frac{\partial v}{\partial x}\right)^2
$$
$$
+ \left.\left(\frac{\partial u}{\partial z} + \frac{\partial w}{\partial x}\right)^2 + \left(\frac{\partial v}{\partial z} + \frac{\partial w}{\partial y}\right)^2\right]
$$
(2.62)

Equation (2.62) is the *conservation* form of the energy equation, written in terms of the internal energy.

Repeating the steps from Eq. (2.57) to Eq. (2.61), except operating on the *total* energy, $(e + V^2/2)$, instead of just the internal energy, e, we obtain

$$\rho\frac{D\left(e + \frac{V^2}{2}\right)}{Dt} = \frac{\partial}{\partial t}\left[\rho\left(e + \frac{V^2}{2}\right)\right] + \nabla\left[\rho\left(e + \frac{V^2}{2}\right)\vec{V}\right]$$
(2.63)

Substituting Eq. (2.63) into the left-hand side of Eq. (2.52), we obtain

$$
\frac{\partial}{\partial t}\left[\rho\left(e + \frac{V^2}{2}\right)\right] + \nabla \cdot \left[\rho\left(e + \frac{V^2}{2}\vec{V}\right)\right]
$$
$$
= \rho\dot{q} + \frac{\partial}{\partial x}\left(k\frac{\partial T}{\partial x}\right) + \frac{\partial}{\partial y}\left(k\frac{\partial T}{\partial y}\right)
$$
$$
+ \frac{\partial}{\partial z}\left(k\frac{\partial T}{\partial z}\right) - \frac{\partial(up)}{\partial x} - \frac{\partial(vp)}{\partial y} - \frac{\partial(wp)}{\partial z} + \frac{\partial(u\tau_{xx})}{\partial x}
$$
$$
+ \frac{\partial(u\tau_{yx})}{\partial y} + \frac{\partial(u\tau_{zx})}{\partial z} + \frac{\partial(v\tau_{xy})}{\partial x} + \frac{\partial(v\tau_{yy})}{\partial y} + \frac{\partial(v\tau_{zy})}{\partial z}
$$
$$
+ \frac{\partial(w\tau_{xz})}{\partial x} + \frac{\partial(w\tau_{yz})}{\partial y} + \frac{\partial(w\tau_{zz})}{\partial z} + \rho\vec{f}\cdot\vec{V}
$$
(2.64)

Equation (2.64) is the *conservation* form of the energy equation, written in terms of the *total* energy, $(e + V^2/2)$.

As a final note in this section, there are many other possible forms of the energy equation; for example, the equation can be written in terms of enthalpy, h, or total enthalpy, $(h + V^2/2)$. We will not take the time to derive these forms here; see Refs. [1–3] for more details.

2.8 Summary of the Governing Equations for Fluid Dynamics: With Comments

By this point in our discussions, you have seen a large number of equations, and they may seem to you at this stage to be 'all looking alike'. Equations by themselves can be tiring, and this chapter would seem to be 'wall-to-wall' equations. However, *all* of theoretical and computational fluid dynamics is based on these equations, and therefore it is absolutely *essential* that you are familiar with them, and that you understand their physical significance. That is why we have spent so much time and effort in deriving the governing equations.

Considering this time and effort, it is important to now summarize the important forms of these equations, and to sit back and digest them.

2.8.1 Equations for Viscous Flow

The equations that have been derived in the preceding sections apply to a *viscous* flow, i.e. a flow which includes the dissipative, transport phenomena of viscosity and thermal conduction. The additional transport phenomenon of mass diffusion has not been included because we are limiting our considerations to a homogenous, non-chemically reacting gas. If diffusion were to be included, there would be additional continuity equations—the species continuity equations involving mass transport of chemical species i due to a concentration gradient in the species. Moreover, the energy equation would have an additional term to account for energy transport due to the diffusion of species. See, for example, Ref. [4] for a discussion of such matters.

With the above restrictions in mind, the governing equations for an unsteady, three-dimensional, compressible, viscous flow are:

Continuity equations
(Non-conservation form—Eq. (2.18))

$$\frac{D\rho}{Dt} + \rho \nabla \cdot \vec{V} = 0$$

(Conservation form—Eq. (2.27))

$$\frac{\partial \rho}{\partial t} + \nabla \cdot (\rho \vec{V}) = 0$$

Momentum equations
(Non-conservation form—Eqs. (2.36a–c))

x-component : $\rho\dfrac{Du}{Dt} = -\dfrac{\partial p}{\partial x} + \dfrac{\partial \tau_{xx}}{\partial x} + \dfrac{\partial \tau_{yx}}{\partial y} + \dfrac{\partial \tau_{zx}}{\partial z} + \rho f_x$

y-component : $\rho\dfrac{Dv}{Dt} = -\dfrac{\partial p}{\partial y} + \dfrac{\partial \tau_{xy}}{\partial x} + \dfrac{\partial \tau_{yy}}{\partial y} + \dfrac{\partial \tau_{zy}}{\partial z} + \rho f_y$

z-component : $\rho\dfrac{Dw}{Dt} = -\dfrac{\partial p}{\partial z} + \dfrac{\partial \tau_{xz}}{\partial x} + \dfrac{\partial \tau_{yz}}{\partial y} + \dfrac{\partial \tau_{zz}}{\partial z} + \rho f_z$

(Conservation form—Eqs. (2.42a–c))

x-component : $\dfrac{\partial(\rho u)}{\partial t} + \nabla \cdot (\rho u \vec{V}) = -\dfrac{\partial p}{\partial x} + \dfrac{\partial \tau_{xx}}{\partial x} + \dfrac{\partial \tau_{yx}}{\partial y} + \dfrac{\partial \tau_{zx}}{\partial z} + \rho f_x$

y-component : $\dfrac{\partial(\rho v)}{\partial t} + \nabla \cdot (\rho v \vec{V}) = -\dfrac{\partial p}{\partial y} + \dfrac{\partial \tau_{xy}}{\partial x} + \dfrac{\partial \tau_{yy}}{\partial y} + \dfrac{\partial \tau_{zy}}{\partial z} + \rho f_y$

z-component : $\dfrac{\partial(\rho w)}{\partial t} + \nabla \cdot (\rho w \vec{V}) = -\dfrac{\partial p}{\partial z} + \dfrac{\partial \tau_{xz}}{\partial x} + \dfrac{\partial \tau_{yz}}{\partial y} + \dfrac{\partial \tau_{zz}}{\partial z} + \rho f_z$

Energy equation
(Non-conservation form—Eq. (2.52))

$$\rho\frac{D}{Dt}\left(e + \frac{V^2}{2}\right) = \rho\dot{q} + \frac{\partial}{\partial x}\left(k\frac{\partial T}{\partial x}\right) + \frac{\partial}{\partial y}\left(k\frac{\partial T}{\partial y}\right) + \frac{\partial}{\partial z}\left(k\frac{\partial T}{\partial z}\right)$$

$$- \frac{\partial(up)}{\partial x} - \frac{\partial(vp)}{\partial y} - \frac{\partial(wp)}{\partial z} + \frac{\partial(u\tau_{xx})}{\partial x}$$

$$+ \frac{\partial(u\tau_{yx})}{\partial y} + \frac{\partial(u\tau_{zx})}{\partial z} + \frac{\partial(v\tau_{xy})}{\partial x} + \frac{\partial(v\tau_{yy})}{\partial y}$$

$$+ \frac{\partial(v\tau_{zy})}{\partial z} + \frac{\partial(w\tau_{xz})}{\partial x} + \frac{\partial(w\tau_{yz})}{\partial y} + \frac{\partial(w\tau_{zz})}{\partial z} + \rho\vec{f}\cdot\vec{V}$$

(Conservation form—Eq. (2.64))

$$\frac{\partial}{\partial t}\left[\rho\left(e + \frac{V^2}{2}\right)\right] + \nabla\cdot\left[\rho\left(e + \frac{V^2}{2}\vec{V}\right)\right]$$

$$= \rho\dot{q} + \frac{\partial}{\partial x}\left(k\frac{\partial T}{\partial x}\right) + \frac{\partial}{\partial y}\left(k\frac{\partial T}{\partial y}\right)$$

$$+ \frac{\partial}{\partial z}\left(k\frac{\partial T}{\partial z}\right) - \frac{\partial(up)}{\partial x} - \frac{\partial(vp)}{\partial y} - \frac{\partial(wp)}{\partial z} + \frac{\partial(u\tau_{xx})}{\partial x}$$

$$+ \frac{\partial(u\tau_{yx})}{\partial y} + \frac{\partial(u\tau_{zx})}{\partial z} + \frac{\partial(v\tau_{xy})}{\partial x} + \frac{\partial(v\tau_{yy})}{\partial y}$$

$$+ \frac{\partial(v\tau_{zy})}{\partial z} + \frac{\partial(w\tau_{xz})}{\partial x} + \frac{\partial(w\tau_{yz})}{\partial y} + \frac{\partial(w\tau_{zz})}{\partial z} + \rho\vec{f}\cdot\vec{V}$$

2.8.2 Equations for Inviscid Flow

Inviscid flow is, by definition, a flow where the dissipative, transport phenomena of viscosity, mass diffusion and thermal conductivity are *neglected*. The governing equations for an unsteady, three-dimensional, compressible inviscid flow are obtained by dropping the viscous terms in the above equations.

Continuity equation
(Non-conservation form)

$$\frac{D\rho}{Dt} + \rho \nabla \cdot \vec{V} = 0$$

(Conservation form)

$$\frac{\partial \rho}{\partial t} + \nabla \cdot (\rho \vec{V}) = 0$$

Momentum equations
(Non-conservation form)

$$x\text{-component}: \quad \rho \frac{Du}{Dt} = -\frac{\partial p}{\partial x} + \rho f_x$$

$$y\text{-component}: \quad \rho \frac{Dv}{Dt} = -\frac{\partial p}{\partial y} + \rho f_y$$

$$z\text{-component}: \quad \rho \frac{Dw}{Dt} = -\frac{\partial p}{\partial z} + \rho f_z$$

(Conservation form)

$$x\text{-component}: \quad \frac{\partial (\rho u)}{\partial t} + \nabla \cdot (\rho u \vec{V}) = -\frac{\partial p}{\partial x} + \rho f_x$$

$$y\text{-component}: \quad \frac{\partial (\rho v)}{\partial t} + \nabla \cdot (\rho v \vec{V}) = -\frac{\partial p}{\partial y} + \rho f_y$$

$$z\text{-component}: \quad \frac{\partial (\rho w)}{\partial t} + \nabla \cdot (\rho w \vec{V}) = -\frac{\partial p}{\partial z} + \rho f_z$$

Energy equation
(Non-conservation form)

$$\rho \frac{D}{Dt}\left(e + \frac{V^2}{2}\right) = p\dot{q} - \frac{\partial (up)}{\partial x} - \frac{\partial (vp)}{\partial y} - \frac{\partial (wp)}{\partial z} + \rho \vec{f} \cdot \vec{V}$$

(Conservation form)

$$\frac{\partial}{\partial t}\left[\rho\left(e + \frac{V^2}{2}\right)\right] + \nabla \cdot \left[\rho\left(e + \frac{V^2}{2}\right)\vec{V}\right] = \rho\dot{q} - \frac{\partial (up)}{\partial x} - \frac{\partial (vp)}{\partial y}$$

$$- \frac{\partial (wp)}{\partial z} + \rho \vec{f} \cdot \vec{V}$$

2.8.3 Comments on the Governing Equations

Surveying the above governing equations, several comments and observations can be made.

(1) They are a coupled system of non-linear partial differential equations, and hence are very difficult to solve analytically. To date, there is no general closed-form solution to these equations.

(2) For the momentum and energy equations, the difference between the non-conservation and conservation forms of the equations is just the left-hand side. The right-hand side of the equations in the two different forms is the same.

(3) Note that the conservation form of the equations contain terms on the left-hand side which include the divergence of some quantity, such as $\nabla \cdot (\rho \vec{V})$, $\nabla \cdot (\rho u \vec{V})$, etc. For this reason, the conservation form of the governing equations is sometimes called the *divergence form*.

(4) The normal and shear stress terms in these equations are functions of the velocity gradients, as given by Eqs. (2.43a, b, c, d, e and f).

(5) The system contains five equations in terms of six unknown flow-field variables, ρ, p, u, v, w, e. In aerodynamics, it is generally reasonable to assume the gas is a perfect gas (which assumes that intermolecular forces are negligible—see Refs. [1, 3]. For a perfect gas, the equation of state is

$$p = \rho R T$$

where R is the specific gas constant. This provides a sixth equation, but it also introduces a seventh unknown, namely temperature, T. A seventh equation to close the entire system must be a thermodynamic relation between state variables. For example,

$$e = e(T, p)$$

For a calorically perfect gas (constant specific heats), this relation would be

$$e = c_v T$$

where c_v is the specific heat at constant volume.

(6) In Sect. 2.6, the momentum equations for a viscous flow were identified as the *Navier–Stokes equations*, which is historically accurate. However, in the modern CFD literature, this terminology has been expanded to include the *entire system* of flow equations for the solution of a viscous flow—continuity and energy as well as momentum. Therefore, when the computational fluid dynamic literature discusses a numerical solution to the 'complete Navier–Stokes equations', it is usually referring to a numerical solution of the *complete system of equations*, say for example Eqs. (2.27), (2.42a, b, c, d, e and c) and (2.64). In this sense, in the CFD literature, a 'Navier–Stokes solution' simply means a solution of a *viscous flow* problem using the *full governing equations*.

2.8.4 Boundary Conditions

The equations given above govern the flow of a fluid. They are the same equations whether the flow is, for example, over a Boeing 747, through a subsonic wind tunnel or past a windmill. However, the flow fields are quite *different* for these cases, although the governing equations are the *same*. Why? Where does the difference enter? The answer is through the *boundary conditions*, which are quite different for each of the above examples. The boundary conditions, and sometimes the initial conditions, dictate the particular solutions to be obtained from the governing equations. For a viscous fluid, the boundary condition on a surface assumes no relative velocity between the surface and the gas immediately at the surface. This is called the *no-slip* condition. If the surface is stationary, with the flow moving past it, then

$$u = v = w = 0 \quad \text{at the surface (for a viscous flow)}$$

For an inviscid fluid, the flow slips over the surface (there is no friction to promote its 'sticking' to the surface); hence, at the surface, the flow must be *tangent* to the surface.

$$\vec{V} \cdot \vec{n} = 0 \quad \text{at the surface (for an inviscid flow)}$$

where $\vec{n}$ is a unit vector perpendicular to the surface. The boundary conditions elsewhere in the flow depend on the type of problem being considered, and usually pertain to inflow and outflow boundaries at a finite distance from the surfaces, or an 'infinity' boundary condition infinitely far from the surfaces.

The boundary conditions discussed above are *physical boundary conditions* imposed by nature. In computational fluid dynamics we have an additional concern, namely, the *proper numerical implementation of the boundary conditions*. In the same sense as the real flow field is dictated by the physical boundary conditions, the computed flow field is driven by the numerical boundary conditions. The subject of proper and accurate boundary conditions in CFD is very important, and is the subject of much current CFD research. We will return to this matter at appropriate stages in these chapters.

2.9 Forms of the Governing Equations Particularly Suited for CFD: Comments on the Conservation Form

We have already noted that all the previous equations in *conservation form* have a divergence term on the left-hand side. These terms involve the divergence of the *flux* of some physical quantity, such as:

(From Eq. (2.27)): $\rho\vec{V}$ — mass flux

(From Eq. (2.42b)): $\rho u\vec{V}$ —flux of x-component of momentum

(From Eq. (2.42b)): $\rho v\vec{V}$ —flux of y-component of momentum

(From Eq. (2.42c)): $\rho w \vec{V}$ —flux of z-component of momentum
(From Eq. (2.62)): $\rho e \vec{V}$ — flux of internal energy
(From Eq. (2.64)): $\rho \left(e + V^2/2\right) \vec{V}$ — flux of total energy

Recall that the conservation form of the equations was obtained *directly* from a control volume that was *fixed in space*, rather than moving with the fluid. When the volume is fixed in space, we are concerned with the *flux* of mass, momentum and energy into and out of the volume. In this case, the *fluxes* themselves become important dependent variables in the equations, rather than just the primitive variables such as p, ρ, $\vec{V}$, etc.

Let us pursue this idea further. Examine the *conservation* form of *all* the governing equations—continuity, momentum and energy. Note that they all have the same generic form, given by

$$\frac{\partial U}{\partial t} + \frac{\partial F}{\partial x} + \frac{\partial G}{\partial y} + \frac{\partial H}{\partial z} = J \qquad (2.65)$$

Equation (2.65) can represent the *entire system* of governing equations in conservation form if U, F, G, H and J are interpreted as column vectors, given by

$$U = \begin{Bmatrix} \rho \\ \rho u \\ \rho v \\ \rho w \\ \rho(e + V^2/2) \end{Bmatrix}$$

$$F = \begin{Bmatrix} \rho u \\ \rho u^2 + p - \tau_{xx} \\ \rho vu - \tau_{xy} \\ \rho wu - \tau_{xz} \\ \rho(e + V^2/2)u + pu - k\dfrac{\partial T}{\partial x} - u\tau_{xx} - v\tau_{xy} - w\tau_{xz} \end{Bmatrix}$$

$$G = \begin{Bmatrix} \rho v \\ \rho uv - \tau_{yx} \\ \rho v^2 + p - \tau_{yy} \\ \rho wv - \tau_{yz} \\ \rho(e + V^2/2)v + pv - k\dfrac{\partial T}{\partial y} - u\tau_{yx} - v\tau_{yy} - w\tau_{yz} \end{Bmatrix}$$

$$H = \begin{Bmatrix} \rho w \\ \rho uw - \tau_{zx} \\ \rho vw - \tau_{zy} \\ \rho w^2 + p - \tau_{zz} \\ \rho(e + V^2/2)w + pw - k\dfrac{\partial T}{\partial z} - u\tau_{zx} - v\tau_{zy} - w\tau_{zz} \end{Bmatrix}$$

$$J = \left\{ \begin{array}{l} 0 \\ \rho f_x \\ \rho f_y \\ \rho f_z \\ \rho(u f_x + v f_y + w f_z) + \rho \dot{q} \end{array} \right\}$$

In Eq. (2.65), the column vectors F, G, and H are called the flux terms (or flux vectors), and J represents a 'source term' (which is zero if body forces are negligible). For an unsteady problem, U is called the solution vector because the elements in U (ρ, ρu, ρv, etc.) are the dependent variables which are usually solved numerically in steps of time. Please note that, in this formalism, it is the elements of U that are obtained computationally, i.e. numbers are obtained for the products ρu, ρv, ρw and $\rho(e + V^2/2)$ rather than for the primitive variables u, v, w and e by themselves. Hence, in a computational solution of an unsteady flow problem using Eq. (2.65), the dependent variables are treated as ρ, ρu, ρv, ρw and $\rho(e + V^2/2)$. Of course, once numbers are known for these dependent variables (which includes ρ by itself), obtaining the primitive variables is simple:

$$\rho = \rho$$
$$u = \frac{\rho u}{\rho}$$
$$v = \frac{\rho v}{\rho}$$
$$w = \frac{\rho w}{\rho}$$
$$e = \frac{\rho(e + V^2/2)}{\rho} - \frac{u^2 + v^2 + w^2}{2}$$

For an *inviscid* flow, Eq. (2.65) remains the same, except that the elements of the column vectors are simplified. Examining the conservation form of the inviscid equations summarized in Sect. 2.8.2, we find that

$$U = \left\{ \begin{array}{l} \rho \\ \rho u \\ \rho v \\ \rho w \\ \rho(e + V^2/2) \end{array} \right\} \quad ; \qquad F = \left\{ \begin{array}{l} \rho u \\ \rho u^2 + p \\ \rho u v \\ \rho u w \\ \rho u(e + V^2/2) + p u \end{array} \right\}$$

$$G = \left\{ \begin{array}{l} \rho v \\ \rho u v \\ \rho v^2 + p \\ \rho w v \\ \rho v(e + V^2/2) + p v \end{array} \right\} \quad ; \qquad H = \left\{ \begin{array}{l} \rho w \\ \rho u w \\ \rho v w \\ \rho w^2 + p \\ \rho w(e + V^2/2) + p w \end{array} \right\}$$

$$J = \begin{Bmatrix} 0 \\ \rho f_x \\ \rho f_y \\ \rho f_z \\ \rho(u f_x + v f_y + w f_z) + \rho \dot{q} \end{Bmatrix}$$

For the numerical solution of an unsteady inviscid flow, once again the solution vector is U, and the dependent variables for which numbers are directly obtained are ρ, ρu, ρv, ρw, and $\rho(e + V^2/2)$. For a steady inviscid flow, $\partial U / \partial t = 0$. Frequently, the numerical solution to such problems takes the form of 'marching' techniques; for example, if the solution is being obtained by marching in the x-direction, then Eq. (2.65) can be written as

$$\frac{\partial F}{\partial x} = J - \frac{\partial G}{\partial y} + \frac{\partial H}{\partial z} \tag{2.66}$$

Here, F becomes the 'solutions' vector, and the dependent variables for which numbers are obtained are $\rho u, (\rho u^2 + p), \rho u v, \rho u w$ and $[\rho u(e + V^2/2) + pu]$. From these dependent variables, it is still possible to obtain the primitive variables, although the algebra is more complex than in our previously discussed case (see Ref. [5] for more details).

Notice that the governing equations, when written in the form of Eq. (2.65), have no flow variables outside the single x, y, z and t derivatives. Indeed, the terms in Eq. (2.65) have everything buried inside these derivatives. The flow equations in the form of Eq. (2.65) are said to be in *strong* conservation form. In contrast, examine the form of Eqs. (2.42a, b and c) and (2.64). These equations have a number of x, y and z derivatives explicitly appearing on the right-hand side. These are the *weak* conservation form of the equations.

The form of the governing equations given by Eq. (2.65) is popular in CFD; let us explain why. In flow fields involving shock waves, there are sharp, discontinuous changes in the primitive flow-field variables p, ρ, u, T, etc., across the shocks. Many computations of flows with shocks are designed to have the shock waves appear naturally within the computational space as a direct result of the overall flow-field solution, i.e. as a direct result of the general algorithm, without any special treatment to take care of the shocks themselves. Such approaches are called *shock-capturing methods*. This is in contrast to the alternate approach, where shock waves are *explicitly* introduced into the flow-field solution, the exact Rankine–Hugoniot relations for changes across a shock are used to relate the flow immediately ahead of and behind the shock, and the governing flow equations are used to calculate the remainder of the flow field. This approach is called the *shock-fitting method*. These two different approaches are illustrated in Figs. 2.8 and 2.9. In Fig. 2.8, the computational domain for calculating the supersonic flow over the body extends both upstream and downstream of the nose. The shock wave is allowed to form within the computational domain as a consequence of the general flow-field algorithm,

Fig. 2.8 Mesh for the
shock-capturing approach

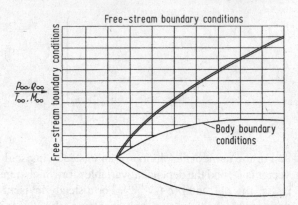

without any special shock relations being introduced. In this manner, the shock
wave is 'captured' within the domain by means of the computational solution of
the governing partial differential equations. Therefore, Fig. 2.8 is an example of the
shock-capturing method. In contrast, Fig. 2.9 illustrates the same flow problem, except that now the computational domain is the flow between the shock and the body.
The shock wave is introduced directly into the solution as an explicit discontinuity,
and the standard oblique shock relations (the Rankine–Hugoniot relations) are used
to fit the freestream supersonic flow ahead of the shock to the flow computed by the
partial differential equations downstream of the shock. Therefore, Fig. 2.9 is an example of the *shock-fitting method*. There are advantages and disadvantages of both
methods. For example, the shock-capturing method is ideal for complex flow problems involving shock waves for which we do not know either the location or number
of shocks. Here, the shocks simply form within the computational domain as nature
would have it. Moreover, this takes place without requiring any special treatment
of the shock within the algorithm, and hence simplifies the computer programming.
However, a disadvantage of this approach is that the shocks are generally smeared
over a number of grid points in the computational mesh, and hence the numerically obtained shock thickness bears no relation what-so-ever to the actual physical
shock thickness, and the *precise* location of the shock discontinuity is uncertain
within a few mesh sizes. In contrast, the advantage of the shock-fitting method is

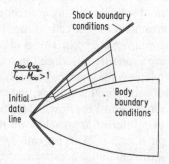

Fig. 2.9 Mesh for the
shock-fitting approach

that the shock is always treated as a discontinuity, and its location is well-defined numerically. However, for a given problem you have to know in advance approximately where to put the shock waves, and how many there are. For complex flows, this can be a distinct disadvantage. Therefore, there are pros and cons associated with both shock-capturing and shock-fitting methods, and both have been employed extensively in CFD. In fact, a combination of these two methods is possible, wherein a shock-capturing approach during the course of the solution is used to predict the formation and approximate location of shocks, and then these shocks are fit with explicit discontinuities midway through the solution. Another combination is to fit shocks explicitly in those parts of a flow field where you know in advance they occur, and to employ a shock-capturing method for the remainder of the flow field in order to generate shocks that you cannot predict in advance.

Again, what does all of this discussion have to do with the conservation form of the governing equations as given by Eq. (2.65)? Simply this. For the *shock-capturing* method, experience has shown that the *conservation form* of the governing equations should be used. When the conservation form is used, the computed flow-field results are generally smooth and stable. However, when the non-conservation form is used for a shock-capturing solution, the computed flow-field results usually exhibit unsatisfactory spatial oscillations (wiggles) upstream and downstream of the shock wave, the shocks may appear in the wrong location and the solution may even become unstable. In contrast, for the shock-fitting method, satisfactory results are usually obtained for either form of the equations—conservation or non-conservation.

Why is the use of the conservation form of the equations so important for the shock-capturing method? The answer can be seen by considering the flow across a normal shock wave, as illustrated in Fig. 2.10. Consider the density distribution across the shock, as sketched in Fig. 2.10(a). Clearly, there is a discontinuous increase in ρ across the shock. If the non-conservation form of the governing equations were used to calculate this flow, where the primary dependent variables are the primitive variables such as ρ and p, then the equations would see a large discontinuity in the dependent variable ρ. This in turn would compound the numerical errors associated with the calculation of ρ. On the other hand, recall the continuity equation for a normal shock wave (see Refs. [1,3]):

$$\rho_1 u_1 = \rho_2 u_2 \tag{2.67}$$

From Eq. (2.67), the *mass flux*, ρu, is *constant* across the shock wave, as illustrated in Fig. 2.10(b). The conservation form of the governing equations uses the product ρu as a dependent variable, and hence the conservation form of the equations see *no* discontinuity in this dependent variable across the shock wave. In turn, the numerical accuracy and stability of the solution should be greatly enhanced. To reinforce this discussion, consider the momentum equation across a normal shock wave [1,3]:

$$p_1 + \rho_1 u_1^2 = p_2 + \rho_2 u_2^2 \tag{2.68}$$

As shown in Fig. 2.10(c), the pressure itself is discontinuous across the shock; however, from Eq. (2.68) the flux variable $(p + \rho u^2)$ is constant across the shock.

Fig. 2.10 Variation of flow
properties through a normal
shock wave

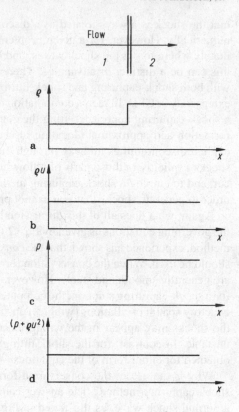

This is illustrated in Fig. 2.10(d). Examining the inviscid flow equations in the conservation form given by Eq. (2.65), we clearly see that the quantity $(p + \rho u^2)$ is one of the dependent variables. Therefore, the conservation form of the equations would see no discontinuity in this dependent variable across the shock. Although this example of the flow across a normal shock wave is somewhat simplistic, it serves to explain why the use of the conservation form of the governing equations are so important for calculations using the shock-capturing method. Because the conservation form uses flux variables as the dependent variables, and because the changes in these flux variables are either zero or small across a shock wave, the numerical quality of a shock-capturing method will be enhanced by the use of the conservation form in contrast to the non-conservation form, which uses the primitive variables as dependent variables.

In summary, the previous discussion is one of the primary reasons why CFD makes a distinction between the two forms of the governing equations—conservation and non-conservation. And this is why we have gone to great lengths in this chapter to derive these different forms, to explain what basic physical models lead to the different forms, and why we should be aware of the differences between the two forms.

References

1. Anderson, John D., Jr., *Fundamentals of Aerodynamics*, 2nd Edition McGraw-Hill, New York, 1991.
2. Liepmann, H.W. and Roshko, A., *Elements of Gasdynamics*, Wiley, New York, 1957.
3. Anderson, J.D., Jr., *Modern Compressible Flow: With Historical Perspective*, 2nd Edition McGraw-Hill, New York, 1990.
4. Bird, R.B., Stewart, W.E. and Lightfoot, E.N., *Transport Phenomena*, 2nd edition, Wiley, 2004.
5. Kutler, P., 'Computation of Three-Dimensional, Inviscid Supersonic Flows,' in H.J. Wirz (ed.), *Progress in Numerical Fluid Dynamics*, Springer-Verlag, Berlin, 1975, pp. 293–374.

Chapter 3
Incompressible Inviscid Flows: Source and Vortex Panel Methods

J.D. Anderson, Jr.

3.1 Introduction

In the present chapter we will consider the numerical analysis of incompressible inviscid flows. In principle, the finite–difference approach discussed later can be used to solve such flows, but there are other approaches which are usually more appropriate solutions for inviscid, incompressible flow. This chapter discusses one such approach, namely, the use of source and vortex panels. *Panel methods*, since the late 1960s, have become standard aerodynamic tools in the aerospace industry. Panel methods are numerical methods which require a high-speed digital computer for their implementation; therefore we include panel methods as part of the overall structure of computational fluid dynamics. For this reason, it is appropriate to spend some time discussing panel methods in our introduction to CFD. Frequently in the literature panel methods will be classified under the title of *computational aerodynamics* which has a slightly more specialized connotation than the more general meaning of CFD. In this author's opinion, computational aerodynamics is simply a sub-speciality under the more general heading of computational fluid dynamics.

3.2 Some Basic Aspects of Incompressible, Inviscid Flow

In this section we briefly review some fundamental aspects of incompressible, inviscid flow. For those readers who are familiar with such flows, this section should serve as a short refresher; for those who have not studied such flows, hopefully this section will give enough background to understand the following sections on the source and vortex panel methods.

Incompressible flow is constant density flow, i.e. ρ = constant. Visualize a fluid element of fixed mass moving along a streamline in an incompressible flow. Because its density is constant, then the *volume* of the fluid element is also constant. In

J.D. Anderson, Jr.
National Air and Space Museum, Smithsonian Institution, Washington, DC
e-mail: AndersonJA@si.edu

J.F. Wendt (ed.), *Computational Fluid Dynamics*, 3rd ed.,
© Springer-Verlag Berlin Heidelberg 2009

Sect. 2.4, we related $\nabla \vec{V}$ to the time rate of change of the volume of a fluid element, per unit volume; see Eq. (2.14). Since the volume is constant for a fluid element in incompressible flow, we have from Eq. (2.14) that

$$\nabla \cdot \vec{V} = 0 \tag{3.1}$$

Furthermore, if the fluid element does not rotate as it moves along the streamline, i.e. if its motion is translational only, then the flow is called *irrotational flow*. For such flow, the velocity can be expressed as the gradient of a scalar function called the velocity potential, denoted by ϕ. (For details, see Ref. [1]).

$$\vec{V} = \nabla \phi \tag{3.2}$$

Combining Eqs. (3.1) and (3.2), we have

$$\nabla \cdot \nabla \phi = 0$$

or,

$$\boxed{\nabla^2 \phi = 0} \tag{3.3}$$

Equation (3.3) is *Laplace's* equation—one of the most famous and extensively studied equations in mathematical physics. From Eq. (3.3), we see that inviscid, irrotational, incompressible flow (sometimes called 'potential flow') is governed by Laplace's equation.

Laplace's equation is linear, and hence any number of particular solutions to Eq. (3.3) can be added together to obtain another solution. This establishes a basic philosophy of the solution of incompressible flows, namely, that a *complicated flow pattern for an irrotational, incompressible flow can be synthesized by adding together a number of elementary flows which are also irrotational and incompressible.*

Let us examine a few of the important *elementary* flows which satisfy Laplace's equation.

3.2.1 Uniform Flow

Consider a uniform flow with velocity V_∞ moving in the x-direction, as sketched in Fig. 3.1. This flow is irrotational, and a solution of Laplace's equation for uniform flow yields:

$$\boxed{\phi = V_\infty x} \tag{3.4}$$

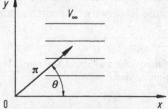

Fig. 3.1 Uniform flow

In polar coordinates, (r, θ), Eq. (3.4) can be expressed as

$$\phi = V_\infty r \cos\theta \qquad (3.5)$$

3.2.2 Source Flow

Consider a flow with straight streamlines emanating from a point, where the velocity along each streamline varies inversely with distance from the point, as shown in Fig. 3.2. Such flow is called *source flow*. This flow is also irrotational, and a solution of Laplace's equation yields (see Ref. [1])

$$\phi = \frac{\Lambda}{2\pi} \ln r \qquad (3.6)$$

where Λ is defined as the *source strength*; Λ is physically the rate of volume flow from the source, per unit depth perpendicular to the page in Fig. 3.2. If Λ is negative, we have *sink flow*, which is the opposite of source flow. In Fig. 3.2, point 0 is the origin of the radial streamlines. We can visualize that point 0 is a *point source or sink* that *induces* the radial flow about it; in this interpretation, the point source or sink is a *singularity* in the flow field (because V becomes infinite there). We can also visualize that point 0 in Fig. 3.2 is simply one point formed by the intersection of the plane of the paper and a *line* perpendicular to the paper. The line perpendicular to the paper is a *line source*, with strength Λ per unit length.

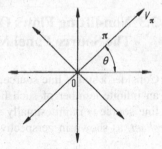

Fig. 3.2 Source flow

3.2.3 Vortex Flow

Consider a flow where all the streamlines are concentric circles about a given point, where the velocity along each streamline is inversely proportional to the distance from the centre, as sketched in Fig. 3.3. Such flow is called *vortex flow*. This flow is irrotational, and a solution of Laplace's equation yields (see Ref. [1])

$$\phi = -\frac{\Gamma}{2\pi}\theta \qquad (3.7)$$

Fig. 3.3 Vortex flow

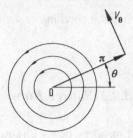

where Γ is the strength of the vortex. In Fig. 3.3, point 0 can be visualized as a *point vortex* that *induces* the circular flow about it; in this interpretation, the point vortex is a *singularity* in the flow field (because V becomes infinite there). We can also visualize that point 0 in Fig. 3.3 is simply one point formed by the intersection of the plane of the paper and a line perpendicular to the paper. This line is called a vortex *filament*, of strength Γ. The strength Γ is the *circulation* around the vortex filament, where circulation is defined as

$$\Gamma = -\oint \vec{V} \cdot d\vec{s}$$

In the above, the line integral of the velocity component tangent to a curve of elemental length ds is taken around a closed curve. This is the general definition of circulation. For a vortex filament, the above expression for Γ (where the closed curves encloses and contains the point vortex) is defined as the vortex strength.

3.3 Non-lifting Flows Over Arbitrary Two-Dimensional Bodies: The Source Panel Method

Consider a single line source, as discussed Sect. 3.2.2. Now imagine that we have an infinite number of such line sources side-by-side, where the strength of each line source is infinitesimally small. These side-by-side line sources form a *source sheet*, as shown in perspective in the upper left of Fig. 3.4. If we look along the

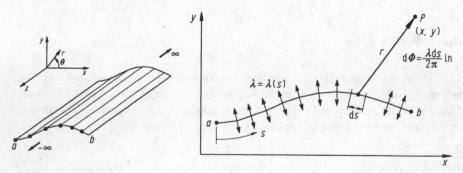

Fig. 3.4 Source sheet

series of line sources (looking along the z-axis in Fig. 3.4), the source sheet will appear as sketched at the lower right of Fig. 3.4. Here, we are looking at an edge view of the sheet; the line sources are all perpendicular to the page. Let s be the distance measured along the source sheet in the edge view. Define $\lambda = \lambda(s)$ to be the *source strength per unit length along s*. [To keep things in perspective, recall from Sect. 3.22 that the strength of a single line source Λ was defined as the volume flow rate per unit depth, i.e. per unit length in the z-direction. Typical units for Λ are square meters per second or square feet per second. In turn, the strength of a source sheet $\lambda(s)$ is the volume flow rate per unit depth (in the z-direction) and per unit length (in the s direction). Typical units for λ are meters per second or feet per second.] Therefore, the strength of an infinitesimal portion ds of the sheet, as shown in Fig. 3.4, is $\lambda\,ds$. This small section of the source sheet can be treated as a distinct source of strength $\lambda\,ds$. Now consider point P in the flow, located a distance r from ds; the cartesian coordinates of P are (x, y). The small section of the source sheet of strength $\lambda\,ds$ induces an infinitesimally small potential, $d\phi$, at point P. From Eq. (3.6), $d\phi$ is given by

$$d\phi = \frac{\lambda\,ds}{2\pi}\ln r \qquad\qquad (3.8)$$

The complete velocity potential at point P, induced by the entire source sheet from a to b, is obtained by integrating Eq. (3.8):

$$\phi(x,y) = \int_a^b \frac{\lambda\,ds}{2\pi}\ln r \qquad\qquad (3.9)$$

Note that, in general, $\lambda(s)$ can change from positive to negative along the sheet, i.e. the 'source' sheet is really a combination of line sources and line sinks.

Next, consider a given body of arbitrary shape in a flow with free-stream velocity V_∞, as shown in Fig. 3.5. Let us cover the surface of the prescribed body with a source sheet, where the strength $\lambda(s)$ varies in such a fashion that the combined action of the uniform flow and the source sheet makes the airfoil surface a streamline of the flow. Our problem now becomes one of finding the appropriate $\lambda(s)$. The solution of this problem is carried out numerically, as follows.

Let us approximate the source sheet by a series of straight panels, as shown in Fig. 3.6. Moreover, let the source strength λ per unit length be constant over a given panel, but allow it to vary from one panel to the next. That is, if there is a total of n panels, the source panel strengths per unit length are $\lambda_1, \lambda_2, \ldots, \lambda_j, \ldots, \lambda_n$. These

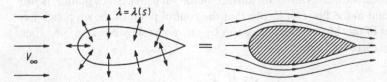

Fig. 3.5 Superposition of a uniform flow and a source sheet on a body of given shape, to produce the flow over the body

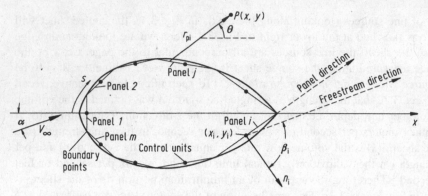

Fig. 3.6 Source panel distribution over the surface of a body of arbitrary shape

panel strengths are unknown; the main thrust of the panel technique is to solve for λ_j, $j = 1$ to n, such that the body surface becomes a streamline of the flow. This boundary condition is imposed numerically by defining the midpoint of each panel to be a *control point* and by determining the λ_j's such that the normal component of the flow velocity is zero at each control point. Let us now quantify this strategy.

Let P be a point located at (x, y) in the flow, and let r_{pj} be the distance from any point on the jth panel to P, as shown in Fig. 3.6. The velocity potential induced at P due to the jth panel $\Delta\phi_j$ is, from Eq. (3.9),

$$\Delta\phi_j = \frac{\lambda_j}{2\pi} \int_j \ln r_{pj} \, ds_j \tag{3.10}$$

In Eq. (3.10), λ_j is constant over the jth panel, and the integral is taken over the jth panel only. In turn, the *potential at P due to all* the panels is Eq. (3.10) summed over all the panels.

$$\phi(P) = \sum_{j=1}^{n} \Delta\phi_j = \sum_{j=1}^{n} \frac{\lambda_j}{2\pi} \int_j \ln r_{pj} \, ds_j \tag{3.11}$$

In Eq. (3.11), the distance r_{pj} is given by

$$r_{pj} = \sqrt{(x - x_j)^2 + (y - y_j)^2} \tag{3.12}$$

where (x_j, y_j) are coordinates along the surface of the jth panel. Since point P is just an arbitrary point in the flow, let us put P at the control point of the ith panel. Let the coordinates of this control point be given by (x_i, y_i) as shown in Fig. 3.6. Then Eqs. (3.11) and (3.12) become

$$\phi(x_i, y_i) = \sum_{j=1}^{n} \frac{\lambda_j}{2\pi} \int_j \ln r_{ij} \, ds_j \tag{3.13}$$

and

$$r_{ij} = \sqrt{(x_i - x_j)^2 + (y_i - y_j)^2} \tag{3.14}$$

Equation (3.13) is physically the contribution of *all* the panels to the potential at the control point of the *i*th panel.

Recall that the boundary condition is applied at the control points, i.e. the normal component of the flow velocity is zero at the control points. To evaluate this component, first consider the component of free-stream velocity perpendicular to the panel. Let $\vec{n}_i$ be the unit vector normal to the *i*th panel, directed out of the body as shown in Fig. 3.6. Also, note that the slope of the *i*th panel is $(dy/dx)_i$. In general, the free-stream velocity will be at some incidence angle α to the *x* axis, as shown in Fig. 3.6. Therefore, inspection of the geometry of Fig. 3.6 reveals that the component of V_∞ normal to the *i*th panel is

$$V_{\infty,n} = \vec{V}_\infty \cdot \vec{n}_i = V_\infty \cos\beta_i \tag{3.15}$$

where β_i is the angle between $\vec{V}_\infty$ and $\vec{n}_i$. Note that $V_{\infty,n}$ is positive when directed away from the body, and negative when directed toward the body.

The normal component of velocity induced at (x_i, y_i) by the source panels is, from Eq. (3.13),

$$V_n = \frac{\partial}{\partial n_i}[\phi(x_i, y_i)] \tag{3.16}$$

where the derivative is taken in the direction of the outward unit normal vector, and hence again, V_n is positive when directed away from the body. When the derivative in Eq. (3.16) is carried out, r_{ij} appears in the denominator. Consequently, a singular point arises on the *i*th panel because when $j = i$, at the control point itself $r_{ij} = 0$. It can be shown that when $j = i$, the contribution to the derivative is simply $\lambda_i/2$. Hence, Eq. (3.16) combined with Eq. (3.13) becomes

$$V_n = \frac{\lambda_i}{2} + \sum_{\substack{j=1 \\ (j \neq i)}}^{n} \frac{\lambda_j}{2\pi} \int_j \frac{\partial}{\partial n_i}(\ln r_{ij})\, ds_j \tag{3.17}$$

In Eq. (3.17), the first term $\lambda_i/2$ is the normal velocity induced at the *i*th control point by the *i*th panel itself, and the summation is the normal velocity induced at the *i*th control point by all the other panels.

The normal component of the flow velocity at the *i*th control point is the sum of that due to the freestream (Eq. (3.15)) and that due to the source panels (Eq. (3.17)). The boundary condition states that this sum must be zero.

$$V_{\infty,n} + V_n = 0 \tag{3.18}$$

Substituting Eqs. (3.15) and (3.17) into Eq. (3.18), we obtain

$$\frac{\lambda_i}{2} + \sum_{\substack{j=1 \\ (j \neq i)}}^{n} \frac{\lambda_j}{2\pi} \int_j \frac{\partial}{\partial n_i}(\ln r_{ij})\, ds_j + V_\infty \cos\beta_i = 0 \tag{3.19}$$

Equation (3.19) is the crux of the source panel method. The values of the integrals in Eq. (3.19) depend simply on the panel geometry; they are not properties of the flow. Let I_{ij} be the value of this integral when the control point is on the ith panel and the integral is over the jth panel. Then, Eq. (3.19) can be written as

$$\frac{\lambda_i}{2} + \sum_{\substack{j=1 \\ (j \neq i)}}^{n} \frac{\lambda_j}{2\pi} I_{i,j} + V_\infty \cos\beta_i = 0 \tag{3.20}$$

Equation (3.20) is a linear *algebraic* equation with n unknowns $\lambda_1, \lambda_2, \ldots, \lambda_n$. It represents the flow boundary condition evaluated at the control points of the ith panel. Now apply the boundary condition to the control points of *all* the panels, i.e. in Eq. (3.20), let $i = 1, 2, \ldots, n$. The results will be a system of n linear algebraic equations with n unknowns $(\lambda_1, \lambda_2, \ldots, \lambda_n)$, which can be solved simultaneously by conventional numerical methods.

Look what has happened! After solving the system of equations represented by Eq. (3.20) with $i = 1, 2, \ldots, n$, we now have the distribution of source panel strengths which, in an approximate fashion, cause the body surface in Fig. 3.6 to be a streamline of the flow. This approximation can be made more accurate by increasing the number of panels, hence more closely representing the source sheet of continuously varying strength $\lambda(s)$ shown in Fig. 3.5. Indeed, the accuracy of the source panel method is amazingly good; a circular cylinder can be accurately represented by as few as 8 panels, and most airfoil shapes by 50–100 panels. (For an airfoil, it is desirable to cover the leading-edge region with a number small panels to accurately represent the rapid surface curvature and to use larger panels over the relatively flat portions of the body. Note that in general, all the panels in Fig. 3.6 can be different lengths.)

Once the λ_i's $(i = 1, 2, \ldots, n)$ are obtained, the velocity *tangent* to the surface at each control point can be calculated as follows. Let s be the distance along the body surface, measured positive from front to rear, as shown in Fig. 3.6. The component of freestream velocity tangent to the surface is

$$V_{\infty,s} = V_\infty \sin\beta_i \tag{3.21}$$

The tangential velocity V_s at the control point of the ith panel induced by all the panels is obtained by differentiating Eq. (3.13) with respect to s.

$$V_s = \frac{\partial\phi}{\partial s} = \sum_{j=1}^{n} \frac{\lambda_j}{2\pi} \int_j \frac{\partial}{\partial s} (\ln r_{ij}) \, ds_j \tag{3.22}$$

[The tangential velocity on a flat source panel induced by the panel itself is zero; hence, in Eq. (3.22), the term corresponding to $j = i$ is zero. This is easily seen by intuition, because the panel can only emit volume flow from its surface in a direction perpendicular to the panel itself.] The total surface velocity at the ith control point V_i is the sum of the contribution from the freestream [Eq. (3.21)] and from the source panels [Eq. (3.22)].

$$V_i = V_{\infty,s} + V_s = V_\infty \sin\beta_i + \sum_{j=1}^{n} \frac{\lambda_j}{2\pi} \int_j \frac{\partial}{\partial s}(\ln r_{ij}) \, ds_j \qquad (3.23)$$

In turn, the pressure coefficient at the ith control point is obtained from Bernoulli's equation as (see Ref. [1])

$$C_{p,i} = 1 - \left(\frac{V_i}{V_\infty}\right)^2$$

In this fashion, the source panel method gives the pressure distribution over the surface of a non-lifting body of arbitrary shape.

When you carry out a source panel solution as described above, the accuracy of your results can be tested as follows. Let S_j be the length of the jth panel. Recall that λ_j is the strength of the jth panel *per unit length*. Hence, the strength of the jth panel itself is $\lambda_j S_j$. For a closed body, such as in Fig. 3.6, the *sum* of all the source and sink strengths must be zero, or else the body itself would be adding or absorbing mass from the flow—an impossible situation for the case we are considering here. Hence, the values of the λ_j's obtained above should obey the relation

$$\sum_{j=1}^{n} \lambda_j S_j = 0 \qquad (3.24)$$

Equation (3.24) provides an independent check on the accuracy of the numerical results.

Let us now demonstrate the above technique with an example; we will calculate the pressure distribution around a circular cylinder using the source panel technique. We choose to cover the body with eight panels of equal length, as shown in Fig. 3.7. This choice is arbitrary; however, experience has shown that, for the case of a circular cylinder, the arrangement shown in Fig. 3.7 provides sufficient accuracy. The panels are numbered from 1 to 8, and the control points are shown by the dots in the centre of each panel.

Let us evaluate the integrals $I_{i,j}$ which appear in Eq. (3.20). Consider Fig. 3.8, which illustrates two arbitrarily chosen panels. In Fig. 3.8, (x_i, y_i) are the coordinates

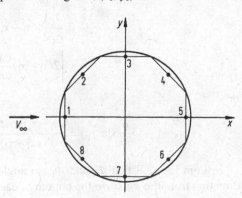

Fig. 3.7 Source panel distribution around a circular cylinder

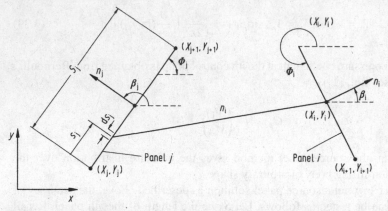

Fig. 3.8 Geometry required for the evaluation of I_{ij}

of the control point of the ith panel, and (x_j, y_j) are the running coordinates over the entire jth panel. The coordinates of the boundary points for the ith panel are (X_i, Y_i) and (X_{i+1}, Y_{i+1}); similarly, the coordinates of the boundary points for the jth panel are (X_j, Y_j) and (X_{j+1}, Y_{j+1}). In this problem, $\vec{V}_\infty$ is in the x-direction; hence, the angles between the x-axis and the unit vectors n_i and n_j are β_i and β_j, respectively. Note that in general both β_i and β_j vary from 0 to 2π. Recall that the integral $I_{i,j}$ is evaluated at the ith control point and the integral is taken over the complete jth panel.

$$I_{i,j} = \int_j \frac{\partial}{\partial n_i} (\ln r_{ij}) \, ds_j \qquad (3.25)$$

Since

$$r_{ij} = \sqrt{(x_i - x_j)^2 + (y_i - y_j)^2}$$

then

$$\frac{\partial}{\partial n_i}(\ln r_{ij}) = \frac{1}{r_{ij}} \frac{\partial r_{ij}}{\partial n_i}$$

$$= \frac{1}{r_{ij}} \frac{1}{2} [(x_i - x_j)^2 + (y_i - y_j)^2]^{-1/2}.$$

$$\left[2(x_i - x_j)\frac{dx_i}{dn_i} + 2(y_i - y_j)\frac{dy_i}{dn_i} \right]$$

or

$$\frac{\partial}{\partial n_i}(\ln r_{ij}) = \frac{(x_i - x_j)\cos\beta_i + (y_i - y_j)\sin\beta_i}{(x_i - x_j)^2 + (y_i - y_j)^2} \qquad (3.26)$$

Note in Fig. 3.8 that Φ_i and Φ_j are angles measured in the counter-clockwise direction from the x-axis to the bottom of each panel. From this geometry,

$$\beta_i = \Phi_i + \frac{\pi}{2}$$

hence,

$$\sin\beta_i = \cos\Phi_i \tag{3.27}$$
$$\cos\beta_i = -\sin\Phi_i \tag{3.28}$$

Also from the geometry of Fig. 3.8, we have

$$x_j = X_j + s_j \cos\Phi_j \tag{3.29}$$

and

$$y_j = Y_j + s_j \sin\Phi_j \tag{3.30}$$

Substituting Eqs. (3.26), (3.27), (3.28), (3.29) and (3.30) into Eq. (3.25), we obtain.

$$I_{i,j} = \int_0^{s_j} \frac{Cs_j + D}{s_j^2 + 2As_j + B} ds_j \tag{3.31}$$

where

$$A = -(x_i - X_j)\cos\Phi_j - (y_i - Y_j)\sin\Phi_j$$
$$B = (x_i - X_j)^2 + (y_i - Y_j)^2$$
$$C = \sin(\Phi_i - \Phi_j)$$
$$D = (y_i - Y_j)\cos\Phi_i - (x_i - X_j)\sin\Phi_i$$
$$S_j = \sqrt{(X_{j+1} - X_j)^2 + (Y_{j+1} - Y_j)^2}$$

Letting

$$E = \sqrt{B - A^2} = (x_i - X_j)\sin\phi_j - (y_i - Y_j)\cos\phi_j$$

we obtain an expression for Eq. (3.31) from any standard table of integrals,

$$I_{i,j} = \frac{C}{2} \ln\left(\frac{S_j^2 + 2AS_j + B}{B}\right) + \frac{D - AC}{E}\left(\tan^{-1}\frac{S_j + A}{E} - \tan^{-1}\frac{A}{E}\right) \tag{3.32}$$

Equation (3.32) is a general expression for two arbitrarily oriented panels; it is not restricted to the case of a circular cylinder.

We now apply Eq. (3.32) to the circular cylinder shown in Fig. 3.7. For purposes of illustration, let us choose panel 4 as the ith panel and panel 2 as the jth panel, i.e. let us calculate $I_{4,2}$. From the geometry of Fig. 3.7, assuming a unit radius for the cylinder, we see that

$$X_j = -0.9239 \quad X_{j+1} = -0.3827 \quad Y_j = 0.3827$$
$$Y_{j+1} = 0.9239 \quad \Phi_i = 315° \quad \Phi_j = 45°$$
$$x_i = 0.6533 \quad y_i = 0.6533$$

Hence, substituting these numbers into the above formulas, we obtain

$$A = -1.3065 \quad B = 2.5607 \quad C = -1 \quad D = 1.3065$$
$$S_j = 0.7654 \quad E = 0.9239$$

Inserting the above values into Eq. (3.32), we obtain

$$I_{4,2} = 0.4018$$

Return to Figs. 3.7 and 3.8. If we now choose panel 1 as the jth panel, keeping panel 4 as the ith panel, we obtain, by means of a similar calculation, $I_{4,1} = 0.4074$. Similarly, $I_{4,3} = 0.3528$, $I_{4,5} = 0.3528$, $I_{4,6} = 0.4018$, $I_{4,7} = 0.4074$, and $I_{4,8} = 0.4084$.

Return to Eq. (3.20), which is evaluated for the ith panel. Written for panel 4, Eq. (3.20) becomes (after multiplying each term by 2 and noting that $\beta_i = 45°$ for panel 4)

$$0.4074\lambda_1 + 0.4018\lambda_2 + 0.3528\lambda_3 + \pi\lambda_4 + 0.3528\lambda_5 + 0.4018\lambda_6$$
$$+ 0.4074\lambda_7 + 0.4084\lambda_8 = -0.7071 \, 2\pi V_\infty \tag{3.33}$$

Equation (3.33) is a linear algebraic equation in terms of the eight unknowns, $\lambda_1, \lambda_2, \ldots, \lambda_8$. If we now evaluate Eq. (3.20) for each of the seven other panels, we obtain a total of eight equations, including Eq. (3.33), which can be solved simultaneously for the eight unknown λ's. The results are

$$\lambda_1/2\pi V_\infty = 0.3765 \qquad \lambda_2/2\pi V_\infty = 0.2662 \qquad \lambda_3/2\pi V_\infty = 0$$
$$\lambda_4/2\pi V_\infty = -0.2662 \qquad \lambda_5/2\pi V_\infty = -0.3765 \qquad \lambda_6/2\pi V_\infty = -0.2662$$
$$\lambda_7/2\pi V_\infty = 0 \qquad \lambda_8/2\pi V_\infty = 0.2662$$

Note the symmetrical distribution of the λ's, which is to be expected for the non-lifting circular cylinder. Also, as a check on the above solution, return to Eq. (3.24). Since each panel in Fig. 3.7 has the same length, Eq. (3.24) can be written simply as

$$\sum_{j=1}^{n} \lambda_j = 0$$

Substituting the values for the λ's obtained above into Eq. (3.24), we see that the equation is identically satisfied.

The velocity at the control point of the ith panel can be obtained from Eq. (3.23). In that equation, the integral over the jth panel is a geometric quantity which is evaluated in a similar manner as before. The result is

$$\int_j \frac{\partial}{\partial s}(\ln r_{ij}) \, ds_j = \frac{D - AC}{E} \ln \frac{S_j^2 + 2AS_j + B}{B}$$
$$- C\left(\tan^{-1}\frac{S_j + A}{E} - \tan^{-1}\frac{A}{E}\right) \tag{3.34}$$

Fig. 3.9 Pressure distribution
over a circular cylinder;
comparison of the source
panel results and theory

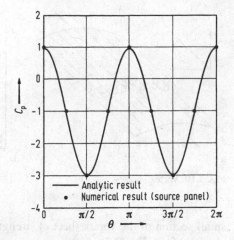

With the integrals in Eq. (3.23) evaluated by Eq. (3.34), and with the values for
λ_1, λ_2,..., λ_8 obtained above inserted into Eq. (3.23), we obtain the velocities
V_1, V_2,..., V_8. In turn, the pressure coefficients $C_{p,1}$, $C_{p,2}$,..., $C_{p,8}$ are obtained
directly from

$$C_{p,i} = 1 - \left(\frac{V_i}{V_\infty}\right)^2$$

Results for the pressure coefficients obtained from this calculation are compared
with the exact analytical result in Fig. 3.9. Amazingly enough, in spite of the relatively crude panelling shown in Fig. 3.7, the numerical pressure coefficient results
are excellent.

3.4 Lifting Flows Over Arbitrary Two-Dimensional Bodies: The Vortex Panel Method

In Sect. 3.3 the concept of a source sheet was introduced. In the present section,
we introduce the analogous concept of a *vortex sheet*. Consider the straight vortex
filament discussed in Sect. 3.2.2. Now imagine an infinite number of straight vortex
filaments side by side, where the strength of each filament is infinitesimally small.
These side-by-side vortex filaments form a *vortex sheet*, as shown in perspective in
the upper left of Fig. 3.10. If we look along the series of vortex filaments (looking
along the y-axis in Fig. 3.10), the vortex sheet will appear as sketched at the lower
right of Fig. 3.10. Here, we are looking at an edge view of the sheet; the vortex
filaments are all perpendicular to the page. Let s be the distance measured along the
vortex sheet in the edge view. Define $\gamma = \gamma(s)$ as the strength of the vortex sheet,
per unit length along s. Thus, the strength of an infinitesimal portion ds of the sheet
is γ ds. This small section of the vortex sheet can be treated as a distinct vortex of
strength γ ds. Now consider point P in the flow, located a distance r from ds. The

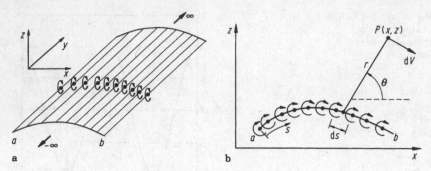

Fig. 3.10 Vortex sheet

small section of the vortex sheet of strength $\gamma \, ds$ induces a velocity potential at P, obtained from Eq. (3.7) as

$$d\Phi = -\frac{\gamma \, ds}{2\pi} \theta \tag{3.35}$$

The velocity potential at P due to the entire vortex sheet from a to b is

$$\Phi = -\frac{1}{2\pi} \int_a^b \theta \gamma \, ds \tag{3.36}$$

In addition, the circulation around the vortex sheet in Fig. 3.10 is the sum of the strengths of the elemental vortices, i.e.

$$\Gamma = \int_a^b \gamma \, ds \tag{3.37}$$

Another property of a vortex sheet is that the component of flow velocity tangential to the sheet experiences a discontinuous change across the sheet, given by

$$\gamma = u_1 - u_2 \tag{3.38}$$

where u_1 and u_2 are the tangential velocities just above and below the sheet respectively. (See Ref. [1] for a derivation of this result). Equation (3.38) is used to demonstrate that, for flow over an airfoil, the value of γ is zero at the trailing edge of the airfoil. This condition, namely

$$\gamma_{TE} = 0 \tag{3.39}$$

is one form of the *Kutta condition* which fixes the precise value of the circulation around an airfoil with a sharp trailing edge. Finally we note that the circulation around the sheet is related to the lift force on the sheet through the Kutta–Joukowski theorem:

$$L = \rho_\infty V_\infty \Gamma \tag{3.40}$$

Fig. 3.11 Simulation of an
arbitrary airfoil by distributing
a vortex sheet over the airfoil
surface

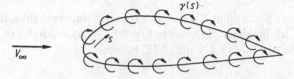

Clearly, a finite value of circulation is required for the existence of lift. In the
present section, we will see that the ultimate goal of the vortex panel method applied
to a given body is to calculate the amount of circulation, and hence obtain the lift on
the body from Eq. (3.40).

With the above in mind, consider an arbitrary two-dimensional body, such as
sketched in Fig. 3.11. Let us wrap a vortex sheet over the complete surface of the
body, as shown in Fig. 3.11. We wish to find $\gamma(s)$ such that the body surface becomes
a streamline of the flow. There exists no closed-form analytical solution for $\gamma(s)$;
rather, the solution must be obtained numerically. This is the purpose of the vortex
panel method.

Let us approximate the vortex sheet shown in Fig. 3.11 by a series of straight
panels, as shown earlier in Fig. 3.6. (In Sect. 3.3, Fig. 3.6 was used to discuss source
panels; here, we use the same sketch for our discussion of vortex panels.) Let the
vortex strength $\gamma(s)$ per unit length be constant over a given panel, but allow it to
vary from one panel to the next. That is, for the n panels shown in Fig. 3.6, the vortex
panel strengths per unit length are $\gamma_1, \gamma_2, \ldots, \gamma_j, \ldots, \gamma_n$. These panel strengths are
unknowns; the main thrust of the panel technique is to solve for γ_j, $j = 1$ to n,
such that the body surface becomes a streamline of the flow and such that the Kutta
condition is satisfied. As explained in Sect. 3.3, the midpoint of each panel is a
control point at which the boundary condition is applied, i.e. at each control point,
the normal component of the flow velocity is zero.

Let P be a point located at (x, y) in the flow, and let r_{pj} be the distance from any
point on the jth panel to P, as shown in Fig. 3.6. The radius r_{pj} makes the angle θ_{pj}
with respect to the x-axis. The velocity potential induced at P due to the jth panel,
$\Delta\phi_j$, is, from Eq. (3.35),

$$\Delta\phi_j = -\frac{1}{2\pi} \int_j \theta_{pj} \gamma_j \, ds_j \tag{3.40a}$$

In Eq. (3.40a), γ_j is constant over the jth panel, and the integral is taken over the
jth panel only. The angle θ_{pj} is given by

$$\theta_{pj} = \tan^{-1}\frac{y - y_j}{x - x_j} \tag{3.41}$$

In turn, the potential at P due to *all* the panels is Eq. (3.40a) summed over all the
panels:

$$\phi(P) = \sum_{j=1}^{n} \phi_j = -\sum_{j=1}^{n} \frac{\gamma_j}{2\pi} \int_j \theta_{pj} \, ds_j \tag{3.42}$$

Since point P is just an arbitrary point in the flow, let us put P at the control point of the ith panel shown in Fig. 3.6. The coordinates of this control point are (x_i, y_i). Then Eqs. (3.41) and (3.42) become

$$\theta_{i,j} = \tan^{-1} \frac{y_i - y_j}{x_i - x_j}$$

and

$$\phi(x_i, y_i) = -\sum_{j=1}^{n} \frac{\gamma_j}{2\pi} \int_j \theta_{ij} \, ds_j \tag{3.43}$$

Equation (3.43) is physically the contribution of *all* the panels to the potential at the control point of the ith panel.

At the control points, the normal component of the velocity is zero; this velocity is the superposition of the uniform flow velocity and the velocity induced by all the vortex panels. The component of V_∞ normal to the ith panel is given by Eq. (3.15):

$$V_{\infty,n} = V_\infty \cos\beta_i \tag{3.44}$$

The normal component of velocity induced at (x_i, y_i) by the vortex panels is

$$V_n = \frac{\partial}{\partial n_i}[\phi(x_i, y_i)] \tag{3.45}$$

Combining Eqs. (3.43) and (3.45), we have

$$V_n = -\sum_{j=1}^{n} \frac{\gamma_j}{2\pi} \int_j \frac{\partial \theta_{ij}}{\partial n_i} \, ds_j \tag{3.46}$$

where the summation is over all the panels. The normal component of the flow velocity at ith control point is the sum of that due to the freestream [Eq. (3.44)] and that due to the vortex panels [Eq. (3.46)]. The boundary condition states that this sum must be zero:

$$V_{\infty,n} + V_n = 0 \tag{3.47}$$

Substituting Eqs. (3.44) and (3.46) into Eq. (3.47), we obtain

$$V_\infty \cos\beta_i - \sum_{j=1}^{n} \frac{\gamma_j}{2\pi} \int_j \frac{\partial \theta_{ij}}{\partial n_i} \, ds_j = 0 \tag{3.48}$$

Equation (3.48) is the crux of the vortex panel method. The values of the integrals in Eq. (3.48) depend simply on the panel geometry; they are not properties of the flow. Let $J_{i,j}$ be the value of this integral when the control point is on the ith panel. Then Eq. (3.48) can be written as

$$V_\infty \cos\beta_i - \sum_{j=1}^{n} \frac{\gamma_j}{2\pi} J_{i,j} = 0 \tag{3.49}$$

Fig. 3.12 Vortex panels at the trailing edge

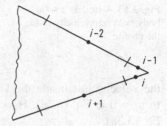

Equation (3.49) is a linear algebraic equation with n unknowns, γ_1, γ_2, ..., γ_n. It represents the flow boundary condition evaluated at the control point of the ith panel. If Eq. (3.49) is applied to the control points of *all* the panels, we obtain a system of n linear equations with n unknowns.

To this point, we have been deliberately paralleling the discussion of the source panel method given in Sect. 3.3; however, the similarity stops here. For the source panel method, the n equations for the n unknown source strengths are routinely solved, giving the flow over a non-lifting body. In contrast, for the lifting case with vortex panels, in addition to the n equations given by Eq. (3.49) applied at all the panels, we must also satisfy the Kutta condition, Eq. (3.39). This can be done in several ways. For example, consider Fig. 3.12, which illustrates a detail of the vortex panel distribution at the trailing edge. Note that the length of each panel can be different; their length and distribution over the body is up to your discretion. Let the two panels at the trailing edge (panels i and $i-1$ in Fig. 3.12) be very small. The Kutta condition is applied *precisely* at the trailing edge and is given by $\gamma(\text{TE}) = 0$. To approximate this numerically, if points i and $i-1$ are close enough to the trailing edge, we can write

$$\gamma_i = -\gamma_{i-1} \tag{3.50}$$

such that the strengths of the two vortex panels i and $i-1$ exactly cancel at the point where they touch at the trailing edge. Thus, in order to impose the Kutta condition on the solution of the flow, Eq. (3.50) (or an equivalent expression) must be included. Note that Eq. (3.49) evaluated at all the panels and Eq. (3.50) constitute an *over-determined* system of n unknowns with $n + 1$ equations. Therefore, to obtain a determined system, Eq. (3.49) is not evaluated at one of the control points on the body. That is, we choose to ignore one of the control points, and we evaluate Eq. (3.49) at the other $n-1$ control points. This, in combination with Eq. (3.50), now gives a system of n linear algebraic equations with n unknowns, which can be solved by standard techniques.

At this stage, we have conceptually obtained the values of $\gamma_1, \gamma_2, ..., \gamma_n$ which make the body surface a streamline of the flow and which also satisfy the Kutta condition. In turn, the flow velocity tangent to the surface can be obtained directly from γ. To see this more clearly, consider the airfoil shown in Fig. 3.13. We are concerned only with the flow outside the airfoil and on its surface. Therefore, let the velocity be zero at every point *inside* the body, as shown in Fig. 3.13. In particular,

Fig. 3.13 Airfoil as a solid
body, with zero velocity inside
the profile

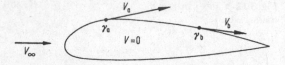

the velocity just inside the vortex sheet on the surface is zero. This corresponds
to $u_2 = 0$ in Eq. (3.38). Hence the velocity just outside the vortex sheet is, from
Eq. (3.38).

$$\gamma = u_1 - u_2 = u_1 - 0 = u_1$$

In Eq. (3.38), u denotes the velocity tangential to the vortex sheet. In terms of
the picture shown in Fig. 3.13, we obtain $V_a = \gamma_a$ at point a, $V_b = \gamma_b$ at point b, etc.
Therefore, *the local velocities tangential to the airfoil surface are equal to the local
values of* γ. In turn, the local pressure distribution can be obtained from Bernoulli's
equation.

The total circulation and the resulting lift are obtained as follows. Let s_j be the
length of the *j*th panel. Then the circulation due to the *j*th panel is $\gamma_j s_j$. In turn, the
total circulation due to all the panels is

$$\Gamma = \sum_{j=1}^{n} \gamma_j s_j \tag{3.51}$$

Hence, the lift per unit span is obtained from

$$L' = \rho_\infty V_\infty \sum_{n=1}^{n} \gamma_j s_j \tag{3.52}$$

The presentation in this section is intended to give only the general flavor of the
vortex panel method. There are many variations of the method is use today, and you
are encouraged to read the moden literature, especially as it appears in the *AIAA
Journal* and the *Journal of Aircraft* since 1970. The vortex panel method as described in this section is termed a 'first-order' method because it assumes a constant
value of γ over a given panel. Although the method may appear to be straightforward, its numerical implementation can sometimes be frustrating. For example, the
results for a given body are sensitive to the number of panels used, their various
sizes and the way they are distributed over the body surface (i.e. it is usually advantageous to place a large number of small panels near the leading and trailing
edges of an airfoil and a smaller number of larger panels in the middle). The need
to ignore one of the control points in order to have a determined system in n equations for n unknowns also introduces some arbitrariness in the numerical solution.
Which control point do you ignore? Different choices sometimes yield different numerical answers for the distribution of γ over the surface. Moreover, the resulting
numerical distributions for γ are not always smooth, but rather they have oscillations
from one panel to the next as a result of numerical inaccuracies. The problems mentioned above are usually overcome in different ways by different groups who have
developed relatively sophisticated panel programs for practical use. Again, you are
encouraged to consult the literature for more information.

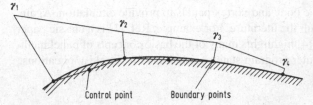

Fig. 3.14 Linear distribution of vortex strength over each panel—a second-order panel method

Such accuracy problems have encouraged the development of higher-order panel techniques. For example, a 'second-order' panel method assumes a *linear* variation of γ over à given panel, as sketched in Fig. 3.14. Here, the value of γ at the edges of each panel is matched to its neighbours, and the values γ_1, γ_2, γ_3, etc., at the *boundary points* become the unknowns to be solved. The flow-tangency boundary condition is still applied at the *control point* of each panel, as before. Some results using a second-order vortex panel technique are given in Fig. 3.15, which shows the distribution of pressure coefficients over the upper and lower surfaces of a NACA 0012 airfoil at a 9° angle of attack. The circles and squares are numerical results from a second-order vortex panel technique developed at the University of Maryland, and the solid lines are from NACA results given in Ref. [2]. Excellent agreement is obtained.

Finally, many groups developing and using panel techniques use a combination of source panels and vortex panels for lifting bodies—source panels to accurately

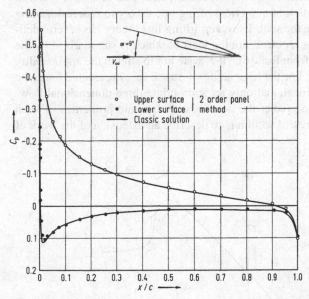

Fig. 3.15 Pressure coefficient distribution over an NACA 0012 airfoil; comparison between second order vortex panel method and theoretical results. The numerical panel results were obtained by one of the author's graduate students, Mr Tae-Hwan Cho

represent the thickness of the body and vortex panels to provide circulation. Again, you are encouraged to consult the literature. For example, Ref. [3] is a classic paper on panel methods, and Ref. [4] highlights many of the basic concepts of panel methods along with actual computer program statement listings for simple applications.

3.5 An Application—The Aerodynamics of Drooped Leading-Edge Wings Below and Above Stall

In this section, in order to illustrate some of the above ideas, we briefly describe an application of a panel method to an applied aerodynamic problem of some interest. Since the late 1970s, low-speed wind tunnel experiments and flight tests (conducted mainly by NASA) have conclusively demonstrated that wings with a discontinuous leading-edge extension and increase in camber (leading-edge droop) exhibit a smoothing of the normally abrupt drop in lift coefficient C_L at stall, and the generation of a relatively large value of C_L at very high post-stall angles of attack. This behaviour is illustrated in Fig. 3.16. As a result, an aeroplane with a properly designed drooped leading edge has increased resistance towards stall/spins—behaviour of great interest to the general aviation community. In response to this interest, an extensive experimental investigation of the fundamental aerodynamic characteristics of drooped leading edge wings is being conducted, an example of which is given in Ref. [5].

Some preliminary theoretical support for such experimental results is given in Ref. [6], which is an application of numerical lifting line theory to drooped leading edge wings below and above the stall. However, lifting line theory has several deficiencies when applied to this problem, not the least of which is summarized by the following statement quoted from Ref. [6]: 'It is wise not to stretch the applicability of lifting-line theory too far. For the high angle-of-attack cases presented here, the flow is highly three-dimensional, and only an appropriate three-dimensional flow-field calculation can hope to predict the detailed aerodynamic properties of such flows.' The purpose of the present section is to describe an extension of the work of

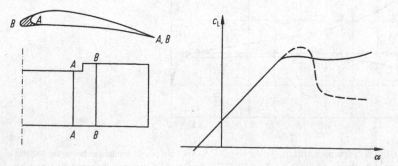

Fig. 3.16 Sketch showing the effect of a drooped leading edge wing on lift coefficient

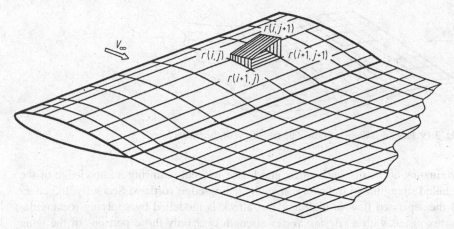

Fig. 3.17 Schematic of the second-order vortex panel

Ref. [6], namely, to present the results of an 'appropriate three-dimensional flowfield calculation' for drooped leading edge wings. The work in this section is patterned after Ref. [7].

In particular, this section presents numerical results obtained with a three-dimensional vortex panel computer program for the calculation of inviscid, incompressible (potential) flow. This program is specially constructed for application to wings with drooped leading-edge discontinuities. The program is essentially a numerical representation of lifting surface theory, involving both spanwise and chordwise distributions of vorticity. Across each panel, the vorticity is assumed to vary linearly in both the spanwise and chordwise directions; hence, this is a second-order panel method. Figure 3.17 illustrates the type of panel used in the present calculations.

The present results also include two approximations of an 'engineering' nature. First, the effect of the leading-edge discontinuities is modelled by assuming that the vortices eminating from these discontinuities aerodynamically divide the wing into three wings of lower aspect ratio, as sketched in Fig. 3.18. Some direct experimental evidence of this effect is discussed in Ref. [8]. Hence, the present calculations were made with three low aspect ratio wings butted against each other (wings A, B and C in Fig. 3.18). In this fashion, the vortex panel analysis is made to 'see' the leading edge discontinuities without explicitly inserting separate vortex filaments

Fig. 3.18 Simulation of the effects of the leading edge discontinuities. Division of the drooped leading edge wing into three wings of lower aspect ratio, butted against each other

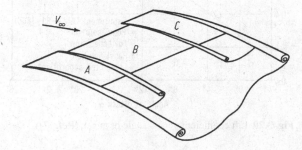

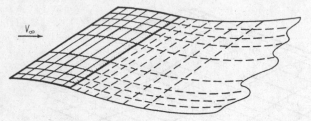

Fig. 3.19 Panel distribution to simulate the effects of flow separation

originating at the discontinuities, and hence without requiring a knowledge of the detailed strength and trajectory of such leading-edge vortices. Secondly, the effect of the separated flow at high angle-of-attack is modelled by applying rectangular vortex panels with a *varying* vortex strength over only those portions of the wing with attached flow, i.e. a wing planform with a scalloped trailing edge as sketched in Fig. 3.19. The separated region of the wing is covered with *constant strength* vortex panels associated with a value of the pressure coefficient, $C_p = -0.6$. This is a reasonable value of C_p in separated regions on wings, in low-speed flow, as shown by numerous experiments. These constant strength panels in the separated region are represented by the shaded region in Fig. 3.19. Obviously, this modelling requires a knowledge of the separation lines on the finite wing. For the present results, these separation lines are obtained from surface oil flow visualization experiments, such as described in Ref. [5].

This modelling of the drooped leading edge discontinuities, and of the separated flow, is a simple engineering approach, and is not meant to be the final theoretical answer to the analysis of such flows. However, this modelling taken in conjunction with the second-order vortex panel program described above yields amazingly good results, as shown in Fig. 3.20. Here, C_L versus angle-of-attack is given for the

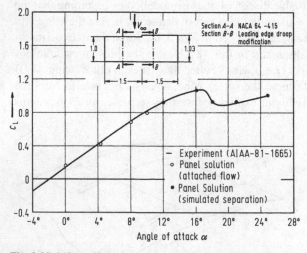

Fig. 3.20 Lift coefficient versus angle of attack (Ref. [7])

drooped leading edge wing sketched in the figure. The solid line represents a curve through the experimental data of Ref. [5]; the open circles give numerical results obtained with the present analysis. The agreement is excellent at all angles-of-attack, both below and above the stall. By taking into account the three-dimensional flow effects, the present results represent a substantial improvement over the C_L versus α results obtained from lifting line theory in Ref. [6].

This example is a good illustration of the usefulness and power of panel programs. A three-dimensional flow has been calculated over a rather complex configuration which could not have been calculated 25 years ago. However, today, with the massive computer power available along with some sophisticated numerical techniques, the calculations of such flows is not only possible, but can be done with great efficiency.

References

1. Anderson, John D., Jr., *Fundamentals of Aerodynamics*, 2nd Edition McGraw-Hill, New York, 1991.
2. Abbott, I.H. and von Doenhoff, A.E., *Theory of Wing-Sections*, McGraw-Hill Book Company, New York, 1949; also, Dover Publications, Inc., New York, 1959.
3. Hess, J.L. and Smith, A.M.O., 'Calculation of Potential Flow about Arbitrary Bodies,' in D. Kucheman (ed.), *Progress in Aeronautical Sciences*, Vol. 8, Pergamon Press, New York, pp. 1–138.
4. Chow, C.Y., *An Introduction to Computational Fluid Dynamics*, John Wiley & Sons, Inc., New York, 1979.
5. Winkelmann, A.E. and Tsao, C.P., 'An Experimental Study of the Flow on a Wing With a Partial Span Dropped Leading Edge,' AIAA Paper No. 81–1665, 1981.
6. Anderson, J.D., Jr., Corda, S. and Van Wie, D.M., 'Numerical Lifting Line Theory Applied to Drooped Leading-Edge Wings Below and Above Stall,' *Journal of Aircraft*, Vol. 17, No. 12, Dec. 1980, pp. 898–904.
7. Cho, T.H. and Anderson, J.D., Jr., 'Engineering Analysis of Drooped Leading-Edge Wings Near Stall,' *Journal of Aircraft*, Vol. 21, No. 6, June 1984, pp. 446–448.
8. Johnson, J.L., Jr., Newsom, W.A. and Satran, D.R., 'Full-Scale Wind Tunnel Investigation of the Effects of Wing Leading-Edge Modifications on the High Angle-of-Attack Aerodynamic Characteristics of a Low-Wing General Aviation Airplane,' AIAA Paper No. 80, 1844, 1980.

Chapter 4
Mathematical Properties of the Fluid Dynamic Equations

J.D. Anderson, Jr.

4.1 Introduction

The governing equations of fluid dynamics derived in Chap. 2 are either integral forms (such as Eq. (2.23) obtained directly from a finite control volume) or partial differential equations (such as Eqs (2.36a–c) obtained directly from an infinitesimal fluid element). The governing equations in the form of partial differential equations are by far the most prevalent form used in computational fluid dynamics. Therefore, before taking up a study of numerical methods for the solution of these equations, it is useful to examine some mathematical properties of partial differential equations themselves. Any valid numerical solution of the equations should exhibit the property of obeying the general mathematical properties of the governing equations.

Examine the governing equations of fluid dynamics as derived in Chap. 2. Note that in all cases the *highest order* derivatives occur *linearly*, i.e. there are no products or exponentials of the highest order derivatives—they appear by themselves, multiplied by coefficients which are functions of the dependent variables themselves. Such a system of equations is called a *quasilinear system*. For example, for inviscid flows, examining the equations in Sect. 2.8.2 we find that the highest order derivatives are first order, and all of them appear linearly. For viscous flows, examining the equations in Sect. 2.8.1 we find the highest order derivatives are second order, and they always occur linearly. For this reason, in the next section, let us examine some properties of a system of quasilinear partial differential equations. In the process, we will establish a classification of three types of partial differential equations—all three of which are encountered in fluid dynamics.

4.2 Classification of Partial Differential Equations

For simplicity, let us consider a fairly simple system of quasilinear equations. They will *not* be the flow equations, but they are similar in some respects. Therefore, this section serves as a simplified example.

J.D. Anderson, Jr.
National Air and Space Museum, Smithsonian Institution, Washington, DC
e-mail: AndersonJA@si.edu

J.F. Wendt (ed.), *Computational Fluid Dynamics*, 3rd ed.,
© Springer-Verlag Berlin Heidelberg 2009

Consider the system of quasilinear equations given below.

$$a_1 \frac{\partial u}{\partial x} + b_1 \frac{\partial u}{\partial y} + c_1 \frac{\partial v}{\partial x} + d_1 \frac{\partial v}{\partial y} = f_1 \tag{4.1a}$$

$$a_2 \frac{\partial u}{\partial x} + b_2 \frac{\partial u}{\partial y} + c_2 \frac{\partial v}{\partial x} + d_2 \frac{\partial v}{\partial y} = f_2 \tag{4.1b}$$

where u and v are the dependent variables, functions of x and y, and the coefficients $a_1, a_2, b_1, b_2, c_1, c_2, d_1, d_2, f_1$ and f_2 can be functions of x, y, u and v.

Consider any point in the xy-plane. Let us seek the lines (or directions) through this point (if any exist) along which the *derivatives* of u and v are indeterminant, and across which may be discontinuous. Such lines are called *characteristic lines*. To find such lines, we assume that u and v are continuous, and hence

$$\text{since } u = u(x,\ y): \ du = \frac{\partial u}{\partial x} dx + \frac{\partial u}{\partial y} dy \tag{4.2a}$$

$$\text{since } v = v(x,\ y): \ dv = \frac{\partial v}{\partial x} dx + \frac{\partial v}{\partial y} dy \tag{4.2b}$$

Equations (4.1a and b) and (4.2a and b) constitute a system of four linear equations with four unknowns ($\partial u/\partial x, \partial u/\partial y, \partial v/\partial x$, and $\partial v/\partial y$). These equations can be written in matrix form as

$$\begin{bmatrix} a_1 & b_1 & c_1 & d_1 \\ a_2 & b_2 & c_2 & d_2 \\ dx & dy & 0 & 0 \\ 0 & 0 & dx & dy \end{bmatrix} \begin{bmatrix} \partial u/\partial x \\ \partial u/\partial y \\ \partial v/\partial x \\ \partial v/\partial y \end{bmatrix} = \begin{bmatrix} f_1 \\ f_2 \\ du \\ dv \end{bmatrix} \tag{4.3}$$

Let $[A]$ denote the coefficient matrix.

$$[A] = \begin{bmatrix} a_1 & b_1 & c_1 & d_1 \\ a_2 & b_2 & c_2 & d_2 \\ dx & dy & 0 & 0 \\ 0 & 0 & dx & dy \end{bmatrix}$$

Moreover, let ;$|A|$ be the determinant of $[A]$. From Cramer's rule, if $|A| \neq 0$, then unique solutions can be obtained for $\partial u/\partial x$, $\partial u/\partial y$, $\partial v/\partial x$, and $\partial v/\partial y$. On the other hand, if $|A| = 0$, then $\partial u/\partial x$, $\partial u/\partial y$, $\partial v/\partial x$ and $\partial v/\partial y$ are, at best, indeterminant. We are seeking the particular directions in the xy-plane along which these derivatives of u and v are indeterminant. Therefore, let us set $|A| = 0$, and see what happens.

$$\begin{vmatrix} a_1 & b_1 & c_1 & d_1 \\ a_2 & b_2 & c_2 & d_2 \\ dx & dy & 0 & 0 \\ 0 & 0 & dx & dy \end{vmatrix} = 0$$

Hence:

$$(a_1c_2 - a_2c_1)(dy)^2 - (a_1d_2 - a_2d_1 + b_1c_2 - b_2c_1)(dx)(dy) + (b_1d_2 - b_2d_1)(dx)^2 = 0$$
(4.4)

Divide Eq. (4.4) by $(dx)^2$.

$$(a_1c_2 - a_2c_1)\left(\frac{dy}{dx}\right)^2 - (a_1d_2 - a_2d_1 + b_1c_2 - b_2c_1)\frac{dy}{dx} + (b_1d_2 - b_2d_1) = 0 \quad (4.5)$$

Equation (4.5) is a quadratic equation in dy/dx. For any point in the xy-plane, the solution of Eq. (4.5) will give the slopes of the lines along which the derivatives of u and v are indeterminant. Why? Because Eq. (4.5) was obtained by setting $|A| = 0$, which from the matrix Eq. (4.3) insures that the solutions for the derivatives $\partial u/\partial x$, $\partial u/\partial y$, $\partial v/\partial x$ and $\partial v/\partial y$ are, at best, indeterminant. These lines in the xy space along which the derivatives of u and v are indeterminant are called the *characteristic lines* for the system of equations given by Eq. (4.1a and b).

In Eq. (4.5), let

$$a = (a_1c_2 - a_2c_1)$$
$$b = -(a_1d_2 - a_2d_1 + b_1c_2 - b_2c_1)$$
$$c = (b_1d_2 - b_2d_1)$$

Then Eq. (4.5) can be written as

$$a\left(\frac{dy}{dx}\right)^2 + b\left(\frac{dy}{dx}\right) + c = 0$$
(4.6)

Hence, from the quadratic formula:

$$\frac{dy}{dx} = \frac{-b \pm \sqrt{b^2 - 4ac}}{2a}$$
(4.7)

Equation (4.7) gives the direction of the characteristic lines through a given xy point. These lines have a different nature, depending on the value of the discriminant in Eq. (4.7). Denote the discriminant by D.

$$D = b^2 - 4ac$$
(4.8)

The characteristic lines may be real and distinct, real and equal, or imaginary, depending on the value of D. Specifically:

If $D > 0$: Two real and distinct characteristics exist through each point in the xy-plane. When this is the case, the system of equations given by Eqs. (4.1a and b) is called *hyperbolic*.

If $D = 0$: One real characteristic exists. Here, the system of Eqs. (4.1a and b) is called *parabolic*.

If $D < 0$: The characteristic lines are imaginary. Here, the system of Eqs. (4.1a and b) is called *elliptic*.

The classification of quasilinear partial differential equations as either *elliptic, parabolic* or *hyperbolic* is common in the analysis of such equations. These three classes of equations have totally different behavior, as will be discussed shortly. The origin of the words elliptic, parabolic or hyperbolic used to label these equations is simply a direct analogy with the case for conic sections. The general equation for a conic section from analytic geometry is

$$ax^2 + bxy + cy^2 + dx + ey + f = 0$$

where, if

$$b^2 - 4ac > 0, \quad \text{the conic is a hyperbola}$$
$$b^2 - 4ac = 0, \quad \text{the conic is a parabola}$$
$$b^2 - 4ac < 0, \quad \text{the conic is an ellipse}$$

We note that, for hyperbolic partial differential equations, the fact that two real and distinct characteristics exist allows the development of a method for the ready solution of these equations. If we return to Eq. (4.3), and actually attempt to solve for, say $\partial u/\partial y$, using Cramer's rule, we have

$$\partial u/\partial y = \frac{|N|}{|A|} = \frac{0}{0}$$

where the numerator determinant is

$$|N| = \begin{vmatrix} a_1 & f_1 & c_1 & d_1 \\ a_2 & f_2 & c_2 & d_2 \\ dx & du & 0 & 0 \\ 0 & dv & dx & dy \end{vmatrix} \tag{4.9}$$

The reason why $|N|$ must be zero is that $\partial u/\partial y$ is indeterminant, of the form $0/0$. Since $|A|$ has already been made equal to zero, then $|N|$ must be zero to allow $\partial u/\partial y$ to be indeterminant. The expansion of Eq. (4.9) will lead to equations involving the flowfield variables which are *ordinary differential equations*, and in some cases are algebraic equations; these equations obtained from Eq. (4.9) are called the *compatibility* equations. They hold only along the characteristic lines. This is the essence of solving the original hyperbolic partial differential equation: simply integrate simpler, ordinary differential equations (the compatibility equations) along the characteristic lines in the xy-plane. This is called the *method of characteristics*. This method is highly developed for the solution of inviscid supersonic flows, for which the system of governing flow equations is hyperbolic. The practical implementation of the method of characteristics requires the use of a high-speed digital computer, and therefore may legitamately be considered a part of CFD. However, the method of characteristics is a well-known classical technique for the solution of inviscid supersonic flows, and therefore we will not consider it in any detail in these notes. For more information, see Ref. [1].

4.3 General Behaviour of the Different Classes of Partial Differential Equations and Their Relation to Fluid Dynamics

In this section we simply discuss, without proof, some of the behaviour of hyperbolic, parabolic and elliptic partial differential equations, and relate this behaviour to the solution of problems in fluid dynamics. For more details on the characteristics of partial differential equations, see any good text on advanced mathematics, such as Ref. [2].

4.3.1 Hyperbolic Equations

For hyperbolic equations, information at a given point P influences only those regions between the advancing characteristics. For example, examine Fig. 4.1, which is sketched for a two-dimensional problem with two independent space variables. Point P is located at a given (x, y). Consider the left- and right-running characteristics through point P, as shown in Fig. 4.1. Information at point P influences only the shaded region—the region labelled I between the two advancing characteristics through P in Fig. 4.1. This has a collorary effect on boundary conditions for hyperbolic equations. Assume that the x-axis is a given boundary condition for the problem, i.e. the dependent variables u and v are known along the x-axis. Then the solution can be obtained by 'marching forward' in the distance y, starting from the given boundary. However, the solution for u and v at point P will depend *only* on that part of the boundary between a and b, as shown in Fig. 4.1. Information at point c, which is outside the interval ab, is propagated along characteristics through c, and influences only region II in Fig. 4.1. Point P is outside region II, and hence does not feel the information from point c. For this reason, point P depends on only that part

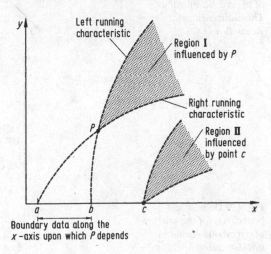

Fig. 4.1 Domain and boundaries for the solution of hyperbolic equations. Two-dimensional steady flow

of the boundary which is intercepted by and included between the two retreating characteristic lines through point P, i.e. interval ab.

In fluid dynamics, the following types of flows are governed by hyperbolic partial differential equations, and hence exhibit the behavior described above:

(1) *Steady, inviscid supersonic flow*. If the flow is two-dimensional, the behaviour is like that already discussed in Fig. 4.1. If the flow is three-dimensional, there are characteristic surfaces in *xyz* space, as sketched in Fig. 4.2. Consider point P at a given (x, y, z) location. Information at P influences the shaded volume within the advancing characteristic surface. In addition, if the $x-y$ plane is a boundary surface, then only that portion of the boundary shown as the cross-hatched area in the $x-y$ plane, intercepted by the retreating characteristic surface, has any effect on P. In Fig. 4.2, the dependent variables are solved by starting with data given in the *xy*-plane, and 'marching' in the *z*-direction. For an inviscid supersonic flow problem, the general flow direction would also be in the *z*-direction.

(2) *Unsteady inviscid compressible flow*. For unsteady one- and two-dimensional inviscid flows, the governing equations are hyperbolic, no matter whether the flow is locally subsonic or supersonic. Here, time is the marching direction. For one-dimensional unsteady flow, consider a point P in the (x, t) plane shown in Fig. 4.3.

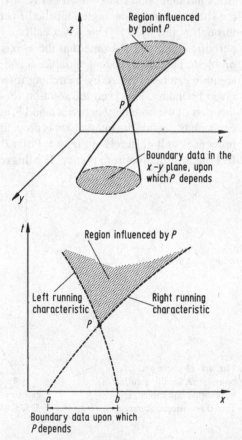

Fig. 4.2 Domain and boundaries for the solution of hyperbolic equations. Three-dimensional steady flow

Fig. 4.3 Domain and boundaries for the solution of hyperbolic equations: one-dimensional unsteady flow

Fig. 4.4 Domain and
boundaries for the solution
of hyperbolic equations:
two-dimensional
unsteady flow

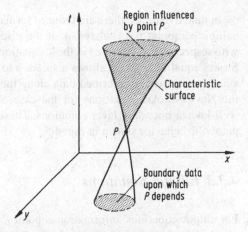

Once again, the region influenced by P is the shaded area between the two advancing characteristics through P, and the interval ab is the only portion of the boundary along the x-axis upon which the solution at P depends. For two-dimensional unsteady flow, consider a point P in the (x, y, t) space as shown in Fig. 4.4. The region influenced by P, and the portion of the boundary in the xy-plane upon which the solution at P depends, are shown in this figure. Starting with known initial data in the xy-plane, the solution 'marches' forward in time.

4.3.2 Parabolic Equations

For parabolic equations, information at point P in the xy-plane influences the entire region of the plane to one side of P. This is sketched in Fig. 4.5, where the single characteristic line through point P is drawn. Assume the x- and y-axes are boundaries; the solution at P depends on the boundary conditions along the entire y axis, as well as on that portion of the x-axis from a to b. Solutions to parabolic equations are also 'marching' solutions; starting with boundary conditions along both the x- and y-axes, the flow-field solution is obtained by 'marching' in the general x-direction.

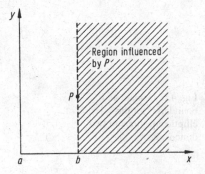

Fig. 4.5 Domain and
boundaries for the solution of
parabolic equations in two
dimensions

In fluid dynamics, there are reduced forms of the Navier–Stokes equations which exhibit parabolic-type behaviour. If the viscous stress terms involving derivatives with respect to x are ignored in these equations, we obtain the 'parabolized' Navier–Stokes equations, which allows a solution to march downstream in the x-direction, starting with some prescribed data along the x- and y-axes. A further reduction of the Navier–Stokes equations for the case of high Reynolds number leads to the well-known boundary layer equations. These boundary layer equations exhibit the parabolic behavior shown in Fig. 4.5.

4.3.3 Elliptic Equations

For elliptic equations, information at point P in the xy-plane influences *all* other regions of the domain. This is sketched in Fig. 4.6, which shows a rectangular domain. Here, the domain is fully closed, surrounded by the closed boundary *abcd*. This is in contrast to the open domains for parabolic and hyperbolic equations discussed earlier, and shown in the previous figures, namely, Figs. 4.1, 4.2, 4.3, 4.4 and 4.5. For elliptic equations, because point P influences all points in the domain, then in turn the solution at point P is influenced by the *entire* closed boundary *abcd*. Therefore, the solution at point P must be carried out *simultaneously* with the solution at all other points in the domain. This is in stark contrast to the 'marching' solutions germaine to parabolic and hyperbolic equations. For this reason, problems involving elliptic equations are frequently called 'equilibrium', or 'jury' problems, because the solution within the domain depends on the *total* boundary around the domain. (See Ref. [3] for more details.)

In fluid dynamics steady, subsonic, inviscid flow is governed by elliptic equations. As a sub-case, this also includes incompressible flow (which theoretically implies that the Mach number is zero). Hence, for such flows, physical boundary conditions must be applied over a closed boundary that totally surrounds the flow, and the flow-field solution at all points in the flow must be obtained simultaneously because the solution at one point influences the solution at all other points. In terms of Fig. 4.6, boundary conditions must be applied over the entire boundary *abcd*. These boundary conditions can take the following forms:

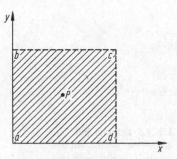

Fig. 4.6 Domain and boundaries for the solution of elliptic equations in two dimensions

(1) A specification of the *dependent variables* u and v along the boundary. This type of boundary conditions is called the *Dirichlet* condition.
(2) A specification of *derivatives* of the dependent variables, such as $\partial u/\partial x$, etc., along the boundary. This type of boundary condition is called the *Neumann* condition.

4.3.4 Some Comments

At this stage it is instructive to return to our discussion of the inviscid flow over a supersonic blunt body in Chap. 1, and in particular to Fig. 1.1. There we pointed out that the locally subsonic steady flow is governed by elliptic partial differential equations, and that the locally supersonic steady flow is governed by hyperbolic partial differential equations. Now we have a better understanding of what this means mathematically; and because of the totally different mathematical behavior of elliptic and hyperbolic equations, we have a new appreciation for the difficulties that were encountered by early researchers in trying to solve the blunt body problem. The sudden change in the nature of the governing equations across the sonic line virtually precluded any practical solution of the steady flow blunt body problem involving a uniform treatment of both the subsonic and supersonic regions. However, recall from Fig. 4.4 that *unsteady* inviscid flow is governed by hyperbolic equations no matter whether the flow is locally subsonic or supersonic. This provides the following opportunity. Starting with rather arbitrary initial conditions for the flow field in the xy-plane in Fig. 1.1, solve the *unsteady*, two-dimensional inviscid flow equations, marching forward in time as sketched in Fig. 4.4. At large times, the solution approaches a steady state, where the time derivatives of the flow variables approach zero. This steady state is the desired result, and what you have when you approach this steady state is a solution for the *entire* flow field *including* both the subsonic and supersonic regions. Moreover, this solution is obtained with the same, uniform method throughout the entire flow. The above discussion gives the elementary philosophy of the *time-dependent technique* for the solution of flow problems. Its practical numerical implementation by Moretti and Abbett [4] in 1966 constituted the major scientific breakthrough for the solution of the supersonic blunt body problem as discussed in Chap. 1.

At this stage, it would be worthwhile for the student to examine the actual, closed-form solution to some linear partial differential equations of the elliptic, parabolic and hyperbolic types. Numerous classical solutions can be found; Refs. [2, 3] are good sources. However, we will not carry out such an examination in these notes; rather, we will use our remaining time and space here to move on to numerical solutions that are germane to fluid flows. Again, the student is referred to Refs. [2,3] for more details.

4.3.5 Well-Posed Problems

In the solution of partial differential equations it is sometimes easy to attempt a solution using incorrect or insufficient boundary and initial conditions. Whether the solution is being attempted analytically or numerically, such an 'ill-posed' problem will usually lead to spurious results.

Therefore, we define a well-posed problem as follows: If the solution to a partial differential equation exists and is unique, and if the solution depends continuously upon the initial and boundary conditions, then the problem is *well-posed*. In CFD, it is important that you establish that your problem is well-posed before you attempt to carry out a numerical solution.

References

1. Anderson, J.D., Jr., *Modern Compressible Flow: With Historical Perspective*, 2nd Edition McGraw-Hill, New York, 1990.
2. Hildebrand, F.B., *Advanced Calculus for Applications*, Prentice-Hall, New Jersey, 1976.
3. Anderson, D.A., Tannehill, John C. and Pletcher, Richard H., *Computational Fluid Mechanics and Heat Transfer*, McGraw-Hill, New York, 1984.
4. Moretti, G. and Abbett, M., 'A Time-Dependent Computational Method for Blunt Body Flows,' *AIAA Journal*, Vol. 4, No. 12, December 1966, pp. 2136–2141.

Chapter 5
Discretization of Partial Differential Equations

J.D. Anderson, Jr.

5.1 Introduction

Analytical solutions of partial differential equations involve closed-form expressions which give the variation of the dependent variables *continuously* throughout the domain. In contrast, numerical solutions can give answers at only *discrete points* in the domain, called *grid points*. For example, consider Fig. 5.1, which shows a section of a discrete grid in the xy-plane. For convenience, let us assume that the spacing of the grid points in the x-direction is uniform, and given by Δx, and that the spacing of the points in the y-direction is also uniform, and given by Δy, as shown in Fig. 5.1. In general, Δx and Δy are different. Indeed, it is not absolutely necessary that Δx or Δy be uniform; we could deal with totally unequal spacing in both directions, where Δx is a different value between each successive pairs of grid points, and similarly for Δy. However, the vast majority of CFD applications involve numerical solutions on a grid which involves uniform spacing in each direction, because this greatly simplifies the programming of the solution, saves storage space and usually results in greater accuracy. This uniform spacing does not have to occur in the physical xy space; as is frequently done in CFD, the numerical calculations are carried out in a transformed computational space which has uniform spacing in the transformed independent variables, but which corresponds to non-uniform spacing in the physical plane. These matters will be discussed in detail in Chap. 6. In any event, in this chapter we will asume uniform spacing in each coordinate direction, but not necessarily equal spacing for both directions, i.e. we will assume Δx and Δy to be constants, but that Δx does not have to equal Δy.

Returning to Fig. 5.1, the grid points are identified by an index i which runs in the x-direction, and an index j which runs in the y-direction. Hence, if $(i,\ j)$ is the index for point P in Fig. 5.1, then the point immediately to the right of P is labeled as $(i+1,\ j)$, the immediately to the left is $(i-1, j)$, the point directly above is $(i,\ j+1)$, and the point directly below is $(i,\ j-1)$.

J.D. Anderson, Jr.
National Air and Space Museum, Smithsonian Institution, Washington, DC
e-mail: AndersonJA@si.edu

J.F. Wendt (ed.), *Computational Fluid Dynamics*, 3rd ed.,
© Springer-Verlag Berlin Heidelberg 2009

Fig. 5.1 Discrete grid points

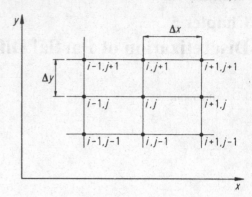

The *method of finite-differences* is widely used in CFD, and therefore most of this chapter will be devoted to matters concerning finite differences. The philosophy of finite difference methods is to replace the partial derivatives appearing in the governing equations of fluid dynamics (as derived in Chap. 2) with algebraic difference quotients, yielding a system of *algebraic equations* which can be solved for the flow-field variables at the specific, discrete grid points in the flow (as shown in Fig. 5.1). Let us now proceed to derive some of the more common algebraic difference quotients used to discretize the partial differential equations.

5.2 Derivation of Elementary Finite Difference Quotients

Finite difference representations of derivatives are based on Taylor's series expansions. For example, if $u_{i,j}$ denotes the x-component of velocity at point (i, j), then the velocity $u_{i+1, j}$ at point $(i + 1, j)$ can be expressed in terms of a Taylor's series expanded about point (i, j), as follows:

$$u_{i+1,j} = u_{i,j} + \left(\frac{\partial u}{\partial x}\right)_{i,j} \Delta x + \left(\frac{\partial^2 u}{\partial x^2}\right)_{i,j} \frac{(\Delta x)^2}{2} + \left(\frac{\partial^3 u}{\partial x^3}\right)_{i,j} \frac{(\Delta x)^3}{6} + \cdots \qquad (5.1)$$

Equation (5.1) is mathematically an exact expression for u_{i+1j} if:

(a) the number of terms is infinite and the series converges,
(b) and/or $\Delta x \to 0$.

For numerical computations, it is impractical to carry an infinite number of terms in Eq. (5.1). Therefore, Eq. (5.1) is *truncated*. For example, if terms of magnitude $(\Delta x)^3$ and higher order are neglected, Eq. (5.1) reduces to

$$u_{i+1,j} \approx u_{i,j} + \left(\frac{\partial u}{\partial x}\right)_{i,j} \Delta x + \left(\frac{\partial^2 u}{\partial x^2}\right)_{i,j} \frac{(\Delta x)^2}{2} \qquad (5.2)$$

We say that Eq. (5.2) is of *second-order accuracy*, because terms of order $(\Delta x)^3$ and higher have been neglected. If terms of order $(\Delta x)^2$ and higher are neglected, we obtain from Eq. (5.1),

$$u_{i+1,j} \approx u_{i,j} + \left(\frac{\partial u}{\partial x}\right)_{i,j} \Delta x \qquad (5.3)$$

where Eq. (5.3) is of *first-order accuracy*. In Eqs. (5.2) and (5.3), the neglected higher-order terms represent the *truncation error* in the finite series representation. For example, the truncation error for Eq. (5.2) is

$$\sum_{n=3}^{\infty} \left(\frac{\partial^n u}{\partial x^n}\right)_{i,j} \frac{(\Delta x)^n}{n!}$$

and the truncation error for Eq. (5.3) is

$$\sum_{n=2}^{\infty} \left(\frac{\partial^n u}{\partial x^n}\right)_{i,j} \frac{(\Delta x)^n}{n!}$$

The truncation error can be reduced by:

(a) Carrying more terms in the Taylor's series, Eq. (5.1). This leads to higher-order accuracy in the representation of $u_{i+1,j}$.
(b) Reducing the magnitude of Δx.

Let us return to Eq. (5.1), and solve for $\left(\frac{\partial u}{\partial x}\right)_{i,j}$

$$\left(\frac{\partial u}{\partial x}\right)_{i,j} = \frac{u_{i+1,j} - u_{i,j}}{\Delta x} \underbrace{- \left(\frac{\partial^2 u}{\partial x^2}\right)_{i,j} \frac{\Delta x}{2} - \left(\frac{\partial^3 u}{\partial x^3}\right)_{i,j} \frac{\Delta x^2}{6} - \cdots}_{\text{Truncation error}}$$

or,

$$\left(\frac{\partial u}{\partial x}\right)_{i,j} = \frac{u_{i+1,j} - u_{i,j}}{\Delta x} + O(\Delta x) \qquad (5.4)$$

In Eq. (5.4), the symbol $O(\Delta x)$ is a formal mathematical notation which represents 'terms of-order-of Δx'. Eq. (5.4) is more precise notation than Eq. (5.3), which involves the 'approximately equal' notation; in Eq. (5.4) the order of magnitude of the truncation error is shown explicitly by the O notation. We now identify the first-order-accurate difference representation for the derivative $(\partial u/\partial x)_{i,j}$ expressed by Eq. (5.4) as a *first-order forward difference*, repeated below

$$\boxed{\left(\frac{\partial u}{\partial x}\right)_{i,j} = \frac{u_{i+1,j} - u_{i,j}}{\Delta x} + O(\Delta x)} \qquad (5.4 \text{ repeated})$$

Let us now write a Taylor's series expansion for $u_{i-1,j}$, expanded about $u_{i,j}$.

$$u_{i-1,j} = u_{i,j} + \left(\frac{\partial u}{\partial x}\right)_{i,j} (-\Delta x) + \left(\frac{\partial^2 u}{\partial x^2}\right)_{i,j} \frac{(-\Delta x)^2}{2}$$

$$+ \left(\frac{\partial^3 u}{\partial x^3}\right)_{i,j} \frac{(-\Delta x)^3}{6} + \cdots$$

or,

$$u_{i-1,j} = u_{i,j} - \left(\frac{\partial u}{\partial x}\right)_{i,j} \Delta x + \left(\frac{\partial^2 u}{\partial x^2}\right)_{i,j} \frac{(\Delta x)^2}{2}$$
$$- \left(\frac{\partial^3 u}{\partial x^3}\right)_{i,j} \frac{(\Delta x)^3}{6} + \cdots \tag{5.5}$$

Solving for $(\partial u/\partial x)_{i,j}$, we obtain

$$\boxed{\left(\frac{\partial u}{\partial x}\right)_{i,j} = \frac{u_{i,j} - u_{i-1,j}}{\Delta x} + O(\Delta x)} \tag{5.6}$$

Equation (5.6) is a *first order rearward difference* expression for the derivative $(\partial u/\partial x)$ at grid point (i, j).

Let us now subtract Eq. (5.5) from (5.1).

$$u_{i+1,j} - u_{i-1,j} = 2\left(\frac{\partial u}{\partial x}\right)_{i,j} \Delta x + \left(\frac{\partial^3 u}{\partial x^3}\right)_{i,j} \frac{(\Delta x)^3}{3} + \cdots \tag{5.7}$$

Solving Eq. (5.7) for $(\partial u/\partial x)_{i,j}$, we obtain

$$\boxed{\left(\frac{\partial u}{\partial x}\right)_{i,j} = \frac{u_{i+1,j} - u_{i-1,j}}{2\Delta x} + O(\Delta x)^2} \tag{5.8}$$

Equation (5.8) is a *second order central difference* for the derivative $(\partial u/\partial x)$ at grid point (i, j).

To obtain a finite-difference expression for the second partial derivative $(\partial^2 u/\partial x^2)_{i,j}$, first recall that the order-of-magnitude term in Eq. (5.8) comes from Eq. (5.7), and that Eq. (5.8) can be written

$$\left(\frac{\partial u}{\partial x}\right)_{i,j} = \frac{u_{i+1,j} - u_{i-1,j}}{2\Delta x} - \left(\frac{\partial^3 u}{\partial x^3}\right)_{i,j} \frac{(\Delta x)^2}{6} + \cdots \tag{5.9}$$

Substituting Eq. (5.9) into (5.1), we obtain

$$u_{i+1,j} = u_{i,j} + \left[\frac{u_{i+1,j} - u_{i-1,j}}{2\Delta x} - \left(\frac{\partial^3 u}{\partial x^3}\right)_{i,j} \frac{(\Delta x)^2}{6} + \cdots \right]\Delta x$$
$$+ \left(\frac{\partial^2 u}{\partial x^2}\right)_{i,j} \frac{(\Delta x)^2}{2} + \left(\frac{\partial^3 u}{\partial x^3}\right)_{i,j} \frac{(\Delta x)^3}{6}$$
$$+ \left(\frac{\partial^4 u}{\partial x^4}\right)_{i,j} \frac{(\Delta x)^4}{24} + \cdots \tag{5.10}$$

Solving Eq. (5.10) for $(\partial^2 u/\partial x^2)_{i,j}$, we obtain

$$\boxed{\left(\frac{\partial^2 u}{\partial x^2}\right)_{i,j} = \frac{u_{i+1,j} - 2u_{i,j} + u_{i-1,j}}{(\varDelta x)^2} + O(\varDelta x)^2} \tag{5.11}$$

Equation (5.11) is a *second-order central second difference* for the derivative $(\partial^2 u/\partial x^2)$ at grid point (i, j).

Difference expressions for the y-derivatives are obtained in exactly the same fashion. The results are directly analogous to the previous equations for the x-derivatives. They are:

$$\left(\frac{\partial u}{\partial y}\right)_{i,j} = \frac{u_{i,j+1} - u_{i,j}}{\varDelta y} + O(\varDelta y) \qquad\qquad \textit{Forward difference}$$

$$\left(\frac{\partial u}{\partial y}\right)_{i,j} = \frac{u_{i,j} - u_{i,j-1}}{\varDelta y} + O(\varDelta y) \qquad\qquad \textit{Rearward difference}$$

$$\left(\frac{\partial u}{\partial y}\right)_{i,j} = \frac{u_{i,j+1} - u_{i,j-1}}{2\varDelta y} + O(\varDelta y)^2 \qquad\quad \textit{Central difference}$$

$$\left(\frac{\partial^2 u}{\partial y^2}\right)_{i,j} = \frac{u_{i,j+1} - 2u_{i,j} + u_{i,j-1}}{(\varDelta y)^2} + O(\varDelta y)^2 \quad \textit{Central second difference}$$

It is interesting to note that the central second difference given for example by Eq. (5.11) can be intepreted as a forward difference of the first derivatives, with rearward differences used for the first derivatives. Dropping the O notation for convenience, we have

$$\left(\frac{\partial^2 u}{\partial x^2}\right)_{i,j} = \left[\frac{\partial}{\partial x}\left(\frac{\partial u}{\partial x}\right)\right]_{i,j} \approx \frac{\left(\frac{\partial u}{\partial x}\right)_{i+1,j} - \left(\frac{\partial u}{\partial x}\right)_{i,j}}{\varDelta x}$$

$$\left(\frac{\partial^2 u}{\partial x^2}\right)_{i,j} \approx \left[\left(\frac{u_{i+1,j} - u_{i,j}}{\varDelta x}\right) - \left(\frac{u_{i,j} - u_{i-1,j}}{\varDelta x}\right)\right]\frac{1}{\varDelta x}$$

$$\left(\frac{\partial^2 u}{\partial x^2}\right)_{i,j} \approx \frac{u_{i+1,j} - 2u_{i,j} + u_{i-1,j}}{(\varDelta x)^2} \tag{5.12}$$

Equation (5.12) is the same difference quotient as Eq. (5.11).

The same philosophy can be used to quickly generate a finite difference quotient for the mixed derivative $(\partial^2 u/\partial x \partial y)$ at grid point (i, j). For example,

$$\frac{\partial^2 u}{\partial x \partial y} = \frac{\partial}{\partial x}\left(\frac{\partial u}{\partial y}\right) \tag{5.13}$$

In Eq. (5.13), write the x-derivative as a central difference of the y-derivatives, and then cast the y-derivatives also in terms of central differences.

$$\frac{\partial^2 u}{\partial x \partial y} = \frac{\partial}{\partial x}\left(\frac{\partial u}{\partial y}\right) = \frac{\left(\frac{\partial u}{\partial y}\right)_{i+1,j} - \left(\frac{\partial u}{\partial y}\right)_{i-1,j}}{2\Delta x}$$

$$\frac{\partial^2 u}{\partial x \partial y} \approx \left[\left(\frac{u_{i+1,j+1} - u_{i+1,j-1}}{2\Delta y}\right) - \left(\frac{u_{i-1,j+1} - u_{i-1,j-1}}{2\Delta y}\right)\right]\frac{1}{2\Delta x}$$

$$\frac{\partial^2 u}{\partial x \partial y} \approx \frac{1}{4\Delta x \Delta y}(u_{i+1,j+1} + u_{i-1,j-1} - u_{i+1,j-1} - u_{i-1,j+1})$$

or

$$\boxed{\begin{aligned}\left(\frac{\partial^2 u}{\partial x \partial y}\right)_{i,j} &= \frac{1}{4\Delta x \Delta y}(u_{i+1,j+1} + u_{i-1,j-1} - u_{i+1,j-1} - u_{i-1,j+1}) \\ &\quad + O[(\Delta x)^2, (\Delta y)^2]\end{aligned}}$$

(5.14)

Many other difference approximations can be obtained for the above derivatives, as well as for derivatives of even higher order. The philosophy is the same. For a detailed tabulation of many forms of difference quotients, see pages 44 and 45 of Ref. [1].

What happens at a boundary? What type of differencing is possible when we have only one direction to go, namely, the direction away from the boundary? For example, consider Fig. 5.2, which illustrates a portion of the boundary, with the y-axis perpendicular to the boundary. Let grid point 1 be on the boundary, with points 2 and 3 a distance Δy and $2\Delta y$ above the boundary respectively. We wish to construct a finite difference approximation for $\partial u/\partial y$ at the boundary. It is easy to construct a forward difference as

$$\left(\frac{\partial u}{\partial y}\right)_1 = \frac{u_2 - u_1}{\Delta y} + O(\Delta y)$$

(5.15)

which is of first-order accuracy. However, how do we obtain a result which is of second-order accuracy? Our central difference in Eq. (5.8) fails us because it requires another point beneath the boundary, such as illustrated as point 2' in Fig. 5.2. Point 2' is outside the domain of computation, and we generally have no information about u at this point. In the early days of CFD, many solutions attempted to side-step this problem by assuming that $u_{2'} = u_2$. This is called the reflection boundary condition. In most cases it does not make physical sense, and is just as inaccurate, if not more so, than the forward difference given by Eq. (5.15).

So we ask the question again, how do we find a second-order accurate finite-difference at the boundary? The answer is simple, and it illustrates another method of deriving finite-difference quotients. Assume that at the boundary u can be expressed by the polynomial

$$u = a + by + cy^2$$

(5.16)

Applied to the grid points in Fig. 5.2, Eq. (5.16) yields

$$u_1 = a$$

$$u_2 = a + b\Delta y + c(\Delta y)^2$$
$$u_3 = a + b(2\Delta y) + c(2\Delta y)^2$$

Solving this system for b:

$$b = \frac{-3u_1 + 4u_2 - u_3}{2\Delta y} \tag{5.17}$$

Returning to Eq. (5.16), and differentiating:

$$\frac{\partial u}{\partial y} = b + 2cy \tag{5.18}$$

Equation (5.18), evaluated at the boundary where $y = 0$, yields

$$\left(\frac{\partial u}{\partial y}\right)_1 = b \tag{5.19}$$

Combining Eqs. (5.18) and (5.19), we obtain

$$\left(\frac{\partial u}{\partial y}\right)_1 = \frac{-3u_1 + 4u_2 - u_3}{2\Delta y} \tag{5.20}$$

It remains to show the order-of-accuracy of Eq. (5.20). Consider a Taylor's series expansion about the point 1.

$$u(y) = u_1 + \left(\frac{\partial u}{\partial y}\right)_1 y + \left(\frac{\partial^2 u}{\partial y^2}\right)_1 \frac{y^2}{2} + \left(\frac{\partial^3 u}{\partial y^3}\right)_1 \frac{y^3}{6} + \cdots \tag{5.21}$$

Compare Eqs. (5.21) and (5.16). Our assumed polynomial expression in Eq. (5.16) is the same as using the first three terms in the Taylor's series. Hence, Eq. (5.16) is of $O(\Delta y)^3$. In forming the derivative in Eq. (5.20), we divided by Δy, which then makes Eq. (5.20) of $O(\Delta y)^2$. Thus, we can write from Eq. (5.20)

$$\left(\frac{\partial u}{\partial y}\right)_1 = \frac{-3u_1 + 4u_2 - u_3}{2\Delta y} + O(\Delta y)^2 \tag{5.22}$$

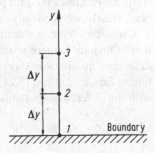

Fig. 5.2 Grid points at a boundary

This is our desired second-order-accurate difference quotient at the boundary.

Both Eqs. (5.15) and (5.22) are called *one-sided differences*, because they express a derivative at a point in terms of dependent variables on *only one side* of the point. Many other one-sided differences can be formed, with higher degrees of accuracy, using additional grid points to one side of the given point. It is not unusual to see four- and five-point one-sided differences applied at a boundary.

5.3 Basic Aspects of Finite-Difference Equations

The essence of finite-difference solutions in CFD is to use the difference quotients derived in Sect. 5.2 (or others that are similar) to replace the partial derivatives in the governing flow equations, resulting in a system of algebraic difference equations for the dependent variables at each grid point. In the present section, we examine some of the basic aspects of a difference equation.

Consider the following model equation, in which we assume that the dependent variable u is a function of x and t.

$$\frac{\partial u}{\partial t} = \frac{\partial^2 u}{\partial x^2} \tag{5.23}$$

We choose this simple equation for convenience; at this stage in our discussions there is no advantage to be obtained by dealing with the much more complex flow equations. The basic aspects of finite-difference equations to be examined in this section can just as well be developed using Eq. (5.23). It should be noted that Eq. (5.23) is parabolic.

If we replace the time derivative in Eq. (5.23) with a forward difference, and the spatial derivative with a central difference, the result is:

$$\frac{u_i^{n+1} - u_i^n}{\Delta t} = \frac{u_{i+1}^n - 2u_i^n + u_{i-1}^n}{(\Delta x)^2} \tag{5.24}$$

In Eq. (5.24), some common notation is used for the difference of the time derivative. The index for time usually appears as a *superscript* in CFD, where n denotes conditions at time t, $(n+1)$ denotes conditions at time $(t+\Delta t)$, and so forth. The subscript still denotes the grid point location; for the one spatial dimension considered here, clearly we need only one index, i.

Question: What is the truncation error for the complete finite-difference equation? Obviously, there must be a truncation error because each one of the finite-difference quotients has its own truncation error. Let us address this question. Combining Eqs. (5.23) and (5.24), and explicitly writing the truncation errors associated with the difference quotients (from Eqs. (5.4) and (5.10)), we have

$$\frac{\partial u}{\partial t} - \frac{\partial^2 u}{\partial x^2} = \frac{u_i^{n+1} - u_i^n}{\Delta t} - \frac{(u_{i+1}^n - 2u_i^n + u_{i-1}^n)}{(\Delta x)^2}$$

$$+ \left[-\left(\frac{\partial^2 u}{\partial t^2}\right)_i^n \frac{\Delta t}{2} + \left(\frac{\partial^4 u}{\partial x^4}\right)_i^n \frac{(\Delta x)^2}{12} + \cdots \right] \tag{5.25}$$

Examining Eq. (5.25), on the left-hand side is the original partial differential equation, the first two terms on the right-hand side are the finite difference representation of this equation and the terms in the square brackets are the *truncation error for the complete equation*. Note that the truncation error for this representation is $O[\Delta t, (\Delta x)^2]$.

Does the finite-difference equation reduce to the original differential equation as the number of grid points goes to infinity, i.e. as $\Delta x \to 0$ and $\Delta t \to 0$? Examining Eq. (5.25), we note that the truncation error approaches zero, and hence the difference equation does indeed approach the original differential equation. When this is the case, the finite-difference representation of the partial differential equation is said to be *consistent*.

The solution of Eq. (5.24) takes the form of a 'marching' solution in steps of time. (Recall from Sect. 4.3.2 that such marching solutions are a characteristic of parabolic equations.) Assume that we know the dependent variable at all x at some instant in time, say from given initial conditions. Examining Eq. (5.24), we see that it contains only one unknown, namely u_j^{n+1}. In this fashion, the dependent variable at time $(t + \Delta t)$ can be obtained *explicitly* from the *known* results at time t, i.e. u_j^{n+1} is obtained directly from the known values u_{j+1}^n, u_j^n, and u_{j-1}^n. This is an example of an *explicit finite-difference solution*.

As a counter example, let us be daring and return to the original partial differential equation given by Eq. (5.23). This time, we write the spatial differences on the right-hand side in terms of *average* properties between n and $(n+1)$, that is

$$\frac{u_i^{n+1} - u_i^n}{\Delta t} = \frac{1}{2}\left[\frac{u_{i+1}^{n+1} + u_{i+1}^n - 2u_i^{n+1} - 2u_i^n + u_{i-1}^{n+1} + u_{i-1}^n}{(\Delta x)^2} \right] \tag{5.26}$$

The differencing shown in Eq. (5.26) is called the *Crank-Nicolson form*. Examine Eq. (5.26) closely. The unknown u_i^{n+1} is not only expressed in terms of the known quantities at time index n, namely u_{i+1}^n, u_i^n, and u_{i-1}^n, but also in terms of unknown quantities at time index $n+1$, namely u_{i+1}^{n+1} and u_{i-1}^{n+1}. Hence, Eq. (5.26) applied at a given grid point i cannot by itself result in the solution for u_i^{n+1}. Rather, Eq. (5.26) must be written at all grid points, resulting in a system of algebraic equations from which the unknown u_i^{n+1} for all i can be solved simultaneously. This is an example of an *implicit finite-difference solution*. Because they deal with the solution of large systems of simultaneous linear algebraic equations, implicit methods are usually involved with the manipulation of large matrices.

The relative major advantages and disadvantages of these two approaches are summarized as follows.

1. *Explicit approach.*

  (a) Advantage. Relatively simple to set up and program.
  (b) Disadvantage. In terms of our above example, for a given Δx, Δt must be less than some limit imposed by stability constraints. In many cases, Δt must be

very small to maintain stability; this can result in long computer running times to make calculations over a given interval of t.

2. *Implicit approach.*

   (a) Advantage. Stability can be maintained over much larger values of Δt, hence using considerably fewer time steps to make calculations over a given interval of t. This results in less computer time.
   (b) Disadvantage. More complicated to set up and program.
   (c) Disadvantage. Since massive matrix manipulations are usually required at each time step, the computer time per time step is much larger than in the explicit approach.
   (d) Disadvantage. Since large Δt can be taken, the truncation error is larger, and the use of implicit methods to follow the exact transients (time variations of the independent variable) may not be as accurate as an explicit approach. However, for a time-dependent solution in which the steady state is the desired result, this relative time-wise inaccuracy is not important.

During the period 1969 to about 1979, the vast majority of practical CFD solutions involving 'marching' solutions (such as in the above example) employed explicit methods. Today, they are still the most straightforward methods for flow field solutions. However, many of the more sophisticated CFD applications—those requiring very closely-spaced grid points in some regions of the flow—would demand inordinately large computer running times due to the small marching steps required. This has made the advantage listed above for implicit methods very attractive, namely the ability to use large marching steps even for a very fine grid. For this reason, implicit methods are today the major focus of CFD applications.

5.3.1 A General Comment

It is clear that finite-difference solutions appear to be philosophically straightforward; just replace the partial derivatives in the governing equations with algebraic difference quotients, and grind away to obtain solutions of these algebraic equations at each grid point. However, this impression is misleading. For any given application, there is no guarantee that such calculations will be accurate, or even stable, under all conditions. Moreover, the boundary conditions for a given problem dictate the solution, and therefore the proper treatment of boundary conditions within the framework of a particular finite-difference technique is vitally important. For these reasons, finite-difference solutions of various aerodynamic flow fields are by no means routine. Indeed, much of computational fluid dynamics today is still more of an art than a science; each different problem usually requires thought and originality in its solution. However, a great deal of research in applied mathematics is now being devoted to CFD, and the next decade should see a major expansion in our understandingof the discipline, as well as the development of more improved, efficient algorithms.[1]

5.4 Errors and an Analysis of Stability

At the end of the last section, we stated that no guarantee exists for the accuracy and stability of a system of finite-difference equations under all conditions. However, for linear equations there is a formal way of examining the accuracy and stability, and these ideas at least provide guidance for the understanding of the behaviour of the more complex non-linear system that is our governing flow equations. In this section we introduce some of these ideas, applied to simple linear equations. The material in this section is patterned somewhat after section 3–6 of the excellent new book on CFD by Dale Anderson, John Tannehill and Richard Pletcher (Ref. [1]), which should be consulted for more details.

Consider a partial differential equation, such as for example Eq. (5.23). The numerical solution of this equation is influenced by two sources of error:

1. *Discretization error*. The difference between the exact analytical solution of the partial differential equation (for example, Eq. (5.23)) and the exact (round-off free) solution of the corresponding difference equation (for example, Eq. (5.24)). From our previous discussion, the discretization error is simply the truncation error for the difference equation plus any errors introduced by the numerical treatment of the boundary conditions.
2. *Round-off error*. The numerical error introduced after a repetitive number of calculations in which the computer is constantly rounding the numbers to some significant figure.

If we let

A = analytical solution of the partial differential equation
D = exact solution of the difference equation
N = numerical solution from a real computer with finite accuracy

then,

$$\text{Discretization error} = A - D$$
$$\text{Round-off} = \varepsilon = N - D \tag{5.27}$$

From Eq. (5.27), we can write

$$N = D + \varepsilon \tag{5.28}$$

---

[1] The author wishes to note in proof that the present text was written in 1985 for use in the first presentation of the VKI short course on Introduction to CFD. Hence, some statements made here are slightly dated. For example, the years since 1985 have seen substantial progress made on sophisticated and advanced algorithm development; please consult the modern CFD literature for such details.

where again ε is the round-off error, which for the remainder of our discussion in this section, we will simply call "error" for brevity. The numerical solution N must satisfy the difference equation. Hence from Eq. (5.24),

$$\frac{D_i^{n+1} + \varepsilon_i^{n+1} - D_i^n - \varepsilon_i^n}{\Delta t} = \frac{D_{i+1}^n + \varepsilon_{i+1}^n - 2D_i^n - 2\varepsilon_i^n + D_{i-1}^n \varepsilon_{i-1}^n}{(\Delta x)^2} \tag{5.29}$$

By definition, D is the *exact* solution of the difference equation, hence it exactly satisfies:

$$\frac{D_i^{n+1} - D_i^n}{\Delta t} = \frac{D_{i+1}^n - 2D_i^n + D_{i-1}^n}{(\Delta x)^2} \tag{5.30}$$

Subtracting Eq. (5.30) from (5.29),

$$\frac{\varepsilon_i^{n+1} - \varepsilon_i^n}{\Delta t} = \frac{\varepsilon_{i+1}^n - 2\varepsilon_i^n + \varepsilon_{i-1}^n}{(\Delta x)^2} \tag{5.31}$$

From Eq. (5.31), we see that the error ε also satisfies the difference equation.

We now consider aspects of the *stability* of the difference equation, Eq. (5.24). If errors ε_i are already present at some stage of the solution of this equation (as they always are in any real computer solution), then the solution will be *stable* if the ε_i's shrink, or at best stay the same, as the solution progresses from step n to $n+1$; on the other hand, if the ε_i's grow larger during the progression of the solution from steps n to $n+1$, then the solution is *unstable*. That is, for a solution to be *stable*,

$$|\varepsilon_i^{n+1} / \varepsilon_i^n| \le 1 \tag{5.32}$$

For Eq. (5.24), let us examine under what conditions Eq. (5.32) holds.

Assume that the distribution of errors along the x-axis is given by a Fourier series in x, and that the time-wise variation is exponential in t, i.e.

$$\varepsilon(x,t) = e^{at} \sum_m e^{ik_m x} \tag{5.33}$$

where k_m is the wave number and where the exponential factor a is a complex number. Since the difference equation is linear, when Eq. (5.33) is substituted into Eq. (5.31) the behaviour of each term of the series is the same as the series itself. Hence, let us deal with just one term of the series, and write

$$\varepsilon_m(x,t) = e^{at} e^{ik_m x} \tag{5.34}$$

Substitute Eq. (5.34) into Eq. (5.31),

$$\frac{e^{a(t+\Delta t)} e^{ik_m x} - e^{at} e^{ik_m x}}{\Delta t} = \frac{e^{at} e^{ik_m(x+\Delta x)} - 2e^{at} e^{ik_m x} + e^{at} e^{ik_m(x-\Delta x)}}{(\Delta x)^2} \tag{5.35}$$

Divide Eq. (5.35) by $e^{at} e^{ik_m x}$.

$$\frac{e^{a\Delta t}-1}{\Delta t} = \frac{e^{ik_m\Delta x}-2+e^{-ik_m\Delta x}}{(\Delta x)^2}$$

or,

$$e^{a\Delta t} = 1 + \frac{\Delta t}{(\Delta x)^2}(e^{ik_m\Delta x}+e^{-ik_m\Delta x}-2) \qquad (5.36)$$

Recalling the identity that

$$\cos(k_m\Delta x) = \frac{e^{ik_m\Delta x}+e^{-ik_m\Delta x}}{2}$$

Equation (5.36) can be written as

$$e^{a\Delta t} = 1 + \frac{2\Delta t}{(\Delta x)^2}[\cos(k_m\Delta x)-1] \qquad (5.37)$$

Recalling another trigonometric identity that

$$\sin^2[(k_m\Delta x)/2] = \frac{1-\cos(k_m\Delta x)}{2}$$

Equation (5.37) finally becomes

$$e^{a\Delta t} = 1 - \frac{4\Delta t}{(\Delta x)^2}\sin^2[(k_m\Delta x)/2] \qquad (5.38)$$

From Eq. (5.34),

$$\frac{\varepsilon_i^{n+1}}{\varepsilon_i^n} = \frac{e^{a(t+\Delta t)}e^{ik_m x}}{e^{at}e^{ik_m x}} = e^{a\Delta t} \qquad (5.39)$$

Combining Eqs. (5.39), (5.38) and (5.32), we have

$$\left|\frac{\varepsilon_i^{n+1}}{\varepsilon_i^n}\right| = |e^{a\Delta t}| = \left|1 - \frac{4\Delta t}{(\Delta x)^2}\sin^2[(k_m\Delta x)/2]\right| \le 1 \qquad (5.40)$$

Equation (5.40) *must* be satisfied to have a *stable* solution, as dictated by Eq. (5.32). In Eq. (5.40) the factor

$$\left|1 - \frac{4\Delta t}{(\Delta x)^2}\sin^2[(k_m\Delta x)/2]\right| \equiv G$$

is called the *amplification* factor, and is denoted by G. Evaluating the inequality in Eq. (5.40), namely $G \le 1$, we have two possible situations which must hold simultaneously:

(1) $1 - \dfrac{4\Delta t}{(\Delta x)^2}\sin^2[(k_m\Delta x)/2] \le 1$

Thus

$$\frac{4\Delta t}{(\Delta x)} \sin^2[(k_m \Delta x)/2] \geq 0$$

Since $\Delta t/(\Delta x)^2$ is always positive, this condition always holds.

(2) $\quad 1 - \dfrac{4\Delta t}{(\Delta x)^2} \sin^2[(k_m \Delta x)/2] \geq -1$

Thus

$$\frac{4\Delta t}{(\Delta x)^2} \sin^2[(k_m \Delta x)/2] - 1 \leq 1$$

For the above condition to hold,

$$\frac{\Delta t}{(\Delta x)^2} \leq \frac{1}{2} \tag{5.41}$$

Equation (5.41) gives the *stability requirement* for the solution of the difference equation, Eq. (5.24), to be *stable*. Clearly, for a given Δx, the allowed value of Δt must be small enough to satisfy Eq. (5.41). Here is a stunning example of the limitation placed on the marching variable by stability considerations for explicit finite difference models. As long as $\Delta t/(\Delta x)^2 \leq \frac{1}{2}$, the error will *not* grow for subsequent marching steps in t, and the numerical solution will proceed in a stable manner. On the other hand, if $\Delta t/(\Delta x)^2 > \frac{1}{2}$, then the error will progressively become larger, and will eventually cause the numerical marching solution to 'blow up' on the computer.

The above analysis is an example of a general method called the *von Neuman stability method*, which is used frequently to study the stability properties of linear difference equations.

Let us quickly examine the stability characteristics of another simple equation, this time a hyperbolic equation. Consider the first order wave equation:

$$\frac{\partial u}{\partial t} + c\frac{\partial u}{\partial x} = 0 \tag{5.42}$$

Let us replace the spatial derivative with a central difference (see Eq. (5.8)).

$$\frac{\partial u}{\partial x} = \frac{u_{i+1}^n - u_{i-1}^n}{2\Delta x} \tag{5.43}$$

Let us replace the time derivative with a first order difference, where $u(t)$ is represented by an average value between grid points $(i+1)$ and $(i-1)$, i.e.

$$u(t) = \frac{1}{2}(u_{i+1}^n + u_{i-1}^n)$$

Then

$$\frac{\partial u}{\partial t} = \frac{u_i^{n+1} - \frac{1}{2}(u_{i+1}^n + u_{i+1}^n)}{\Delta t} \tag{5.44}$$

Substituting Eqs. (5.43) and (5.44) into (5.42), we have

$$u_i^{n+1} = \frac{u_{i+1}^n + u_{i-1}^n}{2} - c\frac{\Delta t}{\Delta x}\left(\frac{u_{i+1}^n - u_{i-1}^n}{2}\right) \tag{5.45}$$

The differencing used in the above equation, where Eq. (5.44) is used to represent the time derivative, is called the *Lax method*, after the mathematician Peter Lax who first proposed it. If we now assume an error of the form $\varepsilon_m(x,\ t) = e^{at}e^{ik_m t}$ as done previously, and substitute this form into Eq. (5.45), the amplification factor becomes

$$G = \cos(k_m\Delta x) - iC\ \sin(k_m\Delta x) \tag{5.46}$$

where $C = c\dfrac{\Delta t}{\Delta x}$. The stability requirement is $|e^{at}| \leq 1$, which when applied to Eq. (5.46) yields

$$C = c\frac{\Delta t}{\Delta x} \leq 1 \tag{5.47}$$

In Eq. (5.47), C is called the *Courant number*. This equation says that $\Delta t \leq \Delta x/c$ for the numerical solution of Eq. (5.45) to be stable. Moreover, Eq. (5.47) is called the *Courant–Friedrichs–Lewy* condition, generally written as the CFL condition. It is an important stability criterion for hyperbolic equations.

Let us examine the physical significance of the CFL condition. Consider the second order wave equation

$$\frac{\partial^2 u}{\partial t^2} = c\frac{\partial^2 u}{\partial x^2} \tag{5.48}$$

The characteristic lines for this equation (see Sect. 4.2) are given by

$$x = ct \quad \text{(right running)}$$

and

$$x = -ct \quad \text{(left running)}$$

and are sketched in Fig. 5.3(a) and (b). In both parts (a) and (b) of Fig. 5.3, let point b be the intersection of the right-running characteristic through grid point $(i-1)$ and the left-running characteristic through grid point $(i+1)$. For Eq. (5.48), the CFL condition as given in Eq. (5.47) holds as the stability criterion. Let $\Delta t_{C=1}$ denote the value of Δt given by Eq. (5.47) when $C = 1$. Then $\Delta t_{C=1} = \Delta x/c$, and the intersection point b is therefore a distance $\Delta t_{C=1}$ above the x-axis, as sketched in Figs. 5.3(a) and (b). Now assume that $C < 1$, which is the case sketched in Fig. 5.3(a). Then from Eq. (5.47), $\Delta t_{C<1} < \Delta t_{C=1}$, as shown in Fig. 5.3(a). Let point d correspond to the grid point at point i, existing at time $(t + \Delta t_{C<1})$. Since properties at point d are calculated numerically from the difference equation using grid points $(i-1)$ and $(i+1)$, the *numerical domain* for point d is the triangle adc shown in Fig. 5.3(a). The *analytical domain* for point d is the shaded triangle in Fig. 5.3(a), defined by the characteristics through point d. Note that in Fig. 5.3(a) the numerical domain of point d *includes* the analytical domain. In contrast, consider the case shown in Fig. 5.3(b). Here, $C > 1$. Then, from Eq. (5.47), $\Delta t_{C>1} > \Delta t_{C=1}$, as shown in Fig. 5.3(b). Let point d

Fig. 5.3 Illustration of the physical significance of the CFL condition

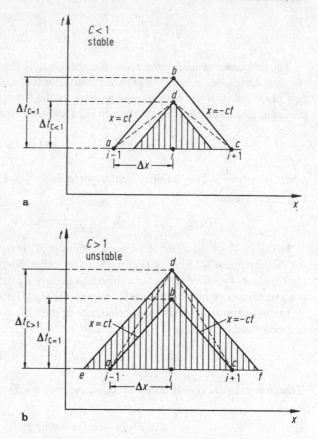

in Fig. 5.3(b) correspond to the grid point i, existing at time $(t + \Delta t_{C>1})$. Since properties at point d are calculated numerically from the difference equation using grid points $(i-1)$ and $(i+1)$, the *numerical domain* for point d is the triangle adc shown in Fig. 5.3(b). The *analytical domain* for point d is the shaded triangle in Fig. 5.3(b), defined by the characteristics through point d. Note that in Fig. 5.3(b), the numerical domain does *not* include all of the analytical domain, and it is this condition which leads to unstable behaviour. Therefore, we can give the following physical interpretation of the CFL condition:

For stability, the computational domain must include all of the analytical domain.

The above considerations dealt with *stability*. The question of *accuracy*, which is sometimes quite different, can also be examined from the point of view of Fig. 5.3. Consider a stable case, as shown in Fig. 5.3(a). Note that the analytic domain of dependence for point d is the shaded triangle in Fig. 5.3(a). From our discussion in Chap. 4, the properties at point d theoretically depend only on those points within the shaded triangle. However, note that the numerical grid points $(i-1)$ and $(i+1)$ are *outside* the domain of dependence, and hence *theoretically* should not influence the properties at point d. On the other hand, the *numerical* calculation of properties at point d takes information from grid points $(i-1)$ and $(i+1)$. This situation is exacerbated when $\Delta t_{C<1}$ is chosen to be very small, $\Delta t_{C<1} \ll \Delta t_{C=1}$. In this case, even though the calculations are stable, the results may be quite *inaccurate* due to the large mismatch between the domain of dependence of point d, and the location of the actual numerical data used to calculate properties at d.

In light of the above discussion, we conclude that the Courant number must be equal to or less than unity for stability, $C \leq 1$, but at the same time it is desirable to have C as close to unity as possible for accuracy.

Reference

1. Anderson, D.A., Tannehill, John C. and Pletcher, Richard H., *Computational Fluid Mechanics and Heat Transfer*, McGraw-Hill, New York, 1984.

Chapter 6
Transformations and Grids

J.D. Anderson, Jr.

6.1 Introduction

If all CFD applications dealt with physical problems where a uniform, rectangular grid could be used in the physical plane, there would be no reason to alter the governing equations derived in Chap. 2. We would simply apply these equations in rectangular (x, y, z, t) space, finite-difference these equations according to the difference quotients derived in Chap. 5, and calculate away, using uniform values of Δx, Δy, Δz and Δt. However, few real problems are ever so accommodating. For example, assume we wish to calculate the flow over an airfoil, as sketched in Fig. 6.1, where we have placed the airfoil in a rectangular grid. Note the problems with this rectangular grid:

(1) Some grid points fall inside the airfoil, where they are completely out of the flow. What values of the flow properties do we ascribe to these points?
(2) There are few, if any, grid points that fall on the surface of the airfoil. This is not good, because the airfoil surface is a vital boundary condition for the determination of the flow, and hence the airfoil surface must be clearly and strongly seen by the numerical solution.

As a result, we can conclude that the rectangular grid in Fig. 6.1 is not appropriate for the solution of the flow field. In contrast, a grid that *is* appropriate is sketched in Fig. 6.2(a). Here we see a *non-uniform, curvilinear* grid which is literally wrapped around the airfoil. New coordinate lines, ξ and η, are defined such that the airfoil surface becomes a coordinate line, $\eta =$ constant. This is called a boundary-fitted coordinate system, and will be discussed in detail later in this chapter. The important point is that grid points *naturally* fall on the airfoil surface, as shown in Fig. 6.2(a). What is equally important is that, in the physical space shown in Fig. 6.2(a), the grid is *not* rectangular, and is not uniformly spaced. As a consequence, the conventional difference quotients are difficult to use. What must be done is to *transform* the curvilinear grid mesh in physical space to a *rectangularmesh* in terms of

J.D. Anderson, Jr.
National Air and Space Museum, Smithsonian Institution, Washington, DC
e-mail: AndersonJA@si.edu

J.F. Wendt (ed.), *Computational Fluid Dynamics*, 3rd ed.,
© Springer-Verlag Berlin Heidelberg 2009

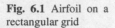

Fig. 6.1 Airfoil on a rectangular grid

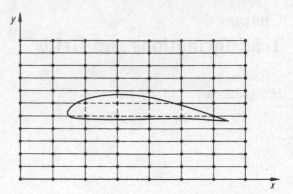

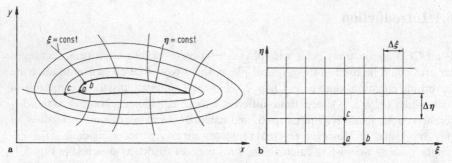

Fig. 6.2 (a) Physical plane. (b) Computational plane

ξ and η. This is shown in Fig. 6.2(b), which illustrates a rectangular grid in terms of ξ and η. The rectangular mesh shown in Fig. 6.2(b) is called the *computational plane*. There is a one-to-one correspondence between this mesh, and the curvilinear mesh in Fig. 6.2(a), called the *physical plane*. For example, points a, b and c in the physical plane (Fig. 6.2a) correspond to points a, b and c in the computational plane, which involves uniform $\Delta\xi$ and uniform $\Delta\eta$. The computed information is then transferred back to the physical plane. Moreover, when the governing equations are solved in the computational space, they must be expressed in terms of the variables ξ and η rather than x and y; i.e., the governing equations must be transformed from (x, y) to (ξ, η) as the new independent variables.

The purpose of this chapter is to first describe the general transformation of the governing flow equations between the physical plane and the computational plane. Following this, various specific grids will be discussed. This material is an example of a very active area of CFD research called *grid generation*.

6.2 General Transformation of the Equations

For simplicity, we will consider a two-dimensional unsteady flow, with independent variables x, y and t; the results for a three-dimensional unsteady flow, with independent variables x, y, z and t, are analogous, and simply involve more terms.

We will transform the variables in physical space (x, y, t) to a transformed space (ξ, η, τ), where

$$\xi = \xi(x,y,t) \tag{6.1a}$$

$$\eta = \eta(x,y,t) \tag{6.1b}$$

$$\tau = \tau(t) \tag{6.1c}$$

In the above transformation, τ is considered a function of t only, and is frequently given by $\tau = t$. This seems rather trivial; however, Eq. (6.1c) must be carried through the transformation in a formal manner, or else certain necessary terms will not be generated. From the chain rule of differential calculus, we have

$$\left(\frac{\partial}{\partial x}\right)_{y,t} = \left(\frac{\partial}{\partial \xi}\right)_{\eta,\tau}\left(\frac{\partial \xi}{\partial x}\right)_{y,t} + \left(\frac{\partial}{\partial \eta}\right)_{\xi,\tau}\left(\frac{\partial \eta}{\partial x}\right)_{y,t}$$
$$+ \left(\frac{\partial}{\partial \tau}\right)_{\xi,\eta}\left(\frac{\partial \tau}{\partial x}\right)_{y,t}^{\!\!\!\!0}$$

The subscripts in the above expression are added to emphasize what variables are being held constant in the partial differentiation. In our subsequent expressions, subscripts will be dropped; however, it is always useful to keep them in your mind. Thus, we will write the above expression as

$$\boxed{\frac{\partial}{\partial x} = \left(\frac{\partial}{\partial \xi}\right)\left(\frac{\partial \xi}{\partial x}\right) + \left(\frac{\partial}{\partial \eta}\right)\left(\frac{\partial \eta}{\partial x}\right)} \tag{6.2}$$

Similarly,

$$\boxed{\frac{\partial}{\partial y} = \left(\frac{\partial}{\partial \xi}\right)\left(\frac{\partial \xi}{\partial y}\right) + \left(\frac{\partial}{\partial \eta}\right)\left(\frac{\partial \eta}{\partial y}\right)} \tag{6.3}$$

Also,

$$\left(\frac{\partial}{\partial t}\right)_{x,y} = \left(\frac{\partial}{\partial \xi}\right)_{\eta,\tau}\left(\frac{\partial \xi}{\partial t}\right)_{x,y} + \left(\frac{\partial}{\partial \eta}\right)_{\xi,\eta}\left(\frac{\partial \eta}{\partial t}\right)_{x,y}$$
$$+ \left(\frac{\partial}{\partial \tau}\right)_{\xi,\eta}\left(\frac{\partial \tau}{\partial t}\right)_{x,y} \tag{6.4}$$

or,

$$\boxed{\frac{\partial}{\partial t} = \left(\frac{\partial}{\partial \xi}\right)\left(\frac{\partial \xi}{\partial t}\right) + \left(\frac{\partial}{\partial \eta}\right)\left(\frac{\partial \eta}{\partial t}\right) + \frac{\partial}{\partial \tau}\frac{d\tau}{dt}} \tag{6.5}$$

Equations (6.2), (6.3) and (6.5) allow the *derivatives* with respect to x, y and t to be transformed into derivatives with respect to ξ, η and τ. The *coefficients* of the derivatives with respect to ξ, η and τ are called metrics, e.g. $\partial\xi/\partial x, \partial\xi/\partial y, \partial\eta/\partial x$ and $\partial\eta/\partial y$ are metric terms which can be obtained from the general transformation given by Eqs. (6.1a, b and c). If Eqs. (6.1a, b and c) are given as closed form analytic expressions, then the metrics can also be obtained in closed form. However, the transformation given by Eqs. (6.1a, b and c) is frequently a purely numerical relationship, in which case the metrics can be evaluated by finite-difference quotients—typically central differences.

Examining the governing equations derived in Chap. 2, we note that the equations for viscous flow involve second derivatives. Therefore, we need a transformation for these derivatives; they can be obtained as follows. From Eq. (6.2), let

$$A = \frac{\partial}{\partial x} = \left(\frac{\partial}{\partial \xi}\right)\left(\frac{\partial \xi}{\partial x}\right) + \left(\frac{\partial}{\partial \eta}\right)\left(\frac{\partial \eta}{\partial x}\right)$$

Then,

$$\frac{\partial^2}{\partial x^2} = \frac{\partial A}{\partial x} = \frac{\partial}{\partial x}\left[\left(\frac{\partial}{\partial \xi}\right)\left(\frac{\partial \xi}{\partial x}\right) + \left(\frac{\partial}{\partial \eta}\right)\left(\frac{\partial \eta}{\partial x}\right)\right]$$

$$= \left(\frac{\partial}{\partial \xi}\right)\left(\frac{\partial^2 \xi}{\partial x^2}\right) + \left(\frac{\partial \xi}{\partial x}\right)\underbrace{\left(\frac{\partial^2}{\partial x \partial \xi}\right)}_{B} + \left(\frac{\partial}{\partial \eta}\right)\left(\frac{\partial^2 \eta}{\partial x^2}\right) + \left(\frac{\partial \eta}{\partial x}\right)\underbrace{\left(\frac{\partial^2}{\partial \eta \partial x}\right)}_{C} \quad (6.6)$$

The mixed derivatives denoted by B and C in Eq. (6.6) can be obtained from the chain rule as follows.

$$B = \frac{\partial^2}{\partial x \partial \xi} = \frac{\partial}{\partial x}\left(\frac{\partial}{\partial \xi}\right)$$

Recalling the chain rule given by Eq. (6.2), we have

$$B = \left(\frac{\partial^2}{\partial \xi^2}\right)\left(\frac{\partial \xi}{\partial x}\right) + \left(\frac{\partial^2}{\partial \eta \partial \xi}\right)\left(\frac{\partial \eta}{\partial x}\right) \quad (6.7)$$

Similarly:

$$C = \frac{\partial^2}{\partial x \partial \eta} = \frac{\partial}{\partial x}\left(\frac{\partial}{\partial \eta}\right) = \left(\frac{\partial^2}{\partial \xi \partial \eta}\right)\left(\frac{\partial \xi}{\partial x}\right) + \left(\frac{\partial^2}{\partial \eta^2}\right)\left(\frac{\partial \eta}{\partial x}\right) \quad (6.8)$$

Substituting B and C from Eqs. (6.7) and (6.8) into Eq. (6.6), and rearranging the sequence of terms, we have

$$\boxed{\begin{aligned} \frac{\partial^2}{\partial x^2} &= \left(\frac{\partial}{\partial \xi}\right)\left(\frac{\partial^2 \xi}{\partial x^2}\right) + \left(\frac{\partial}{\partial \eta}\right)\left(\frac{\partial^2 \eta}{\partial x^2}\right) + \left(\frac{\partial^2}{\partial \xi^2}\right)\left(\frac{\partial \xi}{\partial x}\right)^2 \\ &+ \left(\frac{\partial^2}{\partial \eta}\right)\left(\frac{\partial \eta}{\partial x}\right)^2 + 2\left(\frac{\partial^2}{\partial \eta \partial \xi}\right)\left(\frac{\partial \eta}{\partial x}\right)\left(\frac{\partial \xi}{\partial x}\right) \end{aligned}} \quad (6.9)$$

Equation (6.9) gives the second partial derivative with respect to x in terms of first, second and mixed derivatives with respect to ξ and η, multiplied by various metric terms. Let us now continue to obtain the second partial with respect to y. From Eq. (6.3), let

$$D \equiv \frac{\partial}{\partial y} = \left(\frac{\partial}{\partial \xi}\right)\left(\frac{\partial \xi}{\partial y}\right) + \left(\frac{\partial}{\partial \eta}\right)\left(\frac{\partial \eta}{\partial y}\right)$$

Then,

$$
\begin{aligned}
\frac{\partial^2}{\partial y^2} = \frac{\partial D}{\partial y} &= \frac{\partial}{\partial y}\left[\left(\frac{\partial}{\partial \xi}\right)\left(\frac{\partial \xi}{\partial y}\right) + \left(\frac{\partial}{\partial \eta}\right)\left(\frac{\partial \eta}{\partial y}\right)\right] \\
&= \left(\frac{\partial}{\partial \xi}\right)\left(\frac{\partial^2 \xi}{\partial y^2}\right) + \left(\frac{\partial \xi}{\partial y}\right)\underbrace{\left(\frac{\partial^2}{\partial \xi \partial y}\right)}_{E} + \left(\frac{\partial}{\partial \eta}\right)\left(\frac{\partial^2 \eta}{\partial y^2}\right) + \left(\frac{\partial \eta}{\partial y}\right)\underbrace{\left(\frac{\partial^2}{\partial \eta \partial y}\right)}_{F}
\end{aligned}
\tag{6.10}
$$

Using Eq. (6.3),

$$E = \frac{\partial}{\partial y}\left(\frac{\partial}{\partial \xi}\right) = \left(\frac{\partial^2}{\partial \xi^2}\right)\left(\frac{\partial \xi}{\partial y}\right) + \left(\frac{\partial^2}{\partial \eta \partial \xi}\right)\left(\frac{\partial \eta}{\partial y}\right) \tag{6.11}$$

and

$$F = \frac{\partial}{\partial y}\left(\frac{\partial}{\partial \eta}\right) = \left(\frac{\partial^2}{\partial \eta \partial \xi}\right)\left(\frac{\partial \xi}{\partial y}\right) + \left(\frac{\partial^2}{\partial \eta^2}\right)\left(\frac{\partial \eta}{\partial y}\right) \tag{6.12}$$

Substituting Eqs. (6.11) and (6.12) into (6.10), we have, after rearranging the sequence of terms:

$$
\boxed{
\begin{aligned}
\frac{\partial^2}{\partial y^2} &= \left(\frac{\partial}{\partial \xi}\right)\left(\frac{\partial^2 \xi}{\partial y^2}\right) + \left(\frac{\partial}{\partial \eta}\right)\left(\frac{\partial^2 \eta}{\partial y^2}\right) + \left(\frac{\partial^2}{\partial \xi^2}\right)\left(\frac{\partial \xi}{\partial y}\right)^2 \\
&\quad + \left(\frac{\partial^2}{\partial \eta^2}\right)\left(\frac{\partial \eta}{\partial y}\right)^2 + 2\left(\frac{\partial^2}{\partial \eta \partial \xi}\right)\left(\frac{\partial \eta}{\partial y}\right)\left(\frac{\partial \xi}{\partial y}\right)
\end{aligned}
}
\tag{6.13}
$$

Equation (6.13) gives the second partial derivative with respect to y in terms of first, second and mixed derivatives with respect to ξ and η, multiplied by various metric terms. We now continue to obtain the second partial with respect to x and y.

$$
\begin{aligned}
\frac{\partial^2}{\partial x \partial y} = \frac{\partial}{\partial x}\left(\frac{\partial}{\partial y}\right) = \frac{\partial D}{\partial x} &= \frac{\partial}{\partial x}\left[\left(\frac{\partial}{\partial \xi}\right)\left(\frac{\partial \xi}{\partial y}\right) + \left(\frac{\partial}{\partial \eta}\right)\left(\frac{\partial \eta}{\partial y}\right)\right] \\
&= \left(\frac{\partial}{\partial \xi}\right)\left(\frac{\partial^2 \xi}{\partial x \partial y}\right) + \left(\frac{\partial \xi}{\partial y}\right)\underbrace{\left(\frac{\partial^2}{\partial \xi \partial x}\right)}_{B} + \left(\frac{\partial}{\partial \eta}\right)\left(\frac{\partial^2 \eta}{\partial x \partial y}\right) + \left(\frac{\partial \eta}{\partial y}\right)\underbrace{\left(\frac{\partial^2}{\partial \eta \partial x}\right)}_{C}
\end{aligned}
\tag{6.14}
$$

Substituting Eqs. (6.7) and (6.8) for B and C respectively into Eq. (6.14), and rearranging the sequence of terms, we have

$$
\boxed{
\begin{aligned}
\frac{\partial^2}{\partial x \partial y} &= \left(\frac{\partial}{\partial \xi}\right)\left(\frac{\partial^2 \xi}{\partial x \partial y}\right) + \left(\frac{\partial}{\partial \eta}\right)\left(\frac{\partial^2 \eta}{\partial x \partial y}\right) + \left(\frac{\partial^2}{\partial \xi^2}\right)\left(\frac{\partial \xi}{\partial x}\right)\left(\frac{\partial \xi}{\partial y}\right) \\
&\quad + \left(\frac{\partial^2}{\partial \eta^2}\right)\left(\frac{\partial \eta}{\partial x}\right)\left(\frac{\partial \eta}{\partial y}\right) + \left(\frac{\partial^2}{\partial \eta \partial \xi}\right)\left[\left(\frac{\partial \eta}{\partial x}\right)\left(\frac{\partial \xi}{\partial y}\right) + \left(\frac{\partial \xi}{\partial x}\right)\left(\frac{\partial \eta}{\partial y}\right)\right]
\end{aligned}
}
\tag{6.15}
$$

Equation (6.15) gives the second partial derivative with respect to x and y in terms of first, second and mixed derivatives with respect to ξ and η, multiplied by various metric terms.

Examine all the equations given in the boxes above. They represent all that is necessary to transform the governing flow equations obtained in Chap. 2 with x, y and t as the independent variables to ξ, η and τ as the *new* independent variables. Clearly, when this transformation is made, the governing equations in terms of ξ, η and τ become rather lengthy. Let us consider a simple example, namely that for inviscid, irrotational, steady, incompressible flow, for which Laplace's Equation is the governing equation.

$$\text{Laplace's Equation}: \quad \frac{\partial^2 \phi}{\partial x^2} + \frac{\partial^2 \phi}{\partial y^2} = 0 \qquad (6.16)$$

Transforming Eq. (6.16) from (x, y) to (ξ, η), where $\xi = \xi(x, y)$ and $\eta = \eta(x, y)$, we have from Eqs. (6.9) and (6.13):

$$\left(\frac{\partial^2 \phi}{\partial \xi^2}\right)\left(\frac{\partial \xi}{\partial x}\right)^2 + 2\left(\frac{\partial^2 \phi}{\partial \xi \partial \eta}\right)\left(\frac{\partial \eta}{\partial x}\right)\left(\frac{\partial \xi}{\partial x}\right) + \left(\frac{\partial^2 \phi}{\partial \eta^2}\right)\left(\frac{\partial \eta}{\partial x}\right)^2$$

$$+ \left(\frac{\partial \phi}{\partial \xi}\right)\left(\frac{\partial^2 \xi}{\partial x^2}\right) + \left(\frac{\partial \phi}{\partial \eta}\right)\left(\frac{\partial^2 \eta}{\partial x^2}\right) + \left(\frac{\partial^2 \phi}{\partial \xi^2}\right)\left(\frac{\partial \xi}{\partial y}\right)^2$$

$$+ 2\left(\frac{\partial^2 \phi}{\partial \eta \partial \xi}\right)\left(\frac{\partial \eta}{\partial y}\right)\left(\frac{\partial \xi}{\partial y}\right) + \left(\frac{\partial^2 \phi}{\partial \eta^2}\right)\left(\frac{\partial \eta}{\partial y}\right)^2$$

$$+ \left(\frac{\partial \phi}{\partial \xi}\right)\left(\frac{\partial^2 \xi}{\partial y^2}\right) + \left(\frac{\partial \phi}{\partial \eta}\right)\left(\frac{\partial^2 \eta}{\partial y^2}\right) = 0$$

Rearranging terms, we obtain

$$\frac{\partial^2 \phi}{\partial \xi^2}\left[\left(\frac{\partial \xi}{\partial x}\right)^2 + \left(\frac{\partial \xi}{\partial y}\right)^2\right] + \frac{\partial^2 \phi}{\partial \eta^2}\left[\left(\frac{\partial \eta}{\partial x}\right)^2 + \left(\frac{\partial \eta}{\partial y}\right)^2\right]$$

$$+ 2\frac{\partial^2 \phi}{\partial \xi \partial \eta}\left[\left(\frac{\partial \eta}{\partial x}\right)\left(\frac{\partial \xi}{\partial x}\right) + \left(\frac{\partial \eta}{\partial y}\right)\left(\frac{\partial \xi}{\partial y}\right)\right]$$

$$+ \frac{\partial \phi}{\partial \xi}\left[\frac{\partial^2 \xi}{\partial x^2} + \frac{\partial^2 \xi}{\partial y^2}\right] + \frac{\partial \phi}{\partial \eta}\left[\frac{\partial^2 \eta}{\partial x^2} + \frac{\partial^2 \eta}{\partial y^2}\right] = 0 \qquad (6.17)$$

Examine Eqs. (6.16) and (6.17); the former is Laplace's equation in the physical (x, y) space, and the latter is the transformed Laplace's equation in the computational (ξ, η) space. The transformed equation clearly contains many more terms.

Once again we emphasize that Eqs. (6.1), (6.2), (6.3), (6.5), (6.9), (6.13) and (6.15) are used to transform the governing flow equations from the physical plane (x, y space) to the computational plane (ξ, η space), and that the purpose of the transformation in most CFD applications is to transform a non-uniform grid in physical space (such as shown in Fig. 6.2a) to a *uniform* grid in the computational space (such as shown in Fig. 6.2b). The transformed governing partial differential equations are then finite-differenced in the computational plane, where there exists a uniform $\Delta\xi$ and a uniform $\Delta\eta$, as shown in Fig. 6.2(b). The flow-field variables are calculated at all grid points in the computational plane, such as points a, b and c in Fig. 6.2(b). These are the same flow-field variables which exist in the physical plane at the corresponding points a, b and c in Fig. 6.2(a). The transformation that accomplishes all this is given in general form by Eqs. (6.1a, b and c). Of course, to carry out a solution for a given problem, the transformation given generically by Eqs. (6.1a, b and c) must be explicitly specified. Examples of some specific transformations will be given in subsequent sections.

6.3 Metrics and Jacobians

In Eqs. (6.2), (6.3), (6.4), (6.5), (6.6), (6.7), (6.8), (6.9), (6.10), (6.11), (6.12), (6.13), (6.14), (6.15), the terms involving the geometry of the grids, such as $\partial\xi/\partial x$, $\partial\xi/\partial y$, $\partial\eta/\partial x$, $\partial\eta/\partial y$, etc., are called *metrics*. If the transformation, Eq. (6.1a, b and c), is given analytically, then it is possible to obtain analytic values for the metric terms. However, in many CFD applications, the transformation, Eq. (6.1a, b and c), is given numerically, and hence the metric terms are calculated as finite differences.

Also, in many applications, the transformation may be more conveniently expressed as the *inverse* of Eqs. (6.1a, b), that is, we may have available the *inverse transformation*.

$$x = x(\xi,\eta,\tau) \tag{6.18a}$$

$$y = y(\xi,\eta,\tau) \tag{6.18b}$$

$$t = t(\tau) \tag{6.18c}$$

In Eqs. (6.18a, b and c), ξ, η and τ are the *independent* variables. However, in the derivative transformations given by Eqs. (6.2), (6.3), (6.4), (6.5), (6.6), (6.7), (6.8), (6.9), (6.10), (6.11), (6.12), (6.13), (6.14), and (6.15), the metric terms $\partial\xi/\partial x$, $\partial\eta/\partial y$, etc. are partial derivatives in terms of x, y and t as the independent variables. Therefore, in order to calculate the metric terms in these equations from the inverse transformation in Eqs. (6.18a, b and c), we need to relate $\partial\xi/\partial x$, $\partial\eta/\partial y$, etc. to the inverse forms $\partial x/\partial\xi$, $\partial y/\partial\eta$, etc. These inverse forms of the metrics are the values which can be directly obtained from the inverse transformation, Eqs. (6.18a, b and c). Let us proceed to find such relations.

Consider a dependent variable in the governing flow equations, such as the x-component of velocity, u. Let $u = u(x, y)$, where from Eqs. (6.18a and b), $x = x(\xi, \eta)$ and $y = y(\xi, \eta)$. The total differential of u is given by

$$du = \frac{\partial u}{\partial x}\,dx + \frac{\partial u}{\partial y}\,dy \tag{6.19}$$

It follows from Eq. (6.19) that

$$\frac{\partial u}{\partial \xi} = \frac{\partial u}{\partial x}\frac{\partial x}{\partial \xi} + \frac{\partial u}{\partial y}\frac{\partial y}{\partial \xi} \tag{6.20}$$

and

$$\frac{\partial u}{\partial \eta} = \frac{\partial u}{\partial x}\frac{\partial x}{\partial \eta} + \frac{\partial u}{\partial y}\frac{\partial y}{\partial \eta} \tag{6.21}$$

Equations (6.20) and (6.21) can be viewed as two equations for the two unknowns $\partial u/\partial x$ and $\partial u/\partial y$. Solving the system of equations (6.20) and (6.21) for $\partial u/\partial x$ using Cramer's rule, we have

$$\frac{\partial u}{\partial x} = \frac{\begin{vmatrix} \dfrac{\partial u}{\partial \xi} & \dfrac{\partial y}{\partial \xi} \\[2mm] \dfrac{\partial u}{\partial \eta} & \dfrac{\partial y}{\partial \eta} \end{vmatrix}}{\begin{vmatrix} \dfrac{\partial x}{\partial \xi} & \dfrac{\partial y}{\partial \xi} \\[2mm] \dfrac{\partial x}{\partial \eta} & \dfrac{\partial y}{\partial \eta} \end{vmatrix}} \tag{6.22}$$

In Eq. (6.22), the denominator determinant is identified as the *Jacobian determinant*, denoted by

$$J \equiv \frac{\partial(x,y)}{\partial(\xi,n)} \equiv \begin{vmatrix} \dfrac{\partial x}{\partial \xi} & \dfrac{\partial y}{\partial \xi} \\[2mm] \dfrac{\partial x}{\partial \eta} & \dfrac{\partial y}{\partial \eta} \end{vmatrix}$$

Hence, Eq. (6.22) can be written as

$$\boxed{\frac{\partial u}{\partial x} = \frac{1}{J}\left[\left(\frac{\partial u}{\partial \xi}\right)\left(\frac{\partial y}{\partial \eta}\right) - \left(\frac{\partial u}{\partial \eta}\right)\left(\frac{\partial y}{\partial \xi}\right)\right]} \tag{6.23}$$

Now let us return to Eqs. (6.20) and (6.21), and solve for $\partial u/\partial y$.

$$\frac{\partial u}{\partial y} = \frac{\begin{vmatrix} \dfrac{\partial x}{\partial \xi} & \dfrac{\partial u}{\partial \xi} \\[2mm] \dfrac{\partial x}{\partial \eta} & \dfrac{\partial u}{\partial \eta} \end{vmatrix}}{\begin{vmatrix} \dfrac{\partial x}{\partial \xi} & \dfrac{\partial y}{\partial \xi} \\[2mm] \dfrac{\partial x}{\partial \eta} & \dfrac{\partial y}{\partial \eta} \end{vmatrix}}$$

or,

$$\boxed{\frac{\partial u}{\partial y} = \frac{1}{J}\left[\left(\frac{\partial u}{\partial \eta}\right)\left(\frac{\partial x}{\partial \xi}\right) - \left(\frac{\partial u}{\partial \xi}\right)\left(\frac{\partial x}{\partial \eta}\right)\right]} \tag{6.24}$$

Examine Eqs. (6.23) and (6.24). They express the derivatives of the flow field variables in physical space in terms of the derivatives of the flowfield variables in computational space. Equations (6.23) and (6.24) accomplish the same derivative transformations as given by Eqs. (6.2) and (6.3). However, unlike Eqs. (6.2) and (6.3) where the metric terms are $\partial\xi/\partial x$, $\partial\eta/\partial y$, etc., the new Eqs. (6.23) and (6.24) involve the inverse metrics, $\partial x/\partial\xi$, $\partial y/\partial\eta$, etc. Also notice that Eqs. (6.23) and (6.24) include the Jacobian of the transformation. Therefore, whenever you have the transformation given in the form of Eqs. (6.18a, b and c), from which you can readily obtain the metrics in the form $\partial x/\partial\xi$, $\partial x/\partial\eta$, etc., the transformed governing flow equations can be expressed in terms of these inverse metrics and the Jacobian, J.

A similar but more lengthy set of results can be obtained for a three-dimensional transformation from (x, y, z) to (ξ, η, ζ). Consult Ref. [1] for more details. Our discussion above has been intentionally limited to two dimensions in order to demonstrate the basic principles without cluttering the consideration with details.

6.4 Coordinate Stretching

In the remaining three sections of this chapter, we examine three types of grid transformations. The simplest is discussed here. It consists of stretching the grid in one or more coordinate directions.

For example, consider the physical and computational planes shown in Fig. 6.3(a, b). Assume that we are dealing with the viscous flow over a flat surface, where the velocity varies rapidly near the surface as shown in the velocity profile sketched at the right of the physical plane (Fig. 6.3a). To calculate the details of this flow near the surface, a finely spaced grid in the y-direction should be used, as sketched in the physical plane. However, far away from the surface, the grid can be more coarse. Therefore, a proper grid should be one in which the coordinate lines become progressively more closely spaced as the surface is approached. On the other hand, we wish to deal with a uniform grid in the computational plane, as shown in Fig. 6.3(b). On examination, we see that the grid in the physical space is 'stretched', as if a uniform grid were drawn on a piece of rubber, and then the upper portion of the rubber were stretched upward in the y-direction. A simple analytical transformation which can accomplish this grid stretching is:

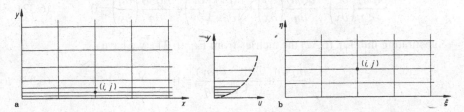

Fig. 6.3 Example of grid stretching. (**a**) Physical plane. (**b**) Computational plane

$$\xi = x \tag{6.25a}$$
$$\eta = \ln(y+1) \tag{6.25b}$$

The *inverse* transformation is

$$x = \xi \tag{6.26a}$$
$$y = e^\eta - 1 \tag{6.26b}$$

from which the inverse metrics are obtained as:

$$\frac{\partial x}{\partial \xi} = 1; \quad \frac{\partial x}{\partial \eta} = 0; \quad \frac{\partial y}{\partial \xi} = 0; \quad \frac{\partial y}{\partial \eta} = e^\eta \tag{6.27}$$

Let us consider the continuity equation, given by Eq. (2.27). For steady, two-dimensional flow, this is

$$\frac{\partial(\rho u)}{\partial x} + \frac{\partial(\rho v)}{\partial y} = 0 \tag{6.28}$$

Equation (6.27) is the continuity equation written in terms of the physical plane. This equation can be formally transformed by means of the general results given by Eqs. (6.23) and (6.24), obtaining

$$\frac{1}{J}\left[\frac{\partial(\rho u)}{\partial \xi}\left(\frac{\partial y}{\partial \eta}\right) - \frac{\partial(\rho u)}{\partial \eta}\left(\frac{\partial y}{\partial \xi}\right)\right] + \frac{1}{J}\left[\frac{\partial(\rho v)}{\partial \eta}\left(\frac{\partial x}{\partial \xi}\right) - \frac{\partial(\rho v)}{\partial \xi}\left(\frac{\partial x}{\partial \eta}\right)\right] = 0 \tag{6.29}$$

Substituting into Eq. (6.29) the inverse metrics from Eq. (6.27), we have

$$e^\eta \frac{\partial(\rho u)}{\partial \xi} + \frac{\partial(\rho v)}{\partial \eta} = 0 \tag{6.30}$$

Equation (6.30) is the continuity equation in the computational plane.

Equation (6.30) can also be obtained from the direct transformation given by Eqs. (6.25a and b). Here, the metrics are:

$$\frac{\partial \xi}{\partial x} = 1; \quad \frac{\partial \xi}{\partial y} = 0; \quad \frac{\partial \eta}{\partial x} = 0; \quad \frac{\partial \eta}{\partial y} = \frac{1}{y+1} \tag{6.31}$$

Using the transformations given by Eqs. (6.2) and (6.3), Eq. (6.28) becomes

$$\frac{\partial(\rho u)}{\partial \xi}\left(\frac{\partial \xi}{\partial x}\right) + \frac{\partial(\rho u)}{\partial \eta}\left(\frac{\partial \eta}{\partial x}\right) + \frac{\partial(\rho v)}{\partial \xi}\left(\frac{\partial \xi}{\partial y}\right) + \frac{\partial(\rho v)}{\partial \eta}\left(\frac{\partial \eta}{\partial y}\right) = 0 \tag{6.32}$$

Substituting into Eq. (6.32) the metrics from Eq. (6.31), we have

$$\frac{\partial(\rho u)}{\partial \xi} + \frac{1}{(y+1)}\frac{\partial(\rho v)}{\partial \eta} = 0 \tag{6.33}$$

However, from Eq. (6.26b), $y + 1 = e^{\eta}$. Therefore, Eq. (6.33) becomes

$$\frac{\partial(\rho u)}{\partial \xi} + \frac{1}{e^{\eta}} \frac{\partial(\rho v)}{\partial \eta} = 0$$

or

$$e^{\eta} \frac{\partial(\rho u)}{\partial \xi} + \frac{\partial(\rho v)}{\partial \eta} = 0 \qquad (6.34)$$

Equation (6.34) is identical to Eq. (6.30). All that we have done here is to demonstrate how the transformed equation can be obtained from either the direct transformation or the inverse transformation; the results are the same.

An example of more complex grid stretching, in both the x- and y-directions, is given in Refs. [2, 3]. Here, the supersonic viscous flow over a blunt base is studied. The physical and computational planes are illustrated in Fig. 6.4. The streamwise stretching is accomplished through a transformation originally used by Holst [4]

$$x = \frac{\xi_0}{A} [\sinh((\xi - x_0)\beta_x) + A]$$

where

$$A = \sinh(\beta_x x_0)$$

and

$$x_0 = \frac{1}{2\beta_x} \ln \left[\frac{1 + (e^{\beta_x} - 1)\xi_0}{1 + (e^{-\beta_x} - 1)\xi_0} \right]$$

where ξ_0 is the location in the computational plane where the maximum clustering is to occur, and β_x is a constant which controls the degree of clustering at ξ_0, with larger values of B_x providing a finer grid in the clustered region. The transverse stretching is accomplished by dividing the physical plane into two sections: (1) the

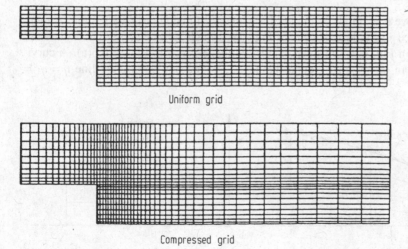

Uniform grid

Compressed grid

Fig. 6.4 Comparison of uniform and compressed grid

space directly behind the step, and (2) the space above (both in front of and behind) the step. The transformation is based on that used by Roberts [5], and is given by

$$y = \frac{(\beta_y + 1) - (\beta_y - 1)e^{-c(\eta-1-\alpha)/(1-\alpha)}}{(2\alpha + 1)(1 + e^{-c(\eta-1-\alpha)/(1-\alpha)})}$$

where

$$c = \log\left(\frac{\beta_y + 1}{\beta_y - 1}\right)$$

and β_y and α are appropriate constants, and are different for the two sections identified above. The algebraic transformations given above result in the grid stretching shown in Fig. 6.4.

6.5 Boundary-Fitted Coordinate Systems

Consider the flow through the divergent duct shown in Fig. 6.5(a). Curve *de* is the upper wall of the duct, and line *fg* is the centreline. For this flow, a simple rectangular grid in the physical plane is not appropriate, for the reasons discussed in Sect. 6.1. Instead, we draw the curvilinear grid in Fig. 6.5(a) which allows both the upper boundary *de* and the centreline *fg* to be coordinate lines, exactly fitting these boundaries. In turn, the curvilinear grid in Fig. 6.5(a) must be transformed to a rectangular grid in the computational plane, Fig. 6.5(b). This can be accomplished as follows. Let $y_s = f(x)$ be the ordinate of the upper surface *de* in Fig. 6.5(a). Then the following transformation will result in a rectangular grid in (ξ, η) space:

$$\xi = x$$
$$\eta = y/y_s \quad \text{where } y_s = f(x)$$

The above is a simple example of a boundary-fitted coordinate system. A more sophisticated example is shown in Fig. 6.6, which is an elaboration of the case illustrated in Fig. 6.2. Consider the airfoil shape given in Figure 6.6(a). A curvilinear system is wrapped around the airfoil, where one coordinate line $\eta = \eta_1 =$

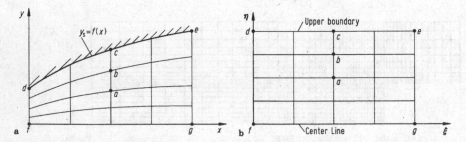

Fig. 6.5 A simple boundary-fitted coordinate system. (**a**) Physical plane. (**b**) Computational plane

Fig. 6.6 (a) Physical plane.
(b) Computational plane

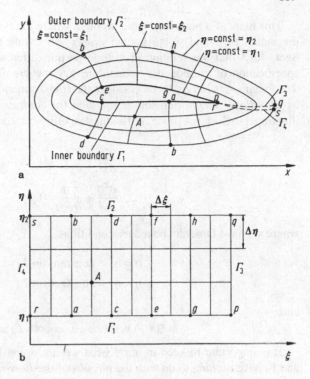

a

b

constant is on the airfoil surface. This is the inner boundary of the grid, designated by Γ_1. The outer boundary of the grid is labelled Γ_2 in Figure 6.6(a), and is given by $\eta = \eta_2 = $ constant. Examining this grid, we see that it clearly fits the boundary, and hence it is a boundary-fitted coordinate system. The lines which fan out from the inner boundary Γ_1 and which intersect the outer boundary Γ_2 are lines of constant ξ, such as line *ef* for which $\xi = \xi_1 = $ constant. (Note that in Fig. 6.6(a) the lines of constant η totally enclose the airfoil, much like elongated circles; such a grid is called an '0' type grid for airfoils. Another related curvilinear grid can have the $\eta = $ constant lines trailing downstream to the right, *not* totally enclosing the airfoil (except on the inner boundary Γ_1). Such a grid is called a 'C' type grid. We will see an example of a 'C' type grid shortly.)

Question: What transformation will cast the curvilinear grid in Fig. 6.6(a) into a uniform grid in the computational plane as sketched in Fig. 6.6(b)? To answer this question, note from Fig. 6.6(a) that along the inner boundary Γ_1, the physical coordinates of the body are known:

$$(x, y) \text{ known along } \Gamma_1$$

Similarly, the physical coordinates of the outer boundary Γ_2 are also known, because Γ_2 is simply a rather arbitrarily drawn loop around the airfoil. Once this loop Γ_2 is specified, then the physical coordinates along it are known:

$$(x, y) \text{ known along } \Gamma_2$$

This hints of a boundary value problem where the boundary conditions (namely the values of x and y) are known *everywhere* along the boundary. Recall from Sect. 4.3.3 that the solution of elliptic partial differential equations requires the specification of the boundary conditions *everywhere* along a boundary enclosing the domain. Therefore, let us consider the transformation in Fig. 6.6 to be defined by an *elliptic partial differential equation* (in contrast to an algebraic relation as illustrated in Sect. 6.4). One of the simplest elliptic equations is Laplace's equation:

$$\frac{\partial^2 \xi}{\partial x^2} + \frac{\partial^2 \xi}{\partial y^2} = 0 \qquad (6.35a)$$

$$\frac{\partial^2 \eta}{\partial x^2} + \frac{\partial^2 \eta}{\partial y^2} = 0 \qquad (6.35b)$$

where we have Dirichlet boundary conditions

$$\eta = \eta_1 = \text{constant on } \Gamma_1$$
$$\eta = \eta_2 = \text{constant on } \Gamma_2$$

and

$$\xi = \xi(x, y) \text{ is specified on both } \Gamma_1 \text{ and } \Gamma_2$$

It is important to keep in mind what we are doing here. The equations (6.35a and b) have *nothing* to do with the physics of the flow field. They are simply elliptic partial differential equations *which we have chosen* to relate ξ and η to x and y, and hence constitute a transformation (a one-to-one correspondence of grid points) from the physical plane to the computational plane. Because this transformation is governed by elliptic equations, it is an example of a general class of grid generation called *elliptic grid generation*. Such elliptic grid generation was first used on a practical basis by Joe Thompson at Mississippi State University, and is described in detail in the pioneering paper given in Ref. [6].

Let us look more closely at the physical and computational planes shown in Fig. 6.6. In order to construct a rectangular grid in the computational plane plane (Fig. 6.6b), a cut must be made in the physical plane (Fig. 6.6a) at the trailing edge of the airfoil. This cut can be visualized as two lines superimposed on each other: the line pq denoted by Γ_3 represents a boundary line for the physical space above pq, and and the line rs denoted by Γ_4 represents a boundary line for the physical space below rs. In the physical plane, the points p and r are the same point, and the points q and s are the same point; in Fig. 6.6(a) they are slightly displaced for clarity. However, in the computational plane, these points are all different. Indeed, the grid in the computational plane is obtained by slicing the physical grid at the cut, and then 'unwrapping' the grid from the airfoil. For example, the airfoil surface in the physical plane, curve *pgecar*, becomes the lower straight line denoted by Γ_1 in the computational plane. Similarly, the outer boundary *ghfdbs* becomes the upper straight line denoted by Γ_2 in the computational plane. The left and right sides of the rectangle in the computational plane are formed from the cut in the physical

Fig. 6.7 Computational plane, illustrating the boundary conditions and an internal point

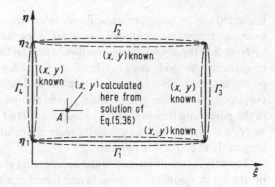

plane; the left side is line *rs* denoted by Γ_4 in Fig. 6.6(b), and the right side is line *pq* denoted by Γ_3 in Fig. 6.6(b).

The computational plane is sketched again in Fig. 6.7. Here we emphasize that values of (x, y) are *known along all four boundaries*, Γ_1, Γ_2, Γ_3 and Γ_4. The key aspect of the elliptic grid generation approach is that, with the given boundary conditions, Eqs. (6.35a and b) are solved for the (x, y) values which apply to *all the internal points*. An example of such an internal point is given by point A in Fig. 6.7, which corresponds to the same point A in Figs. 6.6(a) and (b). In reality, the equations to be solved are the inverse of Eqs. (6.35a and b), that is, equations obtained from Eqs. (6.35a and b) by interchanging the dependent and independent variables. The result is:

$$\alpha\frac{\partial^2 x}{\partial\xi^2} - 2\beta\frac{\partial^2 x}{\partial\xi\partial\eta} + \gamma\frac{\partial^2 x}{\partial\eta^2} = 0 \qquad (6.36a)$$

$$\alpha\frac{\partial^2 y}{\partial\xi^2} - 2\beta\frac{\partial^2 y}{\partial\xi\partial\eta} + \alpha\frac{\partial^2 y}{\partial\eta^2} = 0 \qquad (6.36b)$$

where

$$\alpha = \left(\frac{\partial x}{\partial\eta}\right)^2 + \left(\frac{\partial y}{\partial\eta}\right)^2$$

$$\beta = \left(\frac{\partial x}{\partial\xi}\right)\left(\frac{\partial x}{\partial\eta}\right) + \left(\frac{\partial y}{\partial\xi}\right)\left(\frac{\partial y}{\partial\eta}\right)$$

$$\gamma = \left(\frac{\partial x}{\partial\xi}\right)^2 + \left(\frac{\partial y}{\partial\xi}\right)^2$$

Note in Eqs. (6.36a and b) that x and y are now expressed as the dependent variables. Returning again to Fig. 6.7, Eqs. (6.36a and b) are solved, along with the given boundary conditions for (x, y) on Γ_1, Γ_2, Γ_3 and Γ_4, to obtain the values of (x, y) which correspond to the uniformly spaced grid points in the computational (ξ, η) plane. Thus, a given grid point (ξ_i, η_j) in the computational plane corresponds to the *calculated* grid point (x_i, y_j) in physical space. The solution of Eqs. (6.36a and b) is carried out by an appropriate finite-difference solution for elliptic equations; for example, relaxation techniques are popular for such equations.

Note that the above transformation, using an elliptic partial differential equation to generate the grid, does *not* involve closed-form analytic expressions; rather, it produces a set of *numbers* which locate a grid point (x_i, y_j) in physical space which correspond to a given grid point (ξ_i, η_j) in computational space. In turn, the metrics in the governing flow equations (which are solved in the computational plane), such as $\partial\xi/\partial x$, $\partial\eta/\partial y$, etc. are obtained from finite differences; central differences are frequently used for this purpose.

The curvilinear, boundary-fitted coordinate system shown in Fig. 6.6(a) is simply illustrated in a qualitative sense in that figure, for purposes of instruction. An actual grid generated about an airfoil using the above elliptic grid generation approach is shown in Fig. 6.8, taken from Ref. [7]. Using Thompson's grid generation scheme (Ref. [6]), Wright ([7]) has generated a boundary-fitted coordinate system around a Miley airfoil. (The Miley airfoil is an airfoil specially designed for low Reynolds number applications by Stan Miley at Mississippi State University.) In Fig. 6.6 the white speck in the middle of the figure is the airfoil, and the grid spreads far away from the airfoil in all directions.

In Ref. [7] low Reynolds number flows over airfoils were calculated by means of a time-dependent finite-difference solution of the compressible Navier-Stokes equations (such time-dependent solutions are discussed in Chap. 7). The free stream is subsonic, hence the outer boundary must be placed far away from the airfoil because of the far-reaching propagation of disturbances in a subsonic flow. A detail of the grid in the near vicinity of the airfoil is shown in Fig. 6.9. Note from both Figs. 6.8 and 6.9 that the grid is a 'C' type grid, in contrast to the '0' type grid sketched in Fig. 6.6.

We end this section by emphasizing again that the elliptic grid generation, with its solution of elliptic partial differential equations to obtain the internal grid points, is *completely separate* from the finite-difference solution of the governing equations.

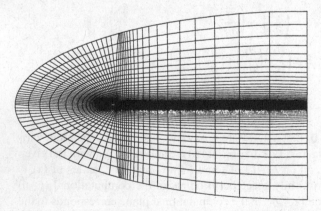

Fig. 6.8 Boundary fitted grid (from Ref. [7])

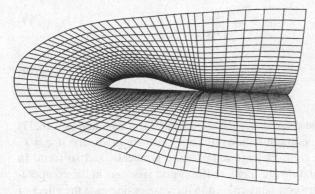

Fig. 6.9 A detail of the boundary fitted grid (from Ref. [7])

The grid is generated first, before any solution of the governing equations is attempted. The use of Laplace's equation (Eq. (6.35a and b)) to obtain this grid has nothing to do whatsoever with the physical aspects of the actual flow field. Here, Laplace's equation is simply used to generate the grid *only*.

6.6 Adaptive Grids

An adaptive grid is a grid network that automatically clusters grid points in regions of high flow field gradients; it uses the solution of the flow field properties to locate the grid points in the physical plane. The adaptive grid evolves in steps of time in conjunction with a time-dependent solution of the governing flow field equations, which computes the flow field variables in steps of time. During the course of the solution, the grid points in the physical plane *move* in such a fashion to 'adapt' to regions of large flow field gradients. Hence, the actual grid points in the physical plane are constantly in motion during the solution of the flow field, and become stationary only when the flow solution approaches a steady state. Therefore, unlike the elliptic grid generation discussed in Sect. 6.5 where the generation of the grid is completely separate from the flow field solution, an adaptive grid is intimately linked to the flow field solution, and changes as the flow field changes. The hoped-for advantages of an adaptive grid are expected because the grid points are clustered in regions where the 'action' is occurring. These advantages are: (1) increased accuracy for a fixed number of grid points, or (2), for a given accuracy, fewer grid points are needed. Adaptive grids are still very new in CFD, and whether or not these advantages are always acheived is not well established.

An example of a simple adaptive grid is that used by Corda [8] for the solution of viscous supersonic flow over a rearward-facing step. Here, the transformation is expressed in the form:

$$\Delta x = \frac{B\Delta\xi}{1 + b\dfrac{\partial g}{\partial x}} \tag{6.37}$$

$$\Delta y = \frac{C\Delta\eta}{1 + c\dfrac{\partial g}{\partial y}} \tag{6.38}$$

where g is a primitive flow field variable, such as p, ρ or T. If $g = p$, then Eqs. (6.37) and (6.38) cluster the grid points in regions of large pressure gradients; if $g = T$, the grid points cluster in regions of large temperature gradients, and so forth. In Eqs. (6.37) and (6.38), $\Delta\xi$ and $\Delta\eta$ are fixed, uniform grid spacings in the computational (ξ, η) plane, b and c are constants chosen to increase or decrease the effect of the gradient in changing the grid spacing in the physical plane, B and C are scale factors and Δx and Δy are the new grid spacings in the physical plane. Because $\partial g/\partial x$ and $\partial g/\partial y$ are changing with time during a time-dependent solution of the flow field, then clearly Δx and Δy change with time, i.e. the grid points move in the physical space. Clearly, in regions of the flow where $\partial g/\partial x$ and $\partial g/\partial y$ are large, Eqs. (6.37) and (6.38) yield small values of Δx and Δy for a given $\Delta\xi$ and $\Delta\eta$; this is the mechanism which clusters the grid points.

In dealing with an adaptive grid, the computational plane consists of fixed points in the (ξ, η) space; these points are fixed in time, i.e. they do *not* move in the computational space. Moreover, $\Delta\xi$ is uniform, and $\Delta\eta$ is uniform. Hence, the computational plane is the same as we have discussed in previous sections. The governing flow equations are solved in the computational plane, where the x, y and t derivatives are transformed according to Eqs. (6.2), (6.3) and (6.5). In particular, examine the transformation given by Eq. (6.5) for the time derivative. In the case of stretched or boundary-fitted grids as discussed in Sects. 6.4 and 6.5 respectively, the metrics $\partial\xi/\partial t$ and $\partial\eta/\partial t$ were zero, and Eq. (6.5) yields $\partial/\partial t = \partial/\partial\tau$. However, for an adaptive grid,

$$\frac{\partial\xi}{\partial t} \equiv \left(\frac{\partial\xi}{\partial t}\right)_{x,y}$$

and

$$\frac{\partial\eta}{\partial t} \equiv \left(\frac{\partial\eta}{\partial t}\right)_{x,y}$$

are finite. Why? Because, although the grid points are fixed in the computational plane, the grid points in the physical plane are moving with time. The physical meaning of $(\partial\xi/\partial t)_{x,y}$ is the time rate of change of ξ at a *fixed* (x, y) location in the physical plane. Similarly, the physical meaning of $(\partial\eta/\partial t)_{x,y}$ is the time rate of change of η at a *fixed* (x, y) location in the physical plane. Imagine that you have your eyes locked to a fixed (x, y) point in the physical plane. As a function of time, the values of ξ and η associated with this *fixed* (x, y) point will change. This is why $\partial\xi/\partial t$ and $\partial\eta/\partial t$ are finite. In turn, when dealing with the transformed flow equations in the computational plane, all three terms on the right-hand side of Eq. (6.5) are finite, and must be included in the transformed equations. In this fashion, the time metrics

$\partial\xi/\partial t$ and $\partial\eta/\partial t$ automatically take into account the movement of the adaptive grid during the solution of the governing flow equations.

The values of the time metrics in the form shown in Eq. (6.5) are a bit cumbersome to evaluate; on the other hand, the related time metrics

$$\left(\frac{\partial x}{\partial t}\right)_{\xi,\eta} \quad \text{and} \quad \left(\frac{\partial y}{\partial t}\right)_{\xi,\eta}$$

are much easier to evaluate, because they come from

$$\left(\frac{\partial x}{\partial t}\right)_{\xi,\eta} \approx \frac{\Delta x}{\Delta t} \tag{6.39}$$

and

$$\left(\frac{\partial y}{\partial t}\right)_{\xi,\eta} \approx \frac{\Delta y}{\Delta t} \tag{6.40}$$

where Δx and Δy are obtained directly from the transformation given in Eqs. (6.37) and (6.38) respectively. Let us find the relationship between these two sets of time metrics. Consider

$$x = x(\xi,\eta,\tau)$$

Hence

$$dx = \left(\frac{\partial x}{\partial \xi}\right)_{\eta,\tau} d\xi + \left(\frac{\partial x}{\partial \eta}\right)_{\xi,\tau} d\eta + \left(\frac{\partial x}{\partial \tau}\right)_{\xi,\eta} d\tau$$

From this result, we write

$$\cancel{\left(\frac{\partial x}{\partial t}\right)_{x,y}}^{\,0} = \left(\frac{\partial x}{\partial \xi}\right)_{\eta,\tau}\left(\frac{\partial \xi}{\partial t}\right)_{x,y} + \left(\frac{\partial x}{\partial \eta}\right)_{\xi,\tau}\left(\frac{\partial \eta}{\partial t}\right)_{x,y} + \left(\frac{\partial x}{\partial \tau}\right)_{\xi,\eta}\cancel{\left(\frac{\partial \tau}{\partial t}\right)_{x,y}}^{\,1}$$

or

$$-\left(\frac{\partial x}{\partial \tau}\right)_{\xi,\eta} = \left(\frac{\partial x}{\partial \xi}\right)_{\eta,\tau}\left(\frac{\partial \xi}{\partial t}\right)_{x,y} + \left(\frac{\partial x}{\partial \eta}\right)_{\xi,\tau}\left(\frac{\partial \eta}{\partial t}\right)_{x,y} \tag{6.41}$$

Note that we are carrying the subscripts on the partial derivatives to avoid any confusion over what variables are held constant. Now consider

$$y = y(\xi,\eta,\tau)$$

Hence:

$$dy = \left(\frac{\partial y}{\partial \xi}\right)_{\eta,\tau} d\xi + \left(\frac{\partial y}{\partial \eta}\right)_{\xi,\tau} d\eta + \left(\frac{\partial y}{\partial \tau}\right)_{\xi,\eta} d\tau$$

Thus, from this result we write

$$\cancel{\left(\frac{\partial y}{\partial t}\right)_{x,y}}^{\,0} = \left(\frac{\partial y}{\partial \xi}\right)_{\eta,\tau}\left(\frac{\partial \xi}{\partial t}\right)_{x,y} + \left(\frac{\partial y}{\partial \eta}\right)_{\xi,\tau}\left(\frac{\partial \eta}{\partial t}\right)_{x,y} + \left(\frac{\partial y}{\partial \tau}\right)_{\xi,\eta}\cancel{\left(\frac{\partial \tau}{\partial t}\right)_{x,y}}^{\,1}$$

or

$$-\left(\frac{\partial y}{\partial \tau}\right)_{\xi,\eta} = \left(\frac{\partial y}{\partial \xi}\right)_{\eta,\tau}\left(\frac{\partial \xi}{\partial t}\right)_{x,y} + \left(\frac{\partial y}{\partial \eta}\right)_{\xi,\tau}\left(\frac{\partial \eta}{\partial t}\right)_{x,y} \tag{6.42}$$

Solve Eqs. (6.41) and (6.42) for $\left(\frac{\partial \xi}{\partial t}\right)_{x,y}$

$$\left(\frac{\partial \xi}{\partial t}\right)_{x,y} = \frac{\begin{vmatrix} -\left(\frac{\partial x}{\partial \tau}\right)_{\xi,\eta} & \left(\frac{\partial x}{\partial \eta}\right)_{\xi,\tau} \\ -\left(\frac{\partial y}{\partial \tau}\right)_{\xi,\eta} & \left(\frac{\partial y}{\partial \eta}\right)_{\xi,\tau} \end{vmatrix}}{\begin{vmatrix} \left(\frac{\partial x}{\partial \xi}\right)_{\eta,\tau} & \left(\frac{\partial x}{\partial \eta}\right)_{\xi,\tau} \\ \left(\frac{\partial y}{\partial \xi}\right)_{\eta,\tau} & \left(\frac{\partial y}{\partial \eta}\right)_{\xi,\tau} \end{vmatrix}}$$

Recognizing that $\tau = t$, and that the denominator is the Jacobian J, the above equation becomes (dropping subscripts)

$$\frac{\partial \xi}{\partial t} = \frac{1}{J}\left[-\left(\frac{\partial x}{\partial t}\right)\left(\frac{\partial y}{\partial \eta}\right) + \left(\frac{\partial y}{\partial t}\right)\left(\frac{\partial x}{\partial \eta}\right)\right] \tag{6.43}$$

Solving Eqs. (6.41) and (6.42) for $\left(\frac{\partial \eta}{\partial t}\right)_{x,y}$, we find a likewise fashion that

$$\frac{\partial \eta}{\partial t} = \frac{1}{J}\left[\left(\frac{\partial x}{\partial t}\right)\left(\frac{\partial y}{\partial \xi}\right) - \left(\frac{\partial y}{\partial t}\right)\left(\frac{\partial x}{\partial \xi}\right)\right] \tag{6.44}$$

Let us recapitulate. For an adaptive grid, the governing flow equations, when transformed for solution in the computational (ξ, η) plane, must contain all the terms in the time transformation given by Eq. (6.5). The time metrics, $\partial \xi/\partial t$ and $\partial \eta/\partial t$, in Eq. (6.5) can in turn be expressed in terms of $\partial x/\partial t$ and $\partial y/\partial t$ through Eqs. (6.43) and (6.44). These new time metrics can in turn be readily calculated from Eqs. (6.39) and (6.40), where Δx and Δy are given by the basic transformation in Eqs. (6.37) and (6.38).

An example of an adapted grid for the supersonic viscous flow over a rearward facing step is given in Fig. 6.10, taken from the work of Corda [8]. Flow is from left to right. Note that the grid points cluster around the expansion wave from the top corner of the step, and around the reattachment shock wave downstream of the step. It is interesting to note that the adapted grid itself is a type of 'flow field visualization method' that helps to identify the location of waves and other gradients in the flow.

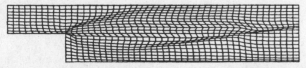

Fig. 6.10 Adapted grid for the rearward-facing step problem (from Corda, Ref. [8])

As a final note, there are many different approaches for the generation of adaptive grids. The above discussion is just one; it is based on ideas presented by Dwyer et al. in Ref. [9]. For a more complete discussion on adaptive grids, as well as grid generation in general, see Ref. [1].

References

1. Anderson, D.A., Tannehill, John C. and Pletcher, Richard H., *Computational Fluid Mechanics and Heat Transfer*, McGraw-Hill, New York, 1984.
2. Sullins, G.A., Anderson, J.D., Jr. and Drummond, J.P., 'Numerical Investigation of Supersonic Base Flow with Parallel Injection,' AIAA Paper No. 82–1001.
3. Sullins, G.A., *Numerical Investigation of Supersonic Base Flow with Tangential Injection*, M.S. Thesis, Department of Aerospace Engineering, University of Maryland, 1981.
4. Holst, T.L., 'Numerical Solution of Axisymmetric Boattail Fields with Plume Simulators,' AIAA Paper No. 77–224, 1977.
5. Roberts, B.O., 'Computational Meshes for Boundary Layer Problems,' *Lecture Notes in Physics*, Springer-Verlag, New York, 1971, pp. 171–177.
6. Thompson, J.F., Thames, F.C. and Mastin, C.W., 'Automatic Numerical Generation of Body-Fitted Curvilinear Coordinate Systems for Fields Containing Any Number of Arbitrary Two-Dimensional Bodies,' *Journal of Computational Physics*, Vol. 15, pp. 299–319, 1974.
7. Wright, Andrew F., *A Numerical Investigation of Low Reynolds Number Flow Over an Airfoil*, M.S. Thesis, Department of Aerospace Engineering, University of Maryland, 1982.
8. Corda, Stephen, *Numerical Investigation of the Laminar, Supersonic Flow over a Rearward-Facing Step Using an Adaptive Grid Scheme*, M.S. Thesis, Department of Aerospace Engineering, University of Maryland, 1982.
9. Dwyer, H.A., Kee, R.J. and Sanders, B.R., 'An Adaptive Grid Method for Problems in Fluid Mechanics and Heat Transfer,' AIAA Paper No. 79–1464, 1979.

Chapter 7
Explicit Finite Difference Methods: Some Selected Applications to Inviscid and Viscous Flows

J.D. Anderson, Jr.

7.1 Introduction

In this chapter we round-out our introductory treatment of computational fluid dynamics by discussing some applications of explicit finite difference methods to selected examples for inviscid and viscous flows. These examples have one thing in common—they are results obtained by either the present author and/or some of his graduate students over the past few years. This is not meant to be chauvinistic; rather this choice is intentionally made to illustrate what can be done by uninitiated students who are new to the ideas of CFD. These examples demonstrate the power and beauty of CFD in the hands of students much like yourselves who may have little or no experience in the field. Moreover, in all cases the applications are carried out with computer programs designed and written completely by each student. This is following the author's educational philosophy that each student should have the experience of starting with paper and pencil, writing down the governing equations, developing the appropriate numerical solution of these equations, writing the FORTRAN program, punching the program into the computer, and then going through all the trials and tribulations of making the program work properly. This is an important aspect of CFD education. No established computer programs ('canned' programs) are used; everything is 'home-grown', with the exception of standard graphics packages which are used to plot the results. Therefore, by examining these examples, you should obtain a reasonable feeling for what you can expect to accomplish when you first jump into the world of CFD applications.

Before we discuss some examples, it is important to describe the mechanism of explicit finite-difference calculations. The distinction between explicit and implicit approaches was made in Sect. 5.3, which should be reviewed before progressing further in this chapter. In the next few sections, we will describe two rather straightforward and popular explicit methods. The treatment and application of implicit methods is given by other lectures in this course, and hence will not be discussed here.

J.D. Anderson, Jr.

National Air and Space Museum, Smithsonian Institution, Washington, DC

e-mail: AndersonJA@si.edu

J.F. Wendt (ed.), *Computational Fluid Dynamics*, 3rd ed.,
© Springer-Verlag Berlin Heidelberg 2009

Finally, the examples discussed in this chapter all incorporate the *time-dependent method*, i.e. forward marching in steps of time. The historic break-through made by this method in the 1960s is discussed in Chap. 1. The vast majority of time-dependent solutions have as their objective the solution of a steady-state flow field which is approached by the solution at large times; here, the time-dependent mechanism is simply a means towards achieving that end. In other applications, the time-dependent method is used to calculate the actual transients in an unsteady flow of interest. Examples of both are given here. We note, however, that although the following sections deal with marching forward in time, the same techniques are easily applied to a steady flow calculation where spatial marching is done along some co-ordinate axis. We have seen in Chap. 4 that such forward marching (in time or space) is appropriate when the governing equations are hyperbolic or parabolic.

7.2 The Lax–Wendroff Method

Let us describe this method by considering a simple gas-dynamic problem, namely the subsonic–supersonic isentropic flow through a convergent–divergent nozzle, as sketched in Fig. 7.1. Here, a nozzle of specified area distribution, $A = A(x)$, is given, and the reservoir conditions are known. Let us consider a quasi-one-dimensional solution where the flow field variables are functions of x (in the steady state). For a calorically perfect gas, the solution of this flow is classical, and can be found in any compressible flow text book (see for example Refs. [1,2]). We use this example here only because it is an excellent vehicle for introducing and describing the time-dependent finite-difference philosophy.

The nozzle is divided into a number of grid points in the x-direction as shown in Fig. 7.1; the spacing between adjacent grid points is Δx. Now *assume* values of the flow field variables at all grid points, and consider this rather arbitrarily assumed flow as an *initial condition* at time $t = 0$. In general, these assumed values will *not* be the exact steady-state results; indeed, the exact steady-state results are what we are trying to calculate. Consider a grid point, say point i. Let g_i denote a flow field variable at this point (g_i might be pressure, density, velocity, etc.). This variable g_i will be a function of time; however, we know g_i at time $t = 0$, i.e. we know $g_i(0)$ because we have *assumed* values for all the flow field variables at all the grid points at the initial time $t = 0$.

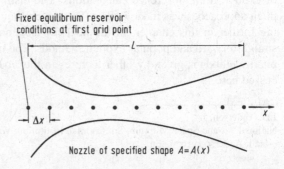

Fig. 7.1 Flow through a convergent-divergent nozzle

We now calculate a *new* value of g_i at time $t + \Delta t$; starting from the initial conditions, the first new time is $t + \Delta t = 0 + \Delta t$. Here, Δt is a small increment in time to be discussed later. The new value of g_i, i.e. $g_i(t + \Delta t)$, is obtained from a Taylor's series expansion in time as

$$g_i(t + \Delta t) = g_i(t) + \left(\frac{\partial g}{\partial t}\right)_i \Delta t + \left(\frac{\partial^2 g}{\partial t^2}\right)_i \frac{(\Delta t)^2}{2} + \cdots$$

or, using the standard notation of time as a superscript,

$$g_i^{t + \Delta t} = g_i^t + \left(\frac{\partial g}{\partial t}_i^t\right) \Delta t + \left(\frac{\partial^2 g}{\partial t^2}\right)_i^t \frac{(\Delta t)}{2} + \cdots \tag{7.1}$$

Here $g_i^{t+\Delta t}$ is the value of g at grid point i and at time $t + \Delta t$; $(\partial g/\partial t)_i^t$ is the first partial of g evaluated at grid point i at time t, etc. In Eq. (7.1), g_i^t is known and Δt is specified. Therefore, we can use Eq. (7.1) to calculate $g_i^{t+\Delta t}$ *if* we have numbers for the derivatives $(\partial g/\partial t)_i^{t+\Delta t}$ and $(\partial^2 g/\partial t^2)_i^{t+\Delta t}$. The numbers for the derivatives are obtained from the physics of the flow as embodied in the governing flow equations. (Note that Eq. (7.1) is simply mathematics, and by itself is certainly not sufficient to solve the problem.) The governing flow equations for the quasi-one-dimensional flow through a nozzle are (14):

$$\text{Continuity}: \quad \frac{\partial \rho}{\partial t} = -\frac{1}{A}\frac{\partial(\rho u A)}{\partial x} \tag{7.2}$$

$$\text{Momentum}: \quad \frac{\partial u}{\partial t} = -\frac{1}{\rho}\left(\frac{\partial p}{\partial x} + \rho u \frac{\partial u}{\partial x}\right) \tag{7.3}$$

$$\text{Energy}: \quad \frac{\partial e}{\partial t} = -\frac{1}{\rho}\left[p\frac{\partial u}{\partial x} + pu\frac{\partial(\ln A)}{\partial x} + \rho u \frac{\partial e}{\partial x}\right] \tag{7.4}$$

Note that Eqs. (7.2), (7.3) and (7.4) are written with the time derivatives on the left-hand side, and spatial derivatives on the right-hand side. For the moment, let us calculate density, i.e. $g \equiv \rho$, and let us consider just the continuity equation, Eq. (7.2).

Expanding the right-hand side of Eq. (7.2), we obtain

$$\frac{\partial \rho}{\partial t} = -\frac{1}{A}\rho u \frac{\partial A}{\partial x} - u\frac{\partial \rho}{\partial x} - \rho \frac{\partial u}{\partial x} \tag{7.5}$$

At time $t = 0$, the flow field variables are assumed; hence we can replace the spatial derivatives with central differences:

$$\left(\frac{\partial \rho}{\partial t}\right)_i^t = -\frac{1}{A}\rho_i^t u_i^t \left(\frac{A_{i+1} - A_{i-1}}{2\Delta x}\right) - u_i^t\left(\frac{\rho_{i+1}^t - \rho_{i-1}^t}{2\Delta x}\right) - \rho_i^t\left(\frac{u_{i+1}^t - u_{i-1}^t}{2\Delta x}\right) \tag{7.6}$$

Equation (7.6) gives us a *number* for $(\partial \rho/\partial t)_i^t$, which is inserted into Eq. (7.1). However, to complete Eq. (7.1), we need a number for the second partial also, namely $(\partial^2 \rho/\partial t^2)_i^t$. To obtain this, differentiate the continuity equation, Eq. (7.5), with respect to time:

$$\frac{\partial^2 \rho}{\partial t^2} = -\frac{1}{A}\left[\frac{\partial A}{\partial x}\left(\rho\frac{\partial u}{\partial t} + u\frac{\partial \rho}{\partial t}\right)\right] - u\frac{\partial^2 \rho}{\partial x \partial t} - \left(\frac{\partial \rho}{\partial x}\right)\left(\frac{\partial u}{\partial t}\right) - \rho\frac{\partial^2 u}{\partial x \partial t} - \left(\frac{\partial u}{\partial x}\right)\left(\frac{\partial \rho}{\partial t}\right) \quad (7.7)$$

Also, differentiate the continuity equation, Eq. (7.5), with respect to x:

$$\frac{\partial^2 \rho}{\partial t \partial x} = -\frac{1}{A}\left[\rho u\frac{\partial^2 A}{\partial x^2} + \left(\frac{\partial A}{\partial x}\right)\left(\rho\frac{\partial u}{\partial x} + u\frac{\partial \rho}{\partial x}\right)\right] - u\frac{\partial^2 \rho}{\partial x^2} - \left(\frac{\partial \rho}{\partial x}\right)\left(\frac{\partial u}{\partial x}\right) - \rho\frac{\partial^2 u}{\partial x^2} - \left(\frac{\partial u}{\partial x}\right)\left(\frac{\partial \rho}{\partial x}\right)$$
$$(7.8)$$

The procedure now works as follows:

(1) In Eq. (7.8), replace all derivatives on the right-hand side with central differences, such as

$$\frac{\partial u}{\partial x} = \frac{u_{i+1}^t - u_{i-1}^t}{2\Delta x}$$
$$\frac{\partial^2 u}{\partial x^2} = \frac{u_{i+1}^t - 2u_i^t + u_{i-1}^t}{(\Delta x)^2}$$
etc.

This now provides a number for $(\partial^2 \rho/\partial t \partial x)_i^t$ from Eq. (7.8).

(2) Insert this *number* for $(\partial^2 \rho/\partial t \partial x)_i^t$ into Eq. (7.7). Also in Eq. (7.7), *numbers* for $\partial u/\partial t$ and $\partial^2 u/\partial x \partial t$ are obtained from a treatment of the momentum equation, Eq. (7.3), in a manner exactly the same as the continuity equation was treated above. The details will not be given here. In Eq. (7.7), a *number* for $(\partial \rho/\partial t)$ is already available, namely from Eq. (7.6). The net result is that we now have a *number* for $(\partial^2 \rho/\partial t^2)_i^t$, obtained from Eq. (7.7).

(3) Insert this number for $(\partial^2 \rho/\partial t^2)_i^t$ into Eq. (7.1), remembering that $g \equiv \rho$ for this case.

(4) Insert the *number* for $(\partial \rho/\partial t)_i^t$, obtained from Eq. (7.6), into Eq. (7.1).

(5) Every quantity on the right-hand side of Eq. (7.1) is now known. This allows the density $\rho_i^{t+\Delta t}$ to be calculated from Eq. (7.1). This is indeed what we wanted. We now have the density at grid point i at the *next* step in time, $t + \Delta t$.

(6) Perform the above procedure at every grid point to obtain $\rho(t + \Delta t)$ everywhere throughout the nozzle.

(7) Perform the above procedure on the momentum and energy equations to obtain $u(t + \Delta t)$ and $e(t + \Delta t)$ everywhere throughout the nozzle. *We now have the complete flowfield at time $(t + \Delta t)$, obtained from the known flowfield at time t.* (Recall that the process is started at $t = 0$ with the assumed initial conditions.)

(8) Repeat the above process for a large number of time steps. At each time step, the flow properties at all grid points will change from one time to the next. However, at large times, these changes become very small, and a steady-state is approached. *This steady-state is the desired result*, and the time-dependent technique is simply a means to that end.

Fig. 7.2 Transient and final steady-state temperature distributions for a calorically perfect gas obtained from the present time-dependent analysis, $\gamma = 1.4$

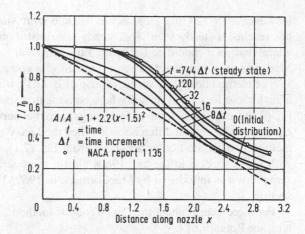

The behaviour of this type of solution is illustrated in Figs. 7.2 and 7.3. In Fig. 7.2, the temperature distribution through a given nozzle is shown. The dashed line labelled $t = 0$ is the initially assumed values for T throughout the nozzle. The curve above it labelled $8\Delta t$ is the temperature distribution after eight time steps following the above procedure. The curves labeled $16\Delta t$ and $32\Delta t$ are similar results after 16 and 32 time steps respectively. Note that the temperature distribution has rapidly changed from the assumed initial distribution at $t = 0$. At later times, the changes become smaller; note that the curve labelled $120\Delta t$ is not too different from that for $32\Delta t$. Finally, after 744 time steps, the changes are so small that the temperature distribution is essentially at a steady state. This steady state is the desired solution. Note that the numerically-obtained steady state agrees virtually perfectly with the classical results, as can be obtained from Refs. [1, 3], and from Ref. [4]. Fig. 7.3 illustrates the variation of mass flow, $\dot{m}$, through the nozzle. The dashed line is the $\dot{m}$ consistent with the assumed initial conditions at $t = 0$. The curves labelled $16\Delta t$ and $32\Delta t$ graphically demonstrate the wild variations in $\dot{m}$ at early times.

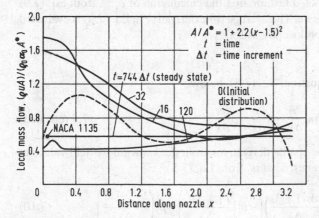

Fig. 7.3 Transient and final steady-state mass-flow distributions for a calorically perfect gas obtained from the present time-dependent analysis, $\gamma = 1.4$

However, after 120 time steps $\dot{m}$ has become more stable, and after 744 time steps has reached a steady state. This steady state distribution for $\dot{m}$ is a straight, horizontal line, as it should be for steady flow, where $\dot{m}$ = constant through the nozzle. Moreover, it is the correct value of mass flow, as compared to results from Ref. [4].

The method described above, utilizing Eq. (7.1), which is the first three terms of a Taylor's series expansion and where both the first and second partial derivatives in Eq. (7.1) are found by finite-differencing the spatial derivatives in the governing flow equations with central differences, is called the Lax-Wendroff method. Note that the method is of second-order accuracy, from Eq. (7.1). This method was employed with much success in the late 1960s until a more straight-forward version of the same idea was introduced by MacCormack in 1969. This is the subject of the next section.

For more details about the Lax-Wendroff method as applied to the nozzle problem, see Refs. [5,6].

7.3 MacCormack's Method

MacCormack's method, first introduced in 1969 (see Ref. [7]), has been the most popular explicit finite-difference method for solving fluid flows. It is closely related to the Lax-Wendroff method, but is easier to apply. Let us use the same nozzle problem discussed in Sect. 7.2 to illustrate MacCormack's method in the present section. MacCormack's method, like the Lax-Wendroff method, is based on a Taylor's series expansion in time. Once again, as in Sect. 7.2, let us consider the density at grid point i.

$$\rho_i^{t+\Delta t} = \rho_i^t + \left(\frac{\partial \rho}{\partial t}\right)_{\text{ave}} \Delta t \tag{7.9}$$

Equation (7.9) is a truncated Taylor's series that looks first-order accurate; however, $(\partial \rho / \partial t)_{\text{ave}}$ is an *average* time derivative taken between time t and $t + \Delta t$. This derivative is evaluated in such a fashion that the calculation of $\rho_i^{t+\Delta t}$ from Eq. (7.9) becomes second-order accurate. The average time derivative in Eq. (7.9) is evaluated from a predictor-corrector philosophy as follows.

Predictor step.
We repeat the continuity equation, Eq. (7.5), below:

$$\frac{\partial \rho}{\partial t} = -\frac{1}{A}\rho u \frac{\partial A}{\partial x} - u \frac{\partial \rho}{\partial x} - \rho \frac{\partial u}{\partial x} \tag{7.5 repeated}$$

In Eq. (7.5), calculate the spatial derivatives from the *known* flow field values at time t *using forward differences*. That is, from Eq. (7.5),

$$\left(\frac{\partial \rho}{\partial t}\right)_i^t = -\frac{1}{A}\left[\rho_i^t u_i^t\left(\frac{A_{i+1} - A_i}{\Delta x}\right)\right] - u_i^t\left(\frac{\rho_{i+1}^t - \rho_i^t}{\Delta x}\right) - \rho_i^t\left(\frac{u_{i+1}^t - u_i^t}{\Delta x}\right) \tag{7.10}$$

Obtain a *predicted* value of density, $\bar{\rho}_i^{t+\Delta t}$, from the first two terms of a Taylor's series, as follows

$$\bar{\rho}_i^{t+\Delta t} = \rho_i^t + \left(\frac{\partial \rho}{\partial t}\right)_i^t \Delta t \tag{7.11}$$

In Eq. (7.11), ρ_i^t is known, and $(\partial \rho / \partial t)_i^t$ is a known number from Eq. (7.10); hence, $\bar{\rho}_i^{t+\Delta t}$ is readily obtained. In a similar fashion, from the momentum and energy equations, predicted values of the other flow variables such as $\bar{u}_i^{t+\Delta t}$, $\bar{e}_i^{t+\Delta t}$, etc. are obtained.

Corrector step

Here, we first obtain a predicted value of the time derivative, $\overline{(\frac{\partial \rho}{\partial t})}_i^{t+\Delta t}$, by substituting the *predicted* values of $\bar{u}_i^{t+\Delta t}$, $\bar{\rho}_i^{t+\Delta t}$, etc. into Eq. 7.5, using *rearward differences*.

$$\overline{\left(\frac{\partial \rho}{\partial t}\right)}_i^{t+\Delta t} = -\frac{1}{A}\bar{\rho}_i^{t+\Delta t}\bar{u}_i^{t+\Delta t}\left(\frac{A_i - A_{i-1}}{\Delta x}\right) - \bar{u}_i^{t+\Delta t}\left(\frac{\bar{\rho}_i^{t+\Delta t} - \bar{\rho}_{i-1}^{t+\Delta t}}{\Delta x}\right) - \bar{\rho}_i^{t+\Delta t}\left(\frac{\bar{u}_i^{t+\Delta t} - \bar{u}_{i-1}^{t+\Delta t}}{\Delta x}\right) \tag{7.12}$$

Now calculate the average time derivative as the arithmetic mean between Eqs. (7.10) and (7.12), i.e.

$$\left(\frac{\partial \rho}{\partial t}\right)_{ave} = \frac{1}{2}\left[\left(\frac{\partial \rho}{\partial t}\right)_i^t + \overline{\left(\frac{\partial \rho}{\partial t}\right)}_i^{t+\Delta t}\right] \tag{7.13}$$

where numbers for the two terms on the right-hand side of Eq. (7.13) come from Eqs (7.10) and (7.12) respectively. Finally, we obtain the *corrected* value of $\rho_i^{t+\Delta t}$ from Eq. (7.9), repeated below:

$$\rho_i^{t+\Delta t} = \rho_i^t + \left(\frac{\partial \rho}{\partial t}\right)_{ave} \Delta t \tag{7.9 repeated}$$

The above predictor–corrector approach is carried out for all grid points throughout the nozzle, and is applied simultaneously to the momentum and energy equations in order to generate $u_i^{t+\Delta t}$ and $e_i^{t+\Delta t}$. In this fashion, the flow field through the entire nozzle at time $t + \Delta t$ is calculated. This is repeated for a large number of time steps until the steady state is achieved, just as in the case of the Lax-Wendroff method described in Sect. 7.2.

MacCormack's technique as described above, because a two-step predictor–corrector sequence is used with forward differences on the predictor and rearward differences on the corrector, is a second-order accurate method. Therefore, it has the same accuracy as the Lax-Wendroff method described in Sect. 7.2. However, the MacCormack method is much easier to apply, because there is no need to evaluate the second time derivatives as was the case for the Lax-Wendroff method. To see this more clearly, recall Eqs. (7.7) and (7.8), which are required for the Lax-Wendroff method. These equations represent a large number of additional calculations. Moreover, for a more complex fluid dynamic problem, the differentiation of the continuity, momentum and energy equations to obtain the second derivatives, first with respect to time, and then the mixed derivatives with respect to time and space, can be very tedious, and provides an extra source for human error. MacCormack's method does *not* require such second derivatives, and hence does *not* deal with equations such as Eqs. (7.7) and (7.8).

A few comments are made with regard to the specific application to the quasi-one-dimensional nozzle flow shown in Fig. 7.1. At the *inflow* boundary (the first grid point at the left), the values of p, T and ρ are fixed, independent of time, and are assumed to be reservoir values. The inflow velocity, which is a very small subsonic value, is calculated from linear extrapolation using the adjacent internal points, or it can be evaluated from the momentum equation applied at the first grid point using one-sided differences. At the *outflow* boundary (the last grid point at the right in Fig. 7.1), all the dependent variables are obtained from linear extrapolation from the adjacent internal points, or by applying the governing equations at this point, using one-sided differences.

Finally, we note that results obtained from the Lax–Wendroff method and from the MacCormack method are virtually identical. For example, these two methods are compared for a vibrationally relaxing, high temperature, non-equilibrium nozzle flow in Ref. [8]; there is no difference between the two sets of results.

7.4 Stability Criterion

Examine Eq. (7.1), which is vital to the Lax–Wendroff method. Note that it requires the specification of a time increment, Δt. Examine Eqs. (7.9) and (7.11), which are vital to the MacCormack method. They too require the specification of a time increment, Δt. For explicit methods, the value of Δt *cannot* be arbitrary, rather it must be less than some maximum value allowable for stability. The time-dependent applications described in Sects. 7.2 and 7.3 are dealing with governing flow equations which are hyperbolic with respect to time. Recall our discussion in Sect. 5.4 dealing with the stability criteria for such equations. There, it was stated that Δt must obey the Courant–Friedrichs–Lewy criterion—the so-called CFL criterion. This is embodied in Eq. (5.47), which was derived from the simple model equation given by Eq. (5.42). This is the linear wave equation, where c is the wave propagation speed. If the wave were propagating through a gas which already has a velocity u, then the wave will travel at the velocity $(u + c)$ relative to the stationary surroundings. For such a case, Eq. (5.47) becomes

$$\Delta t = C\left(\frac{\Delta x}{u+c}\right); \quad C \le 1 \tag{7.14}$$

where C is the Courant number, and c is the speed of sound, $c = (\partial p/\partial \rho)_s$. Eq. (7.14) is the appropriate CFL criterion for the one-dimensional, explicit solutions of nozzle flows discussed in Sects. 7.2 and 7.3. The CFL criterion given by Eq. (7.14) says *physically* that the explicit time step must be no greater than the time required for a sound wave to propagate from one grid point to the next. This author's experience has been that C should be as close to unity as possible, but depending upon the actual application, the maximum allowable value of C for stability in explicit time-dependent finite difference calculations can vary from approximately 0.5–1.0. Keep in mind that the stability criteria exemplified by Eqs. (5.47) and (7.14) are based on analysis of *linear* equations. On the other hand, the governing equations for a general fluid flow are highly non-linear. Therefore, we would not expect the CFL criteria to apply *exactly* to such cases; instead, it provides a reasonable *estimate* of Δt for a given non-linear problem, and as a result the value of the Courant number in Eq. (7.14) can be viewed as an adjustable parameter to compensate for such non-linearities.

Return for a moment to the nozzle flow application discussed in Sects. 7.2 and 7.3. Here, at any given time t, Eq. (7.14) is evaluated at each grid point throughout the flow. Because u and c vary with x, then the local value of Δt associated with each grid point will be different from one point to the next. The value of Δt actually employed in Eqs. (7.1) and (7.9) to advance the flow field through the next step in time should be the *minimum* Δt calculated over all the grid points.

[Some CFD applications have employed the 'local time step method', wherein the local values of Δt are used at each grid point in Eqs. (7.1) and (7.9). In this case, the transient variations calculated over many time steps do not hold physically; a type of 'time-warped' flow field is developed, where all the new flow variables calculated for a subsequent time step actually pertain to different total values of time. This 'local time step method' frequently results in a faster convergence to the steady state, that is, fewer total time steps are required to obtain the steady state. On the other hand, the calculated transients have no physical meaning, and some CFD experts wonder openly about the overall *accuracy* of such a method, even for the final steady state results.]

Finally, we note that for a two or three-dimensional flow, an extension of Eq. (7.14) is:

$$\Delta t = \text{Min}(\Delta t_x, \Delta t_y) \tag{7.15a}$$

where

$$\Delta t_x = C \frac{\Delta x}{u + c} \tag{7.15b}$$

and

$$\Delta t_y = C \frac{\Delta y}{v + c} \tag{7.15c}$$

7.5 Selected Applications of the Explicit Time-Dependent Technique

The purpose of this section is to illustrate various applications of the explicit, time-dependent technique described in the previous sections of this chapter. These applications contain many of the CFD features that have been discussed throughout these notes.

7.5.1 Non-equilibrium Nozzle Flows

References [5,6,8] represent the first application of the time-dependent technique to vibrational and chemical non-equilibrium nozzle flows. A purely steady flow analysis of such flows, which involves forward marching from the reservoir to the exit of the nozzle, encounters a saddle-point singularity at the nozzle throat. This singularity greatly complicates steady-state numerical solutions of the flow. On the other hand, as first demonstrated in Refs. [5,6], the time-dependent numerical solution circumvents such problems in the throat region, and therefore constitutes a relatively straightforward numerical solution of such flows.

The analysis of vibrational non-equilibrium nozzle flows requires the inclusion of a *vibrational rate equation*, such as

$$\frac{\partial e_{vib}}{\partial t} = \frac{1}{\tau}[(e_{vib})_{eq} - e_{vib}] - u\frac{\partial e_{vib}}{\partial x} \qquad (7.16)$$

where e_{vib} is the local non-equilibrium value of molecular vibrational energy per unit mass of gas, $(e_{vib})_{eq}$ is the local equilibrium value, and τ is the vibrational relaxation time which is a function of local p and T. The analysis of chemical non-equilibrium nozzle flows requires the inclusion of species continuity equations—one for each chemical species present in the gas—which are of the form

$$\frac{\partial \eta_i}{\partial t} = \dot{w}_i - u\frac{\partial \eta_i}{\partial x} \qquad (7.17)$$

where η_i is the mole–mass ratio (moles of species i per unit mass of mixture), and $\dot{w}_i$ is the rate of formation (or extinction of species i) due to finite-rate chemical reactions. The form of $\dot{w}_i$ involves chemical rate constants and the local concentrations of the chemical species. For an introductory development of Eqs. (7.16) and (7.17), see Chaps. 13 and 14 of Ref. [3]. Note that, in the same vein as Eqs. (7.2), (7.3) and (7.4), Eqs. (7.16) and (7.17) are written in the form of a time derivative on the left-hand side, and spatial derivatives on the right-hand side. In turn, the non-equilibrium variables e_{vib} and η_i are calculated in steps of time in the same fashion as ρ, u and e from Eqs. (7.2), (7.3) and (7.4). Indeed, for the time-dependent solution of non-equilibrium nozzle flows, Eqs (7.2), (7.3) (7.4), (7.16) and (7.17) are *coupled*, and are solved in the same coupled fashion at each time step as described in Sects. 7.2 and 7.3. However, there is one additional stability restriction brought about by the non-equilibrium phenomena. For explicit solutions of non-equilibrium flows, in addition to the CFL criterion discussed in Sect. 7.4, the value of Δt must also be less than the characteristic time for the fastest finite rate taking place in the system. That is

$$\Delta t < B\Gamma$$

where $\Gamma = \tau$ for vibrational non-equilibrium, and $\Gamma = (\partial \dot{w}_i/\partial \eta_i)^{-1}$ which is an effective chemical relaxation time. (See Refs. [5,6] for more details.)

For this problem, no grid transformation is necessary; the physical and computational planes are one-in-the-same.

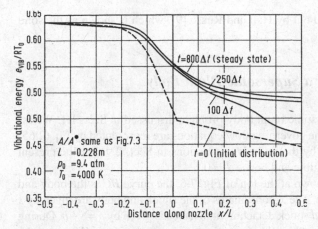

Fig. 7.4 Transient and final steady-state e_{vib} distributions for the non-equilibrium expansion of N_2 obtained from the present time-dependent analysis

Typical results obtained with the Lax–Wendroff time-dependent technique are shown in Figs. 7.4 and 7.5, from Ref. [5]. The case of the vibrational non-equilibrium expansion of pure N_2 is illustrated in Fig. 7.4. Here, the time-dependent nature of the non-equilibrium value of e_{vib} as a function of distance through the nozzle is shown. The dashed line represents the assumed initial distribution at $t = 0$. Several intermediate distributions, after 100 and 250 time steps, are shown, along with the final steady state after 800 time steps. A different case, namely that of the non-equilibrium chemically reacting expansion of dissociated oxygen, is illustrated in Fig. 7.5. Here, the dashed line represents the initially assumed variation of the mass fraction of atomic oxygen through the nozzle at $t = 0$. Several intermediate curves after 100 and 400 time steps are shown, along with the final, converged steady state after 2800 time steps. This final steady state distribution agrees well with an earlier

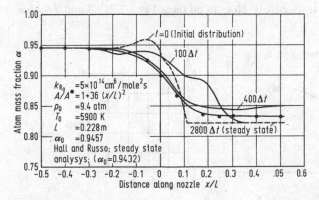

Fig. 7.5 Transient and final steady-state atom mass fraction distributions for the non-equilibrium expansion of dissociating oxygen obtained from the present time-dependent method; the steady-state distribution is compared with the steady-flow analysis of Ref. [9]

steady flow solution carried out by Hall and Russo [9], which is shown as the solid circles in Fig. 7.5.

7.5.2 Flow Field Over a Supersonic Blunt Body

Here we return to the supersonic blunt body problem discussed in Sect. 1.1. We assume inviscid flow, hence the governing flow equations are represented by Eq. (2.65) with U, F, G, and H given by the *inviscid* expressions in Sect. 2.9. For the present case, body forces are negligible and hence $J = 0$.

The physical plane is shown at the top of Fig. 7.6; the curve BC is the body and curve AD is the shock wave. The x-coordinates of the shock and body are given by s and b respectively. The *local* shock detachment distance is given by $\delta = s - b$. During the time-dependent solution, the body is stationary, hence $b = b(y)$. However, the shock wave will change shape and location with time, hence $s = s(y, t)$. Therefore,

$$\delta(y,t) = s(y,t) - b(y) \tag{7.18}$$

The computational plane (ξ, η) is shown in Fig. 7.6b, and is obtained from the transformation

$$\xi = \frac{x-b}{\delta}; \qquad \eta = y; \ \tau = t \tag{7.19}$$

where δ is obtained from Eq. (7.18). Note that this transformation is an example of a *boundary-fitted coordinate system* as discussed in Sect. 5.5.

Typical results, obtained from Ref. [10], are shown in Figs. 7.7, 7.8 and 7.9. These results were obtained using the Lax–Wendroff method. In Fig. 7.7, the time-dependent wave motion is illustrated, starting from its initially assumed value of $t = 0$, and progressing to its steady state shape and location after 500 time steps. The time variations of the centreline wave velocity and the stagnation point pressure are shown in Figs. 7.8 and 7.9 respectively. Note in all three Figs. 7.7, 7.8 and 7.9, that

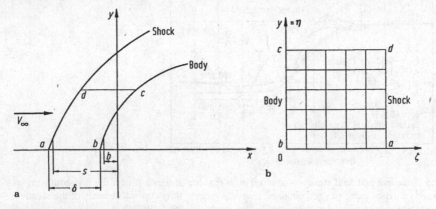

Fig. 7.6 Coordinate system for the blunt body problem

Fig. 7.7 Time-dependent shock wave motion, parabolic cylinder, $M_\infty = 4$

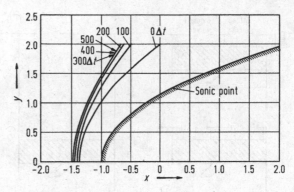

Fig. 7.8 Time variation of wave velocity; parabolic cylinder, $M_\infty = 4$

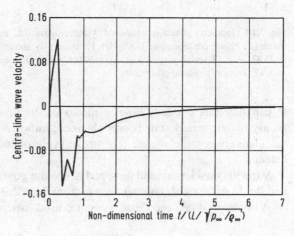

Fig. 7.9 Time variation of stagnation point pressure; parabolic cylinder, $M_\infty = 4$

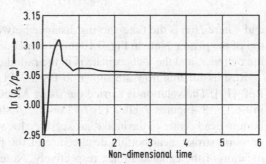

the most rapid changes occur at early times, and the steady state is approached rather asymptotically at large times.

7.5.3 Internal Combustion Engine Flows

Consider the flow inside an internal combustion engine as modelled by the piston-cylinder geometry shown in Fig. 7.10. The piston moves up and down inside the cylinder, and the flow enters through the intake valve and exits through the exhaust valve. The flow field in this problem is truly unsteady, and the objective is

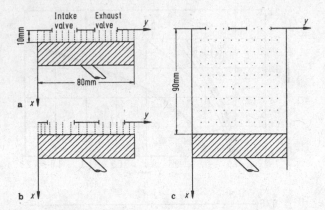

Fig. 7.10 Geometry of two-dimensional cylinder-piston I.C. engine model showing grid arrangement. (**a**) Piston positioned at TDC, 10×17 uniformly spaced grid points; (**b**) Piston positioned at TDC, 10×17 variably spaced grid points (only in y-direction); (**c**) Piston positioned at BDC, 10×17 uniformly spaced grid points

to calculate this unsteady flow by means of the time-dependent technique. Here, *no* asymptotic steady state is ever obtained; rather, a repeatable cyclic flow field is calculated over the complete four-stroke cycle of intake, compression, power and exhaust.

We will consider inviscid flow, and hence the governing equations are Eq. (2.65) and the U, F, G, and H column vectors from Sect. 2.9 for an inviscid flow.

A boundary-fitted coordinate system is used, where the transformation is

$$\xi = x/H(t); \quad \eta - y, \ \tau = t$$

and where $H(t)$ is the time-varying distance between the top of the cylinder and the top of the piston. Note in Fig. 7.10 that the x-coordinate is along the vertical axis of the cylinder, and the y-coordinate is in the radial direction across the cylinder.

Results for this flow are shown in Figs. 7.11, 7.12, 7.13 and 7.14, taken from Ref. [11]. The solution is carried out using MacCormack's technique as described in Sect. 7.3. Figures 7.11, 7.12, 7.13 and 7.14 show the flow field associated with bottom dead centre of the intake stroke, three locations of the piston during the compression stroke, near bottom dead centre of the power stroke, and an intermediate location of the exhaust stroke, respectively. Note that a circulatory flow is created during the intake stroke, and that this circulatory flow persists throughout the four-stroke cycle.

7.5.4 Supersonic Viscous Flow Over a Rearward-Facing Step With Hydrogen Injection

Consider the two-dimensional supersonic viscous flow over a rearward facing step, where H_2 is injected into the flow downstream of the step as sketched in Fig. 7.15.

Fig. 7.11 Velocity pattern on the intake stroke. $X_p^* = 8.78$, CA = 161°, $t = 8.95\,\text{msec} = 3080\,\Delta t$, $22 \times 30\,\text{mesh}$

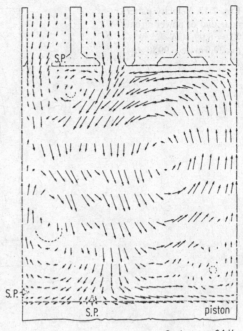

Scale: ⊢——⊣ = 0.1 V_r

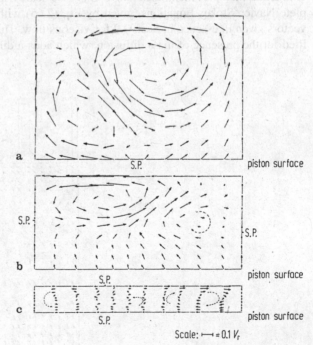

Fig. 7.12 Velocity distributions on compression stroke for the manifold-valve-engine model, $12 \times 12\,\text{mesh}$.
(**a**) $X_p^* = 5.63$, CA = 261°, $t = 14.5\,\text{msec} = 3970\,\Delta t$;
(**b**) $X_p^* = 3.56$, CA = 291°, $t = 16.2\,\text{msec} = 4250\,\Delta t$;
(**c**) $X_p^* = 1.0$, CA = 359°, $t = 19.9\,\text{msec} = 6300\,\Delta t$

Scale: ⊢——⊣ = 0.1 V_r

Fig. 7.13 Velocity pattern
near end of power stroke;
$X_p^* = 8.99$, CA = 539°,
$t = 29.9$ msec = 9950 Δt

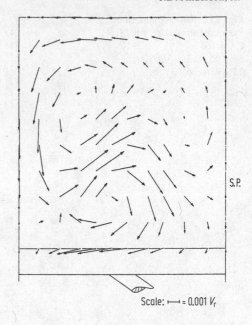

S.P.

Scale: ⊢——⊣ = 0.001 V_r

Unlike the examples mentioned above, this case deals with the solution of the complete Navier–Stokes Equations, given by Eq. (2.65) with the U, F and G column vectors given in essence in Sect. 2.9 for viscous flow. This system is slightly modified for the presence of mass diffusion, which adds a diffusion term in the energy

Fig. 7.14 Velocity
distribution on exhaust stroke;
$X_p^* = 6.99$, CA = 600°,
$t = 33.3$ msec = 11560 Δt,
30×22 mesh

S.P.

Scale: ⊢——⊣ = 0.1 V_r

Fig. 7.15 Rearward facing
step geometry

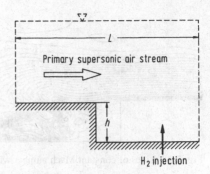

Primary supersonic air stream

h

H_2 injection

equation, and adds another equation, namely, the species continuity equation with diffusion terms. (See Refs. [12, 13] for more details.) The numerical technique used here is MacCormack's method discussed in Sect. 7.3. The present calculations were made on a uniform grid throughout the physical space. In combination with the rectangular geometry already existing in the physical plane (as can be seen by examining Fig. 7.15), this means that no grid transformation is needed.

Typical results obtained from Refs. [12, 13] are given in Figs. 7.16, 7.17, 7.18 and 7.19. In Fig. 7.16, a velocity vector diagram is shown for the case with no H_2 injection. The external Mach number is 2.19, and the Reynolds number based on step height is 70,000. These calculations also include a turbulence model patterned after that of Baldwin and Lomax [14]. Note the recirculating separated flow just downstream of the step. Figure 7.17 is a velocity vector diagram with H_2 injection.

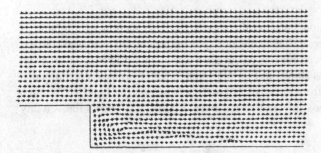

Fig. 7.16 Velocity vectors with no H_2 injection

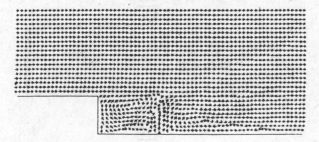

Fig. 7.17 Velocity vectors with H_2 injection

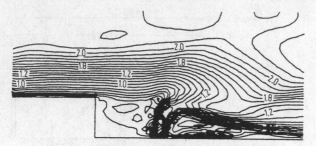

Fig. 7.18 Lines of constant Mach number with H_2 injection

Fig. 7.19 Lines of constant H_2 mass fraction

Recirculating separated flows are now seen between the step and the H_2 jet, as well as downstream of the jet. Figure 7.18 shows a Mach number contour plot of the flow (lines of constant Mach number). Figure 7.19 illustrates the contours of constant H_2 mass fraction; this figure serves to define the extent and shape of the jet flow.

7.5.5 Supersonic Viscous Flow Over a Base

In a somewhat related fashion, consider the supersonic viscous flow over a base, as illustrated in Fig. 7.20. Here, the same viscous flow equations are used as discussed in Sect. 7.5.4 above. However, for this calculation a stretched grid is used, as given in detail in Sect. 6.4, and as shown in Fig. 6.4. Again, MacCormack's technique is used. Some sample results from Refs. [15,16] are given in Figs. 7.21 and 7.22, which deal with *no* secondary mass injection at the base. Figure 7.21 shows the velocity

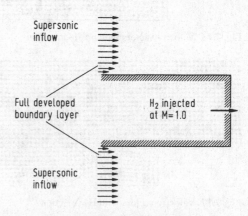

Fig. 7.20 Base flow with mass injection

Fig. 7.21 Velocity vectors
with no base injection

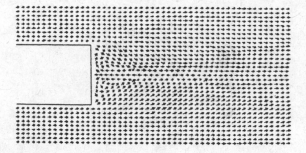

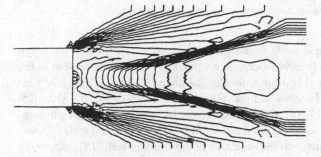

Fig. 7.22 Lines of constant pressure with no base injection

vector diagram for the case with an external Mach number of 2.25 and a Reynolds
number of 477 000 based on the height of the base. Note the recirculating separated flow downstream of the base. Figure 7.22 illustrates the contours of constant
pressure in the flow; the expansion wave around the corner and the recompression
shock downstream of the base are clearly seen. Figures 7.23 and 7.24 show the same
type of results, except now for the case of air injection from the centre of the base.
Note that injection greatly changes the flow field, as can be seen in comparison with
Figs. 7.21 and 7.22.

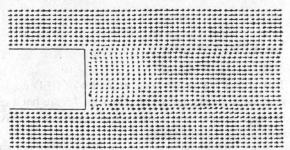

Fig. 7.23 Velocity vectors
with injection from the center
of the base

7.5.6 Compressible Viscous Flow Over an Airfoil

Consider the subsonic compressible, viscous two-dimensional flow over an airfoil.
The governing equations are the Navier–Stokes equations discussed in Chap. 2. For

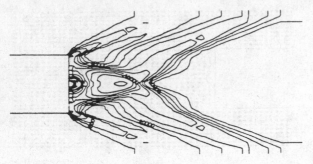

Fig. 7.24 Lines of constant pressure with injection from the center of the base

this application, the choice is made to use the non-conservation form of the equations, namely, Eqs. 2.36(a, b and c), because no shock waves will be present in the flow. MacCormack's method is used. Consider the airfoil and the elliptically generated boundary-fitted grid shown in Figs. 6.8 and 6.9, as discussed in Sect. 6.5, and as taken from Refs. [17, 18]. Calculated results for a free stream Mach number of 0.5 and a Reynolds number based on chord length of 100 000 (this is a low Reynolds number flow, which was the objective of the study in Ref. [18]) are shown in Figs. 7.25, 7.26 and 7.27. The angle-of-attack in these figures is zero. These figures illustrate the instantaneous flow over a Wortmann airfoil at different times. In Figs. 7.25 and 7.26, the flow is laminar, and it separates over the top surface of the airfoil at about the maximum thickness point. The flow is clearly unsteady, as can be seen by comparing Fig. 7.25(a, b and c); there is a rather periodic flow fluctuation over the rearward portion of the airfoil, as well as downstream of the trailing edge. The calculation of such unsteady flows, especially in situations where they may be unexpected, is one of the major advantages of the time-dependent method in comparison to steady-state analyses. In Fig. 7.27, the flow is treated as turbulent; note that in this case the flow is attached.

7.6 Final Comment

This author has many more examples of CFD applications from the work of his graduate students; those listed in Sect. 7.5 are but a small fraction. They are picked for discussion in these notes on a rather arbitrary basis. Time and space do not allow further listing and discussion.

Also, this brings to an end our introduction to CFD. It is the author's hope that these notes have been a reasonable beginning for the unitiated reader, and that he or she can now greatly expand his or her horizons by reading the more advanced literature on CFD. If such advanced reading is indeed more easy after studying the present notes, then this author has accomplished his goal.

In recent years, some modern texts on CFD have been published (Refs. [19–23]); these texts are recommended for advanced studies of the subject. In

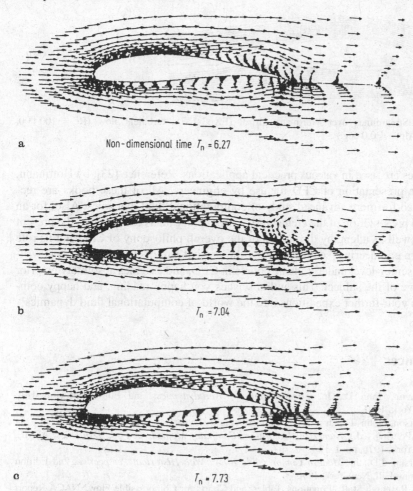

a   Non-dimensional time T_n = 6.27

b T_n = 7.04

c T_n = 7.73

Fig. 7.25 Velocity vector diagrams at three different non-dimensional times for purely laminar flow (Re = 1 000 000, M = 0.5, Alpha = 0.0 deg.). (**a**) Non-dimensional time T_n = 6.27. (**b**) T_n = 7.04. (**c**) T_n = 7.73

particular, Fletcher's two volumes (Refs. [19, 20]) contain a nice theoretical discussion of the subject. Of special note are the two volumes by Hirsch (Refs. [21, 22]); these volumes represent an authoritative presentation of the mathematical and numerical fundamentals of CFD, the modern techniques used in CFD, and how these

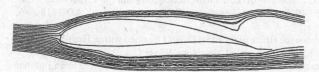

Fig. 7.26 Instantaneous streamlines over Wortmann airfoil (FX63-137)—laminar flow (unsteady results) (Re = 100 000, M = 0.5, Alpha = 0.0 deg.) Non-dimensional time T_n = 7.04

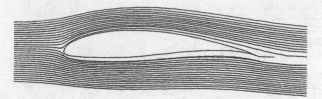

Fig. 7.27 Streamlines over Wortmann airfoil (FX63-137)—turbulent flow (Re = 100 000, M = 0.5, Alpha = 0.0 deg.)

techniques are used in various practical applications. Reference [23], by Hoffmann, is a crisp presentation of CFD for use by engineers. All of these books are recommended for more advanced study of computational fluid dynamics. Also, for an extended presentation of the elementary, introductory ideas contained in the present book, as well as a lengthy discussion of the overall philosophy of CFD and its role in modern engineering, see the book by the present author (Ref. [24]); this is written for a senior-level undergraduate course in CFD, and assumes absolutely no prior knowledge of the subject. This author wishes you happy reading, and happy computing in your further expeditions into the world of computational fluid dynamics!

References

1. Anderson, John D., Jr., *Fundamentals of Aerodynamics*, 2nd Edition McGraw-Hill, New York, 1991.
2. Anderson, John D., Jr., 'Computational Fluid Dynamics—An Engineering Tool?' in A.A. Pouring (ed.), *Numerical Laboratory Computer Methods in Fluid Dynamics*, ASME, New York, 1976, pp. 1–12.
3. Anderson, J.D., Jr., *Modern Compressible Flow: With Historical Perspective*, 2nd Edition McGraw-Hill, New York, 1990.
4. Ames Research Staff, 'Equations, Tables, and Charts for Compressible Flow,' NACA Report 1135, 1953.
5. Anderson, J.D. Jr., 'A Time-Dependent Analysis for Quasi-One-Dimensional Nozzle Flows with Vibrational and Chemical Nonequilibrium,' NOLTR 69-52, Naval Ordnance Laboratory, White Oak, MD, 1969.
6. Anderson, J.D., Jr., 'A Time-Dependent Analysis for Vibrational and Chemical Nonequilibrium Nozzle Flows,' *AIAA Journal*, Vol. 8, No. 3, March 1970, pp. 545–550.
7. MacCormack, R.W., 'The Effect of Viscosity in Hypervelocity Impact Cratering,' AIAA Paper No. 69-354, 1969.
8. Anderson, J.D., Jr., 'Time-Dependent Solutions of Nonequilibrium Nozzle Flow—A Sequel,' *AIAA Journal*, Vol. 5, No. 12, Dec. 1970. pp. 2280–2282.
9. Hall, J.G. and Russo, A.L., 'Studies of Chemical Nonequilibrium in Hypersonic Nozzle Flows,' AFOSR TN 59-1090, Cornell Aeronautical Laboratory Report AD-1118-A-6, November 1969.
10. Anderson, J.D., Jr., 'On Hypersonic Blunt Body Flow Fields Obtained with a Time-Dependent Technique,' NOLTR 68–129, Naval Ordnance Laboratory, White Oak, MD, August 1968.
11. Dallospedale, C.L., 'A Numerical Solution for the Two-Dimensional Flowfield in an Internal Combustion Engine with Realistic Valve-Geometry,' M.S. Thesis, Department of Aerospace Engineering, University of Maryland, College Park, MD, 1978.

12. Berman, H.A., Anderson, J.D., Jr. and Drummond, J.P., 'A Numerical Solution of the Super-sonic Flow Over a Rearward Facing Step with Transverse Non-Reacting Hydrogen Injection,' AIAA Paper No. 82–1002, 1982.

13. Berman, H.A., Anderson, J.D., Jr. and Drummond, J.P., 'Supersonic Flow over a Rearward Facing Step with Transverse Nonreacting Hydrogen Injection,' *AIAA Journal*, Vol. 21, No. 12, December 1983, pp. 1707–1713.

14. Baldwin, B.S. and Lomax, H., 'Thin Layer Approximations and Algebraic Model for Sepa-rated Turbulent Flows,' AIAA Paper 78–257, 1978.

15. Sullins, G.A., Anderson, J.D., Jr. and Drummond, J.P., 'Numerical Investigation of Supersonic Base Flow with Parallel Injection,' AIAA Paper No. 82–1001.

16. Sullins, G.A., *Numerical Investigation of Supersonic Base Flow with Tangential Injection*, M.S. Thesis, Department of Aerospace Engineering, University of Maryland, 1981.

17. Wright, Andrew F., *A Numerical Investigation of Low Reynolds Number Flow Over an Airfoil*, M.S. Thesis, Department of Aerospace Engineering, University of Maryland, 1982.

18. Kothari, A.P. and Anderson, J.D., Jr., 'Flows Over Low Reynolds Number Airfoils—Compressible Navier–Stokes Numerical Solutions', AIAA Paper No. 85–0107, 1985.

19. Fletcher, C.A., Computational Techniques for Fluid Dynamics, Vol. I: *Fundamental and Gen-eral Techniques*, Springer-Verlag, Berlin, 1988.

20. Fletcher, C.A., Computational Techniques for Fluid Dynamics, Vol. II: *Specific Techniques for Different Flow Categories*, Springer-Verlag, Berlin, 1988.

21. Hirsch, C., Numerical Computation of Internal and External Flows, Vol. I: *Fundamentals of Numerical Discretization*, Wiley, New York, 1988.

22. Hirsch, C., Numerical Computation of Internal and External Flows, Vol. II: *Computational Methods for Inviscid and Viscous Flows*, Wiley, New York, 1990.

23. Hoffmann, K.A., *Computational Fluid Dynamics for Engineers*, Engineering Education Sys-tem, Austin, Tex., 1989.

24. Anderson, J.D., Jr., *Computational Fluid Dynamics: The Basis with Applications*, McGraw-Hill, New York, 1995.

Chapter 8
Boundary Layer Equations and Methods of Solution

R. Grundmann

8.1 Introduction

The objective of computational fluid dynamics is to calculate an entire flow field either around an arbitrary obstacle or through a channel of any shape. The flow may be unsteady, three dimensional, compressible and turbulent. At hypersonic speeds, regions of reacting flow (dissociation, ionization, etc.) might also be considered. The equations to describe this task, as derived in Chap. 2, are the Navier–Stokes equation, the energy equation, the global and partial continuity equations and other closure model equations describing turbulence and reacting gas effects. It can easily be shown that, at present, no computer could provide either the capacity or the necessary calculation speed to fulfil this task.

Thereby, coming back to the reality of today, the governing equations have to be simplified such that the properties of the remaining set of equations still describe the flow to be considered. For instance, when the viscous terms in the Navier–Stokes equations are neglected, one arrives at the Euler equations. They can be used to determine far-field flows where the interaction with the viscous layer is not dominant. However, a separation bubble on the surface of a wing cannot be detected without providing viscous flow calculations near the body surface; separation is a matter of viscous effects.

As indicated already, different types of flows can be treated by examining their physical characteristics in detail and in this way establishing the appropriate governing equations by the correct reduction of the general set of fluid mechanical equations. This is what Prandtl [1] did in 1904 concerning a thin layer near walls where the influence of viscosity normal to the wall is dominant. He called this layer a 'boundary layer'. The important detail of the physical meaning of this kind of flow is that the main flow velocity tends to zero approaching the wall. The gradient of this velocity component in the direction normal to the surface is large compared to the gradient of this component in the downstream direction.

This observation leads to an important change in the character of the governing equations from the elliptic to the parabolic type which makes a numerical

R. Grundmann
Technische Universität Dresden, Dresden, Germany
e-mail: grundman@tfd.mw.tu-dresden.de

J.F. Wendt (ed.), *Computational Fluid Dynamics*, 3rd ed.,
© Springer-Verlag Berlin Heidelberg 2009

downstream marching procedure applicable. Reverse flow therefore cannot be calculated with a simple boundary layer method, but the flow field very near to the separation point, where the reverse flow starts, can be detected very well. Reference to the parabolic nature of differential equation is already given in Sect. 4.3.

The boundary layer theory will be the subject of this chapter. Prandtl's idea will be described in detail as an introduction. The hierarchy of the boundary layer equations will be discussed; that is, the relationship of the different types of boundary layer equations to the Navier–Stokes equations will be demonstrated. Furthermore, it will be pointed out that there are transformation techniques to reduce the problems of solution. A generalized discretization scheme will be applied to a set of laminar compressible boundary layer equations and a numerical solution scheme for calculating the remaining tridiagonal linear difference equations will be shown. A sample calculation of a laminar boundary layer along the symmetry line of a highly inclined ellipsoid will conclude the discussion.

This chapter deals only with laminar boundary layer theory. The description of turbulence needs additional effort, especially in seeking suitable turbulence models for specific purposes, but it does not affect the principal solution procedure. As these notes are meant to give an introduction to the boundary layer theory and its methods of solution, turbulent boundary layers will not be considered, but overviews on recent turbulence models are given in Refs. [2–5].

8.2 Description of Prandtl's Boundary Layer Equations

In 1904 Prandtl [1] made an important contribution to the calculation of a specific type of flow for which the Reynolds number is very large. The Reynolds number has the form of a non-dimensional parameter

$$\text{Re} = \frac{LV}{\nu} = \frac{\rho LV}{\mu} \tag{8.1}$$

where L is a characteristic length, usually the length of the considered body, V is the velocity of the flow where it is well-defined and undisturbed. The kinematic and dynamic viscosity are denoted by ν and μ, respectively. The density of the fluid is ρ. The Reynolds number is the ratio of inertia to friction forces following the 'principle of similarity':

$$\text{Re} = \frac{\rho u \partial u/\partial x}{\mu \partial^2 u/\partial x^2} \equiv \frac{\text{inertia force}}{\text{friction force}} \tag{8.2}$$

The velocity u at some point in the velocity field is proportional to the free stream velocity V. The velocity gradient $\partial u/\partial x$ is proportional to V/L and similarly $\partial^2 u/\partial x^2$ is proportional to V/L^2. Hence the ratio, Eq. (8.2) yields

$$\text{Re} = \frac{\rho V^2/L}{\mu V/\text{d}^2} = \frac{\rho LV}{\mu} \tag{8.3}$$

Fig. 8.1 Boundary layer flow
along a wall

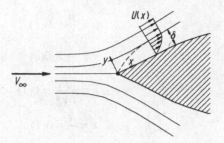

Two flows are similar from the point of view of the relative importance of iner-
tial and viscous effects if the Reynolds number is constant. Now the physical phenomenon of a flow with high Reynolds number is considered for the example of a
cylindrical body shown in Fig. 8.1.

With the exception of the immediate neighbourhood of the surface the flow velocity is comparable to the free stream velocity V. This flow region is nearly free
of friction; it is a potential flow. Considering the region near the surface there is
friction in the flow which means that the fluid is retarded until it adheres at the surface. The transition from zero velocity at the surface to the full magnitude at some
distance from it takes place in a very thin layer, the so-called 'boundary layer'. Its
thickness is δ, which is a function of the downstream coordinate x and is assumed to
be very small compared to the length of the body L. In the normal direction y inside
the thin layer it is clear that the gradient $\partial u/\partial y$ is very large compared to gradients
in the streamwise direction $\partial u/\partial x$. Although the viscosity was meant to be very
small in this flow the shear stress $\tau = \mu(\partial u/\partial y)$ may assume large values. Outside
the boundary layer the velocity gradients are negligibly small and the influence of
the viscosity is unimportant. The flow is frictionless and potential.

The above assumptions are now used to simplify the Navier–Stokes equations
for steady two-dimensional, laminar and incompressible flows, resulting from the
non-conservation form in Sect. 2.8, by a formal procedure. Including the continuity
equation they have the following dimensional form in cartesian coordinates:

$$\bar{u}\frac{\partial \bar{u}}{\partial \bar{x}} + \bar{v}\frac{\partial \bar{u}}{\partial \bar{y}} = -\frac{1}{\bar{\rho}}\frac{\partial \bar{p}}{\partial \bar{x}} + \frac{\bar{\mu}}{\bar{\rho}}\left(\frac{\partial^2 \bar{u}}{\partial \bar{x}^2} + \frac{\partial^2 \bar{u}}{\partial \bar{y}^2}\right) \tag{8.4}$$

$$\bar{u}\frac{\partial \bar{v}}{\partial \bar{x}} + \bar{v}\frac{\partial \bar{v}}{\partial \bar{y}} = -\frac{1}{\bar{\rho}}\frac{\partial \bar{p}}{\partial \bar{y}} + \frac{\bar{\mu}}{\bar{\rho}}\left(\frac{\partial^2 \bar{v}}{\partial \bar{x}^2} + \frac{\partial^2 \bar{v}}{\partial \bar{y}^2}\right) \tag{8.5}$$

$$\frac{\partial \bar{u}}{\partial \bar{x}} + \frac{\partial \bar{v}}{\partial \bar{y}} = 0 \tag{8.6}$$

Here the velocity components $\bar{u}$ and $\bar{v}$ are directed towards the downstream $\bar{x}$ and
the normal $\bar{y}$-direction, respectively. The static pressure is denoted by $\bar{p}$, $\bar{\rho}$ is the
density and $\bar{\mu}$ is the dynamic viscosity of the fluid.

For convenience, this set of second order differential equations is non-dimensionalized which involves the Reynolds number necessary for the following reduction
of the equations. The prescriptions for non dimensionalization are:

$$u = \frac{\bar{u}}{V} = 0(1)$$

$$v = \frac{\bar{v}}{V} = 0(\varepsilon)$$

$$p = \frac{\bar{p}}{\bar{\rho}V^2} = 0(1) \qquad \text{Re} = \frac{\bar{\rho}LV}{\bar{\mu}} = 0\left(\frac{1}{\varepsilon^2}\right) \tag{8.7}$$

$$x = \frac{\bar{x}}{L} = 0(1)$$

$$y = \frac{\bar{y}}{L} = 0(\varepsilon)$$

V is the dimensional free stream velocity and the pressure is non-dimensionalized by twice the dynamic pressure, $\bar{q} = \frac{1}{2}\bar{\rho}V^2$.

Using these definitions, Eqs. (8.4), (8.4) and (8.6) become:

$$u\frac{\partial u}{\partial x} + v\frac{\partial u}{\partial y} = -\frac{\partial p}{\partial x} + \frac{1}{\text{Re}}\left(\frac{\partial^2 u}{\partial x^2} + \frac{\partial^2 u}{\partial y^2}\right)$$

$$(1)\frac{(1)}{(1)}(\varepsilon)\frac{(1)}{(\varepsilon)} \qquad (1) \qquad (\varepsilon^2)\left(\frac{(1)}{(1)} \quad \frac{(1)}{(\varepsilon^2)}\right) \tag{8.8}$$

$$u\frac{\partial v}{\partial x} + v\frac{\partial v}{\partial y} = -\frac{\partial p}{\partial y} + \frac{1}{\text{Re}}\left(\frac{\partial^2 v}{\partial x^2} + \frac{\partial^2 v}{\partial y^2}\right)$$

$$(1)\frac{(\varepsilon)}{(1)} \quad (\varepsilon)\frac{(\varepsilon)}{(\varepsilon)} \qquad (\varepsilon^2)\left(\frac{(\varepsilon)}{(1)} \quad \frac{(\varepsilon)}{(\varepsilon^2)}\right) \tag{8.9}$$

$$\frac{\partial u}{\partial x} + \frac{\partial v}{\partial y} = 0$$

$$\frac{(1)}{(1)} \quad \frac{(\varepsilon)}{(\varepsilon)} \tag{8.10}$$

Now the question is, what order of magnitude do the dimensionless substitutions Eqs. (8.7) have? As stated above, the boundary layer thickness δ is very small, so is the distance y compared to the length of the body L. Consequently y is of the order ε which describes a value much smaller than 1. The u-velocity component can reach the maximum value of V, therefore it is of the order 1. But the v-velocity component also has to be of the order ε as can be seen from the continuity equation, Eq. (8.10). If the derivative $\partial u/\partial x$ is of the order 1 because x becomes, at its maximum, the length L, then the second term in the continuity equation $\partial v/\partial y$ has also to be of the order 1. Consequently, v is not greater than ε. Now, with these assumptions the order of magnitude analysis can be done. It follows from the first equation of motion, Eq. (8.8), that the viscous forces in the boundary layer can become of the same order of magnitude as the inertia forces only if the Reynolds number is of the order of $1/\varepsilon^2$.

The equation of continuity remains unaltered for very large Reynolds numbers. The downstream momentum equations can be reduced by the second derivative of the u-velocity component $\partial^2 u/\partial x^2$ and multiplied by 1/Re because it has the smallest order of magnitude in this equation. It only holds that the forcing function term $(-\mathrm{d}p/\mathrm{d}x)$ will not exceed the order of 1 to be in balance with the other remaining terms.

All terms of the normal momentum equation, Eq. (8.9), are of a smaller magnitude than those of Eq. (8.8). This equation can only be in balance if the pressure term is of the same order of magnitude. Therefore, this equation delivers the information of negligible pressure gradient in the normal direction, i.e.

$$\frac{\partial p}{\partial y} = 0(\varepsilon) \tag{8.11}$$

The meaning of this result is that the pressure is practically constant; it is 'impressed' on the boundary layer by the outer flow. Therefore, the pressure p is only a function of x.

The derivation of Eq. (8.8) at the outer edge of the boundary layer gives, if the inviscid velocity distribution $U(x) = \bar{u}(x)/V$ is known:

$$U \frac{\mathrm{d}U}{\mathrm{d}x} = -\frac{1}{\rho} \frac{\mathrm{d}p}{\mathrm{d}x} \tag{8.12}$$

The other terms involving $\partial u/\partial y$ are zero since there remains no large velocity gradient. After integration of Eq. (8.12) the well known Bernoulli equation is found:

$$p + \frac{1}{2}\rho U^2 = \text{const.} \tag{8.13}$$

Summing up, by the order of magnitude analysis the Navier–Stokes equations, Eqs. (8.8) and (8.9), and the continuity Eq. (8.10), have been simplified. They are known as 'Prandtl's boundary layer equations':

$$u \frac{\partial u}{\partial x} + v \frac{\partial u}{\partial y} = -\frac{\partial p}{\partial x} + \frac{1}{\text{Re}} \frac{\partial^2 u}{\partial y^2} \tag{8.14}$$

$$\frac{\partial p}{\partial y} = 0 \tag{8.15}$$

$$\frac{\partial u}{\partial x} + \frac{\partial v}{\partial y} = 0 \tag{8.16}$$

The boundary conditions are:
On the surface:

$$y = 0 \qquad u = 0, \qquad v = 0 \tag{8.17}$$

On the outer edge of the boundary layer:

$$y = \delta = \frac{\bar{\delta}}{L} \qquad u = U(x) \tag{8.18}$$

This set of equations is reduced by the unknown pressure p, which is, because of Bernoulli's equation, Eq. (8.13), a known value now, if only the inviscid velocity distribution at the surface $U(x)$ is provided. It is still a coupled, non-linear, second-order set of differential equations.

The order of magnitude analysis also described by Schlichting [6] is well suited to analyse the more complicated surface-oriented Navier–Stokes equations with additional surface curvature created Coriolis and centrifugal forces. At least the order of magnitude analysis gives an impression where the boundary layer equations and their more complicated extensions are situated in their level of approximation to the full Navier–Stokes equations. This overview will be given in the next section.

8.3 Hierarchy of the Boundary Layer Equations

To develop a hierarchy of the fluid mechanical equations, the steady, compressible, laminar, two-dimensional Navier–Stokes equations should be written for the Euclidian space in a layer close to the surface. This will say that a coordinate system, which may be surface oriented for a better adaption to the flow problem considered, is related to the cartesian coordinate system. Both systems must be transferable from one to the other. The cartesian and the polar coordinate system, for example, are matched together following this demand of Euclidian space. In other words, the Jacobian matrix must exist.

If the Navier–Stokes equations can be formulated for such a surface-oriented coordinate system, they will contain many additional terms due to the surface curvature. These terms can be understood as Coriolis and centrifugal force terms caused by the change of the streamlines in downstream as well as in the cross flow direction depending on the curvature of the surface. Curvature-induced terms will have different orders of magnitude. Some are important and others can be neglected depending on the specific flow problems.

Now the question is to set the boundary layer equations including curvature terms in relation to Prandtl's boundary layer equations developed in the foregoing chapter.

A simple two-dimensional surface-oriented coordinate system is fixed on an airfoil-like contour sketched in Fig. 8.2. The relations between the new coordinate system and the cartesian one are:

$$x = \int_0^s \cos\theta(x) \, \mathrm{d}s - n\sin\theta(x) \tag{8.19}$$

$$y = \int_0^s \sin\theta(x) \, \mathrm{d}s + n\cos\theta(x) \tag{8.20}$$

The resultant set of differential equations due to the coordinate transformation consists of two equations of motion in the downstream direction s and the perpendicular direction n, the energy and the continuity equations. (For convenience this lengthy dimensional set of equations is not barred for reasons of clarity.)

Momentum equation in tangential direction:

Fig. 8.2 Surface oriented coordinate system

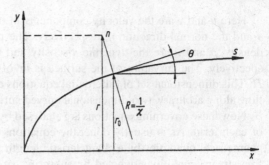

$$\rho\left[u\frac{\partial u}{\partial s}+Hv\frac{\partial u}{\partial n}+\kappa uv\right]=-\frac{\partial p}{\partial s}+\frac{\partial}{\partial s}\left[\frac{4}{3}\frac{\mu}{H}\frac{\partial u}{\partial s}+\frac{4}{3}\frac{\kappa\mu v}{H}-\frac{2}{3}\mu\frac{\partial v}{\partial n}\right]$$

$$+H\frac{\partial}{\partial n}\left[\frac{\mu}{H}\frac{\partial v}{\partial s}+\mu\frac{\partial u}{\partial n}-\frac{\mu\kappa u}{H}\right] \tag{8.21}$$

Momentum equation in normal direction:

$$\rho\left[u\frac{\partial v}{\partial s}+Hv\frac{\partial v}{\partial n}-\kappa u^2\right]=-H\frac{\partial p}{\partial n}+H\frac{\partial}{\partial n}\left[\frac{4}{3}\mu\frac{\partial v}{\partial n}-\frac{2}{3}\frac{\mu}{H}\frac{\partial u}{\partial s}-\frac{2}{3}\frac{\mu\kappa v}{H}\right]$$

$$+\frac{\partial}{\partial s}\frac{\mu}{H}\frac{\partial v}{\partial s}+\mu\frac{\partial u}{\partial n}-\frac{\mu\kappa u}{H}$$

$$+2\kappa\left[\mu\frac{\partial v}{\partial n}-\frac{\mu}{H}\frac{\partial u}{\partial s}-\frac{\mu\kappa v}{H}\right] \tag{8.22}$$

Energy equation:

$$c_p\rho\left[u\frac{\partial T}{\partial s}+Hv\frac{\partial T}{\partial n}\right]=u\frac{\partial p}{\partial s}+Hv\frac{\partial p}{\partial n}+\frac{\partial}{\partial s}\left[\lambda\frac{\partial T}{\partial s}\right]+H\frac{\partial}{\partial n}\left[\lambda\frac{\partial T}{\partial n}\right]$$

$$+H\frac{\mu}{2}\left\{\left[\frac{2}{H}\frac{\partial u}{\partial s}+2\frac{\kappa v}{H}\right]^2\right.$$

$$\left.+\left[2\frac{\partial v}{\partial n}\right]^2+2\left[\frac{1}{H}\frac{\partial v}{\partial s}+\frac{\partial u}{\partial n}-\frac{\kappa u}{H}\right]^2\right\}$$

$$-\frac{2}{3}H\mu\left[\frac{1}{H}\frac{\partial u}{\partial s}+\frac{\kappa v}{H}+\frac{\partial v}{\partial n}\right]^2 \tag{8.23}$$

Continuity equation:

$$\frac{\partial(\rho u)}{\partial s}+\frac{\partial(H\rho v)}{\partial n}=0 \tag{8.24}$$

with

$$H=1+\kappa n=\frac{R+n}{R}$$

Here u and v are the velocity components in the tangential direction of the flow s and the normal direction n, respectively. The pressure is denoted by p, ρ is the density, μ and λ are the dynamic viscosity and the thermal heat conductivity, respectively. The curvature of the surface is involved in the geometrical coefficient H. This dimensional set of differential equations describe the laminar, compressible flow along arbitrary, two-dimensional curved surfaces.

Now these governing equations are analysed by predicting the order of magnitude of each term. As is usually done, the equations will be non-dimensionalized, the geometrical quantities by a characteristic length L and the flow properties by their free stream conditions denoted by subscript ∞. The order of magnitude of these quantities is defined as has been done in the case of a simple boundary layer without curvature in the preceding chapters.

$$s = \frac{\bar{s}}{L} = 0(1), \qquad n = \frac{\bar{n}}{L} = 0(\varepsilon), \qquad \kappa = \bar{\kappa}L = 0(1)$$

$$H = 1 + \bar{\kappa}\bar{\eta} = 0(1), \qquad u = \frac{\bar{u}}{u_\infty} = 0(1), \qquad v = \frac{\bar{v}}{u_\infty} = 0(\varepsilon)$$

$$T = \frac{\bar{T}}{T_\infty} = 0(1) \qquad p = \frac{\bar{p}}{\rho_\infty u_\infty^2} = 0(1) \qquad \rho = \frac{\bar{\rho}}{\rho_\infty} = 0(1)$$

$$\mu = \frac{\bar{\mu}}{\mu_\infty} = 0(1) \qquad \lambda = \frac{\bar{\lambda}}{\lambda_\infty} = 0(1) \qquad c_p = \frac{\bar{c}_p}{c_{p\infty}} = 0(1)$$

$$Re = \frac{\rho_\infty u_\infty L}{\mu_\infty} = 0\left(\frac{1}{\varepsilon^2}\right) \qquad \text{Reynolds number}$$

$$Pr = \frac{c_{p\infty}\mu_\infty}{\lambda_\infty} = 0(1) \qquad \text{Prandtl number} \qquad (8.25)$$

$$Ec = \frac{u_\infty^2}{c_{p\infty}T_\infty} = 0(1) \qquad \text{Eckert number}$$

It is to be mentioned that the radius of curvature R is not allowed to be much larger than the characteristic length L, otherwise κ would belong to another order of magnitude. The radius of curvature R is related to the curvature as follows

$$\kappa = \bar{\kappa}L = \frac{L}{R} \qquad (8.26)$$

When the radius R becomes very small compared to the length, H can exceed the order demanded above.

The combination of Eq. (8.25) with the governing equations, Eqs. (8.21), (8.22), (8.23) and (8.24), provides the order of magnitude of each term. A detailed development of the order of magnitude analysis applied to this set of equations seems not to be necessary here because in the preceding chapter an example was already presented. But in order to give an insight into the origin of the hierarchy of the boundary layer equations, the equations will be shown that retain terms only of the order $0(1)$ and $0(\varepsilon)$. The chosen equation is the tangential and normal momentum equation in dimensional unbarred form.

Order 0(1):

$$\rho\left(u\frac{\partial u}{\partial s} + Hv\frac{\partial u}{\partial n}\right) = -\frac{\partial p}{\partial s} + H\frac{\partial}{\partial n}\left(\mu\frac{\partial u}{\partial n}\right) \tag{8.27}$$

$$\frac{\partial p}{\partial n} = 0 \tag{8.28}$$

These equations, including the continuity equation, are called the 'first order boundary layer equations'. Curvature effects are included in the quantity H defined in Eq. (8.24). These equations become identical to Prandtl's boundary layer equations when the curvature goes to zero. Hence, Prandtl's equations are the lowest level of the hierarchy and therefore they should be called 'zeroth order boundary layer equations'.

Now terms of the order $0(1)$ and $0(\varepsilon)$ are retained.

Order of magnitude $0(\varepsilon)$:

$$\rho\left(u\frac{\partial u}{\partial s} + Hv\frac{\partial u}{\partial n} + \kappa uv\right) = -\frac{\partial p}{\partial s} + II\frac{\partial}{\partial n}\left(\mu\frac{\partial u}{\partial n}\right) - \kappa\frac{\partial u}{\partial n}u + \kappa\mu\frac{\partial u}{\partial n} \tag{8.29}$$

$$\frac{\partial p}{\partial n} = \frac{\kappa\rho u^2}{H} \tag{8.30}$$

These equations show a significant extension of the foregoing ones. In Eq. (8.29) an additional centrifugal term κuv appears as well as dissipative terms due to curvature on the right-hand side; but the most important extension appears in the normal momentum equation, Eq. (8.30). The pressure gradient normal to the flow is no longer zero. Eq. (8.30) is an integral equation for the pressure which is no longer impressed on the boundary layer from the inviscid flow. These equations are the so-called 'second order boundary layer equations' and take into account that, even in the outer inviscid flow normal to the surface, there exist velocity gradients due to the streamline curvature. The outer edge of the boundary layer is matched to this gradient which is no longer equal to zero as the first order of boundary layer theory prescribes.

Consequently terms of higher order than $0(\varepsilon)$ will be retained now. The result is summarized in Table 8.1.

A decisive development takes place proceeding from the second to the third-order set. The mathematical character of the equation changes from parabolic to elliptic. Elliptic differential equations are pure boundary value problems while parabolic equations are initial-boundary value problems. The latter can be solved by the so-called 'marching procedure', but the former require the calculation of the entire flow field surrounded by the boundaries which implies a greater numerical effort.

The conclusion of this discussion is that a boundary layer theory of order higher than second order immediately leads to elliptic equations. This complicates the method of solution because the parabolic approach of the original idea of boundary layer theory no longer holds.

The subject of the following chapter will be to give an impression as to how transformations of the governing first-order boundary layer equations influence the solution techniques positively.

Table 8.1 Hierarchy of the boundary layer equations

Theory	Equation of motion		Energy equation
Navier–Stokes 5th order	Elliptic	Elliptic	Elliptic
Theory 4th order	Elliptic N–S	Parabolic	Elliptic
Theory 3rd order	Elliptic	Parabolic	Elliptic
Boundary layer theory 2nd order	Parabolic	Integral equation	Parabolic
Boundary layer theory 1st order	Parabolic	Constant	Parabolic
Boundary layer theory 0th order	Parabolic	Constant	Parabolic
Prandtl boundary layer equation			

8.4 Transformation of the Boundary Layer Equations

In this section the boundary equations for the flow around a body of revolution without inclination are considered. This flow is two dimensional because it does not vary in the circumferential direction. The coordinate system indicated in Fig. 8.2 was applied to derive the first order laminar compressible boundary layer equations in dimensional form.

Continuity equation:

$$\frac{\partial}{\partial s}(r^j \rho u) + \frac{\partial}{\partial n}(r^j H \rho v) = 0 \tag{8.31}$$

Downstream momentum equation:

$$\frac{\rho u}{H}\frac{\partial u}{\partial s} + \rho v\frac{\partial u}{\partial n} = -\frac{1}{H}\frac{\partial p}{\partial s} + \frac{\partial}{\partial n}\left(\mu\frac{\partial u}{\partial n}\right) \tag{8.32}$$

Energy equation:

$$\frac{c_p \rho u}{H}\frac{\partial T}{\partial s} + c_p \rho v\frac{\partial T}{\partial n} = \frac{u}{H}\frac{\partial p}{\partial s} + \mu\left(\frac{\partial u}{\partial n}\right)^2 + \frac{\partial}{\partial n}\left(\lambda\frac{\partial T}{\partial n}\right) \tag{8.33}$$

with $H = 1 + \kappa n$.

The meaning of the symbols was already explained in the preceding chapter. For $j = 0$ the flow is purely two dimensional and for $j = 1$ it is axisymmetric.

These equations present a system of coupled partial differential equations depending on the spatial directions s and n. Their character is parabolic and therefore one has an initial-boundary value problem that can be solved by a marching procedure.

Often transformations help to simplify the governing equations concerning the solution technique. If the equations are handled as those for incompressible flow, the density ρ would not cause additional effort. Also, when the curvature of the surface in the downstream direction is not too extreme, the shape of the velocity profiles u and v does not change too much, except in their growth following the hierarchy of the boundary layer. These kinds of profiles are called 'quasi-similar'. By a transformation the boundary layer in the transformed plane can be kept at a nearly uniform thickness for many flow situations. Such a transformation, which is called a 'compressibility and similarity transformation', was first proposed by Levy–Lees and is often cited in the literature (Ref. [7]). It reads for axisymmetric bodies:

$$\xi(s) = \int_0^{\mathscr{S}} \rho_e \mu_e u_e R^{2j} \, ds \tag{8.34}$$

$$\eta(s,n) = \frac{\rho_e \mu_e}{\sqrt{2\xi}} \int_0^{\mathscr{N}} \frac{\rho}{\rho_e} r^j dn \tag{8.35}$$

The index e denotes the values at the outer edge of the boundary layer flow and R denotes the local radius of a body of revolution. Introducing the transformation rules, Eqs. (8.31), (8.32) and (8.33) become:

Continuity equation:

$$2\xi \frac{\partial F}{\partial \xi} + \frac{\partial V}{\partial \eta} + F = 0 \tag{8.36}$$

Downsteam momentum equation:

$$\frac{2\xi}{H} F \frac{\partial F}{\partial \xi} + \frac{V}{H} \frac{\partial F}{\partial \eta} = -\frac{2\xi F^2}{Hu_e} \frac{du_e}{d\xi} + \frac{\partial}{\partial n} \left[\left(\frac{r}{R} \right)^{2j} \frac{\rho \mu}{\rho_\infty \mu_\infty} \frac{\partial F}{\partial \eta} \right] \tag{8.37}$$

Energy equation:

$$\frac{2\xi}{H} F \frac{\partial S}{\partial \xi} + \frac{V}{H} \frac{\partial S}{\partial \eta} = \frac{2\xi F}{HT_e} \frac{dT_e}{d\xi} S + \left(\frac{r}{R} \right)^{2j} \frac{\rho \mu}{\rho_\infty \mu_\infty} \frac{u_e^2}{c_p T_e} \left(\frac{\partial F}{\partial \eta} \right)^2$$
$$+ \frac{\partial}{\partial n} \left[\left(\frac{r}{R} \right)^{2j} \frac{\rho \mu}{\rho_\infty \mu_\infty} \frac{1}{Pr} \frac{\partial S}{\partial \eta} \right] \tag{8.38}$$

with $F = u/u_e$, $S = T/T_e$, V represents the transformed velocity component

$$V = \frac{2\xi}{\rho_e u_e \mu_e R^{2j}} \left[F \left(\frac{\partial \eta}{\partial s} + \rho v \frac{r^j}{\sqrt{2\xi}} \right) \right] \tag{8.39}$$

For the case of vanishing ξ near the sharp tip of the body a singularity arises in Eq. (8.35) but this makes the initial conditions for the transformed equations easy to calculate because the ξ-derivatives drop out of the set of equations. Once an initial rough guess of F and S is done, the transformed normal velocity V can easily be integrated from the continuity equation. Iterating steps will correct the initial guesses of F and S.

The purpose of this section is to become acquainted with the transformed boundary layer equations. Their merits have already been mentioned. The disadvantage is that, apart from their complicated form, the total number of grid points in the normal direction n has to be calculated starting right away from the initial profile although here the boundary layer is very thin.

On the contrary, the calculation in the physical plane using Eqs. (8.31), (8.32) and (8.33) needs very few grid points at the beginning. The numbers must be continuously increased due to the growth of the boundary layers. The disadvantage of this method is to correctly overcome the singularity at the sharp tip where S equals zero.

More details concerning transformations of boundary layer equations can be found in Refs. [8–12].

8.5 Numerical Solution Method

8.5.1 Choice of Discretization Model

To come to a numerical solution of a set of partial differential equations it is usual to replace the differential quotients by finite difference quotients taking into account that a truncation error of a certain order of magnitude will now be induced to the set of equations. By rearranging the finite difference equations a system of algebraic equations is obtained which can be solved by means of the known methods. The techniques of the discretization are detailed in Chap. 5. It is stated there that the choice of the computational discretization grid is important as it affects the truncation error, the stability and the consistency. The form of these grids and the solution methods to which they lead will be summarized briefly.

Parabolic equations as observed here have a first order differential in the marching direction. As the flow is not allowed to reverse, the values of each quantity at the last upstream grid line normal to the surface are known. If we consider a grid as shown in Fig. 8.3, where Δx and Δy are the step sizes in the tangential and normal direction to the surface, the known points are on the left-hand side and the unknown on the right. Also the boundary conditions at the wall are given. Therefore, it is easy to calculate the flow quantities at the point with the open circle using discretization models as already given in Chap. 5. Because of the direct calculation of only one point on the grid line, this is called an 'explicit method'. The explicit method causes strong restrictions in the choice of the downstream step size as will briefly be repeated later, so the scheme is slow.

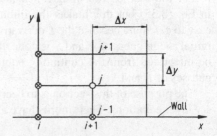

Fig. 8.3 Grid for an explicit method

Fig. 8.4 Grid for a fully
implicit method

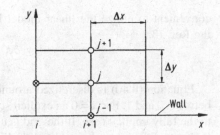

Figure 8.4 shows another extreme choice of a computational grid; the so-called 'fully implicit method'.

Only one known grid point from the preceding step is used, while on the actual one all points are unknown except the boundary values. That leads to an implicit form of the set of algebraic equations as will be shown later. This method is, concerning the choice of the step size, unconditionally stable but may lead to a poor accuracy. If there is no restriction on the step size in the downstream direction it becomes a fast calculation method which is desirable.

Now it is obviously possible to formulate something in between these extremes which will result in both a fast and accurate solution method. Figure 8.5 gives the computational mesh proposed by Crank–Nicholson [13] but in a more general form, so that the discretization methods described before are contained within it as special cases. Here, all points of the known and unknown grid lines are involved, but now the centre of discretization is located at the point $i + \lambda$. $\lambda = 1/2$ was originally proposed by Crank–Nicholson. Although the pure Crank–Nicholson scheme was described in detail in Part I, an example of a linear model equation is utilized to show its discrimination by the more generalized Crank–Nicholson scheme. In a following section the application to the two dimensional, rotational compressible boundary layer equations will be given.

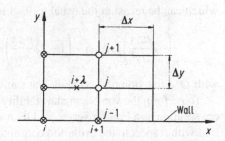

Fig. 8.5 Grid for a
generalized implicit method

8.5.2 Generalized Crank–Nicholson Scheme

This section is taken directly from Arina & Benocci [5]. In order to analyse the stability and accuracy of a generalization of the Crank–Nicholson scheme, it is convenient to utilize the linear model Eq. (8.40), which is rewritten here following Ref. 13.

$$\frac{\partial \phi}{\partial x} = a\frac{\partial^2 \phi}{\partial y^2} \tag{8.40}$$

Equation (8.40) is discretized around the mesh point $(i + \lambda, \ j)$, with λ ranging between 0 and 1. For $\lambda = 0$ an explicit scheme is recovered, while $\lambda = 1$ corresponds to the fully implicit case. If the grid is uniform, the x-derivative is approximated by the finite difference relation developed in Sect. 5.2.1.

$$\left(\frac{\partial \phi}{\partial x}\right)_{1+\lambda,j} = \frac{\phi_{i+1,j} - \phi_{i,j}}{\Delta x} + \left(\lambda - \frac{1}{2}\right)0(\Delta x) + 0(\Delta x^2) \tag{8.41}$$

and the y-derivative is replaced by the weighted mean

$$\left(\frac{\partial^2 \phi}{\partial y^2}\right)_{i+\lambda,j} = \lambda\left(\frac{\partial^2 \phi}{\partial y^2}\right)_{i+1,j} + (1-\lambda)\left(\frac{\partial^2 \phi}{\partial y^2}\right)_{i,j} \tag{8.42}$$

Each second-order derivative is then replaced by the usual three-point centred finite difference relation:

$$\left(\frac{\partial^2 \phi}{\partial y^2}\right)_{i,j} = \frac{\phi_{i,j+1} - 2\phi_{i,j} + \phi_{i,j-1}}{\Delta y^2} + \left(\Delta y^2\right) \tag{8.43}$$

Substituting Eqs. (8.41, 8.42, 8.43) into equation (8.40), a linear difference equation is obtained

$$\frac{\phi_{i+1,j} - \phi_{i,j}}{\Delta x} = \frac{a}{\Delta y^2}\Big[\lambda(\phi_{i+1,j+1} - 2\phi_{i+1,j} + \phi_{i+1,j-1}) \\ + (1-\lambda)\big(\phi_{i,j+1} - 2\phi_{i,j} + \phi_{i,j-1}\big)\Big] \tag{8.44}$$

which can be recast in the usual tridiagonal form

$$\left(-\lambda\frac{a\Delta x}{\Delta y^2}\right)\phi_{i+1,j-1} + \left(1 - 2\lambda\frac{a\Delta x}{\Delta y^2}\right)\phi_{i+1,j} + \left(\lambda\frac{a\Delta x}{\Delta y^2}\right)\phi_{i+1,j+1} = D_j \tag{8.45}$$

with D_j a function of ϕ computed at station i.

To perform the von Neumann stability analysis it is useful to express the numerical solution as a Fourier series, and then verify that none of the harmonics is amplified with respect to the evolution coordinate x. This stability analysis is described in detail in Sect. 4.4; Part I, and is repeated here as a reminder. Hence putting

$$\phi_{i,j} = \rho^i e^{I\omega(j\Delta y)} \tag{8.46}$$

where I in the exponent is the unit complex number, and ρ^i is the amplification factor at level i, and then substituting inside Eq. (8.45), actualizing the indices in Eq. (8.46), we have

$$G = \frac{\rho^{i+1}}{\rho^i} = \frac{1 + 2a(1-\lambda)\frac{\Delta x}{\Delta y^2}[\cos(\omega \Delta y) - 1]}{1 + 2a\lambda\frac{\Delta x}{\Delta y^2}[\cos(\omega \Delta y) - 1]} \qquad (8.47)$$

To have stability, $|G| \le 1$ for all harmonics $\omega \Delta y$; this inequality together with Eq. (8.47), leads to the following stability condition for $0 \le \lambda < 1/2$

$$C \le \frac{1}{2(1-2\lambda)}$$

where $C = a\Delta x/\Delta y^2$. For $1/2 \le \lambda \le 1$ no stability restriction is imposed on C. Hence the scheme presented is unconditionally stable for values of λ equal or larger than $1/2$. In the case of the explicit scheme ($\lambda = 0$), there is a strong limitation to Δx if Δy is chosen rather small for accuracy requirements.

The consistency of the scheme can easily be verified expanding in Taylor series all other terms of Eq. (8.45) about the point $(i + \lambda, \ j)$. The discretization error can be proved to be of $0(\Delta x, \Delta y^2)$ if $\lambda \ne 1$ (Ref. [14]). The scheme is therefore second-order accurate with respect to y and first-order with respect to x. To obtain second-order accuracy with respect to x, λ should be taken equal to $1/2$ (Crank–Nicholson scheme), or slightly different to $1/2$ (e.g. $= 1/2 + 0(\Delta x)$). However, for practical, non-linear problems it is often necessary to increase λ in order to avoid non-linear instabilities. For instance, the full implicit scheme is often very stable, but leads to a worse accuracy.

Equation (8.40) is a linear partial differential equation employed as a model to demonstrate the widely used generalized implicit Crank–Nicholson solution code. Now this will be applied to the boundary layer Eqs. (8.31), (8.32) and (8.33) of Sect. 8.4.

8.5.3 Discretization of the Boundary Layer Equations

The boundary layer equations (8.31), (8.32) and (8.33) now will be non-dimensio nalized by applying the prescriptions of equations (8.25):

Continuity equation:

$$\frac{\partial}{\partial s}(r^j \rho u) + \frac{\partial}{\partial n}(r^j H \rho v) = 0 \qquad (8.48)$$

with $H = 1 + \kappa n$.

Downstream momentum equation:

$$\frac{\rho u}{H}\frac{\partial u}{\partial s} + \rho v \frac{\partial u}{\partial n} = -\frac{1}{H}\frac{\partial p}{\partial s} + \frac{1}{Re}\frac{\partial}{\partial n}\left(\mu \frac{\partial u}{\partial n}\right) \qquad (8.49)$$

* *Energy equation:*

$$\frac{c_p \rho u}{H}\frac{\partial T}{\partial s} + c_p \rho v \frac{\partial T}{\partial n} = Ec\frac{u}{H}\frac{\partial p}{\partial s} + \frac{Ec}{Pr}\mu\left(\frac{\partial u}{\partial n}\right)^2 + \frac{1}{RePr}\frac{\partial}{\partial n}\left(\lambda \frac{\partial T}{\partial n}\right) \qquad (8.50)$$

For convenience the geometrical and physical quantities are not barred to show that they have no dimension.

The boundary conditions at the outer edge of the boundary are the same as described in Chap. 2. Since the velocity u_e is given from measurements or inviscid flow calculations Eq. (8.49) can be developed at the point $n = \delta$:

$$\rho_e u_e \frac{\partial u_e}{\partial s} = -\frac{\partial p}{\partial s} \qquad (8.51)$$

The demand of constant total enthalpy in the outer flow yields the boundary conditions for the temperature T_e:

$$T_e = 1 + Ec(1 - u_e^2) \qquad (8.52)$$

The conditions for the velocity at the wall are the no-slip assumption

$$u_w = 0 \qquad (8.53)$$

and the zero normal velocity statement

$$v_w = 0. \qquad (8.54)$$

For the wall temperature T_w, according to the demands, a distribution can be prescribed or the wall may be adiabatic.

$$T_w = T_w(s) \qquad \text{(prescribed) or} \qquad (8.55)$$

$$\left.\frac{\partial T}{\partial n}\right|_w = 0 \qquad \text{(adiabatic wall)} \qquad (8.56)$$

Since the set of three equations (8.48), (8.49) and (8.50) contains four unknowns, an additional equation is prescribed to close this system; it is the equation for perfect gases

$$p = \rho R T \qquad (8.57)$$

where R is the ideal gas constant. The solution of these equations could be effected as follows:

1. solving the downstream momentum equation for the velocity component u simultaneously by solving the energy equation for the temperature T;
2. hence, the pressure p is constant in the normal direction of the boundary layer, the gas equation delivers a simple connection between the known temperature T and the density ρ;
3. knowing the density and obviously the surface curvature, the continuity equation can be solved for the normal velocity component v.

Since this system of non-linear, partial differential equations is coupled, meaning that each fluid mechanical property is directly or indirectly involved in each of these equations, care has to be taken by solving one equation after the other. The normal velocity, v, following the procedure mentioned above is found only in the last step but takes an important part in the solution of the downstream momentum equation for the velocity component, u, which was calculated first. It is clear that iterations of the total solution process have to take place. Inner iteration loops are followed by outer ones, but finally the correct application of a loop system is a question of experience. One example will be given later.

Now the discretization of the downstream momentum equation will be given as an example using the finite difference relations presented in Eqs. (8.41), (8.42) and (8.43), noting that these were evaluated from the generalized Crank–Nicholson scheme shown in Fig. 8.5.

First-order derivative in downstream direction:

$$\left(\frac{\partial u}{\partial s}\right)_{i+\lambda,j} = \frac{u_{i+1,j} - u_{i,j}}{\Delta S} \tag{8.58}$$

First-order derivative in normal direction:

$$\left(\frac{\partial u}{\partial n}\right)_{i+\lambda,j} = \lambda\left(\frac{\partial u}{\partial n}\right)_{i+1,j} + (1-\lambda)\left(\frac{\partial u}{\partial n}\right)_{i,j} \tag{8.59}$$

where for example the normal derivative at $i+1$, j is a second-order accurate centred difference solution

$$\left(\frac{\partial u}{\partial n}\right)_{i+1,j} = \frac{u_{i+1,j+1} - u_{i+1,j-1}}{2\Delta n} \tag{8.60}$$

The second term on the right-hand side of Eq. (8.49) at first has to be differentiated in detail before the discretization.

$$\frac{\partial}{\partial n}\left(\mu\frac{\partial u}{\partial n}\right) = \mu\frac{\partial^2 u}{\partial n^2} + \frac{\partial \mu}{\partial n}\frac{\partial u}{\partial n} \tag{8.61}$$

Because the dynamic viscosity is an analytical function of T and for high pressures also of p, and not essentially on the spatial location, Eq. (8.61) can be rewritten:

$$\frac{\partial}{\partial n}\left(\mu\frac{\partial u}{\partial n}\right) = \mu\frac{\partial^2 u}{\partial n^2} + \frac{\partial \mu}{\partial T}\frac{\partial T}{\partial n}\frac{\partial u}{\partial n} \tag{8.62}$$

Now the second derivative becomes, in finite difference form following Eqs. (8.42) and (8.43)

$$\left(\frac{\partial^2 u}{\partial n^2}\right)_{i+\lambda,j} = \frac{1}{\Delta n^2}\Big[\lambda(u_{i+1,j+1} - 2u_{i+1,j} + u_{i+1,j-1})$$
$$+ (1-\lambda)\big(u_{i,j+1} - 2u_{i,j} + u_{i,j-1}\big)\Big] \tag{8.63}$$

The non-linear coefficient of $\partial u/\partial s$ for instance is discretized as:

$$\left(\frac{\rho u}{H}\right)_{i+\lambda,j} = \frac{\rho_{i,j}u_{i,j}}{(\lambda H_{i+1,j} + (1-\lambda)H_{i,j})} = A \tag{8.64}$$

The geometry is known in advance but the density ρ and the downstream velocity u are not known until the set of equations is solved, so the opportunity is offered to take these results from the previous step which involves errors. One can overcome this by replacing these results after each iteration step by taking again the weighted mean value of the properties right after the first iteration since now the value at $i+1$ is known in a first approximation. But this causes additional effort to be done which is not presented here. The other coefficients are

$$(\rho v)_{i+\lambda,j} = \rho_{i,j} v_{i,j} = B \tag{8.65}$$

$$(\mu)_{i+\lambda,j} = \mu_{i,j} = C \tag{8.66}$$

$$\left(\frac{\partial \mu}{\partial T}\frac{\partial T}{\partial n}\right)_{i+\lambda,j} = \left(\frac{\partial u}{\partial T}\right)_{i,j}\left(\frac{T_{i,j+1}-T_{i,j-1}}{2\Delta n}\right) = D \tag{8.67}$$

Finally, the pressure gradient terms $\partial p/\partial s$ has to be discussed. First-order boundary layer theory indicates no pressure variation normal to the surface; that means pressure is only a function of the downstream coordinate s and it is known from the inviscid flow calculation

$$-\frac{1}{H}\frac{\partial p}{\partial s} = -\left[\lambda\left(\frac{1}{H}\frac{\partial p}{\partial s}\right)_{i+1,j} + (1-\lambda)\left(\frac{1}{H}\frac{\partial p}{\partial s}\right)_{i,j}\right] = E \tag{8.68}$$

The momentum Eq. (8.49) now can be rewritten in abbreviated form:

$$A\frac{\partial u}{\partial s} + B\frac{\partial u}{\partial n} - C\frac{\partial^2 u}{\partial n^2} - D\frac{\partial u}{\partial n} + E = 0 \tag{8.69}$$

Substituting Eqs. (8.58), (8.59), (8.60), (8.61), (8.62), (8.63), (8.64), (8.65), (8.65), (8.66), (8.67) and (8.68) into Eq. (8.69) and sorting in order of $u_{i+1,j+1}$, $u_{i+1,j}$ and $u_{i+1,j-1}$ yields

$$\frac{\lambda}{2\Delta n}\left[B - \frac{2C}{Re\Delta n} - D\right]u_{i+1,j+1} + \left[\frac{A}{\Delta S} + \frac{2\lambda C}{Re\Delta n^2}\right]u_{i+1,j}$$

$$+ \frac{\lambda}{2\Delta n}\left[-B - \frac{2C}{Re\Delta n} + D\right]u_{i+1,j-1} = F \tag{8.70}$$

On the left-hand side of Eq. (8.70) the unknown values of u are arranged while the known ones from the previous calculation step are on the right-hand side, hidden in F, which is:

$$F = -\frac{(1-\lambda)}{2\Delta n}\left(B - \frac{2C}{Re\Delta n} - D\right)u_{i,j+1} - \left[\frac{A}{\Delta S} + \frac{2(1-\lambda)C}{Re\Delta n^2}\right]u_{i,j}$$

$$- \frac{(1-\lambda)}{2\Delta n}\left[-B - \frac{2C}{Re\Delta n} + D\right]u_{i,j-1} - E \tag{8.71}$$

Eq. (8.70) is now written in abbreviated form:

$$a_j u_{i+1,j+1} + b_j u_{i+1,j} + c_j u_{i+1,j-1} = d_j \qquad 2 \le j \le M-1 \qquad (8.72)$$

Following the same lengthy method of discretization, the energy Eq. (8.50) takes the corresponding form:

$$e_j T_{i+1,j+1} + f_j T_{i+1,j} + g_j T_{i+1,j-1} = h_j \qquad 2 \le j \le M-1 \qquad (8.73)$$

The coefficients a_j to h_j are known from previous spatial or iteration steps. Eqs. (8.72) and (8.73) are tridiagonal matrices that can be solved by common recurrence formulas, for example with the 'Thomas algorithm' (Ref. [13]). It will be briefly described in the following section.

Now the continuity Eq. (8.48) will be rewritten and discretized.

$$\frac{\partial v}{\partial n} + \left(\frac{1}{\rho} \frac{\partial \rho}{\partial n} + \frac{1}{H} \frac{\partial H}{\partial n} + \frac{1}{r^j} \frac{\partial r^j}{\partial n} \right) v = -\frac{1}{r^j H \rho} \frac{\partial}{\partial s} (r^j \rho u) \qquad (8.74)$$

If the coefficient of the second left-hand term and the right-hand term were considered to be known from the previous spatial or iteration step, Eq. (8.74) becomes:

$$\frac{\partial v}{\partial n} + Gv = H \qquad (8.75)$$

The discretization of the reduced continuity Eq. (8.75) yields, with the help of Eq. (8.41), and by using the relation

$$v_{i+\lambda,j} = \lambda v_{i+1,j} + (1 - \lambda) v_{i,j}, \qquad (8.76)$$

$$\left[\frac{\lambda}{2\Delta n} \right] v_{i+1,j+1} + [\lambda G] v_{i+1,j} + \left[-\frac{\lambda}{2\Delta n} \right] v_{i+1,j-1} = I \qquad (8.77)$$

with

$$I = -\left[\frac{(1-\lambda)}{2\Delta n} \right] v_{i,j+1} - [(1-\lambda)G] v_{i,j} - \left[-\frac{(1-\lambda)}{2\Delta n} \right] v_{i,j-1} + H \qquad (8.78)$$

Equation (8.78) corresponds to Eq. (8.71). All known values are collected in this equation, while again Eq. (8.77) keeps all unknown values. One easily can see that this equation has the same form as Eq. (8.70):

$$p_j v_{i+1,j+1} + q_j v_{i+1,j} + r_j v_{i+1,j-1} = S_j \qquad (8.79)$$

The same tridiagonal algorithm as applied for the solution of the downstream momentum Eq. (8.72) and the energy Eq. (8.73) can be used for the continuity Eq. (8.79) to result in the values of the normal velocity component in the boundary layer.

The only unknown which is not yet treated is the density, ρ. As mentioned above, the equation for a perfect gas (8.57) fills this gap. First order boundary layer theory does not know a pressure variation across the boundary layer. The dynamic pressure is constant. Thus, Eq. (8.57) becomes:

$$\rho T = \text{const.} \qquad (8.80)$$

In the discretized form it yields:

$$\rho_{i+1,j+1} = \rho_{i+1,j} \frac{T_{i+1,j}}{T_{i+1,j+1}} \tag{8.81}$$

Since the temperature T and the density ρ at the previous step and the temperature T at the actual step are already calculated the density at the new step will be the result of Eq. (8.81).

As mentioned previously, the governing Eqs. (8.48), (8.49) and (8.50) form a set of coupled, nonlinear differential equations. Firstly, the result of each equation is needed to calculate the other equations and vice-versa, and secondly, since the equations are non-linear, the solution becomes iterative. Therefore the solution of this set of equations will include, so to say, two iterative steps: one for the solution of each single equation and the other for the coupling of the total set of equations. The procedure may be described as follows:

1. Solution of the momentum and the energy equation until convergence is achieved.
2. Solution of the continuity equation.
3. Calculation of the density.
4. Repeat of the previous steps until convergence of the equations is achieved.
5. Start the calculations at the next downstream position.

The above procedure is only one of the many possibilities to effect a solution of these equations.

The following section shortly describes a widely applied solution scheme to solve the discretized fluid mechanical Eqs. (8.72), (8.73) and (8.79) which form a linear system with tridiagonal matrices. This scheme is called the Thomas algorithm (Ref. [13]).

8.5.4 Solution of a Tridiagonal System of Linear Algebraic Equations

An efficient technique, sometimes called the Thomas algorithm (Ref. [13]) is described, that can be used to solve a linear system with a tridiagonal matrix defined by the following equations, where the x_i are the unknowns and a_i, b_i, c_i, d_i are known:

$$\left.\begin{array}{l} b_1 x_1 + c_1 x_2 = d_1 \\ a_i x_{i-1} + b_i x_i + c_i x_{i+1} = d_i \\ a_I x_{I-1} + b_I x_I = d_I \end{array}\right\} 2 \leq i \leq -1 \tag{8.82}$$

It is possible to calculate the unknowns x_i using the following recurrence formula

$$x_i = \alpha_i x_{i+1} + \beta_i \qquad i = (I-1), (I-2), \cdots, 1 \tag{8.83}$$

The expression of α_i and β_i can be derived by substituting Eq. (8.83) into (8.82). We obtain:

$$\left(a_i\alpha_{i-1} + b_i + \frac{c_i}{\alpha_i}\right)x_i + \left(\alpha_i\beta_{i-1} - c_i\frac{\beta_i}{\alpha_i} - d_i\right) = 0 \tag{8.84}$$

This equation can be satisfied for any value of x_i if α_i and β_i are chosen as follows:

$$\left.\begin{array}{l} \alpha_i = -\dfrac{c_i}{b_i + \alpha_{i-1}a_i} \\[2mm] \beta_i = \dfrac{d_i - a_i\beta_{i-1}}{b_i + \alpha_{i-1}a_i} \end{array}\right\} \quad i = 2,\ldots,I \tag{8.85}$$

Writing Eq. (8.83) for $i = 1$, we obtain:

$$x_1 = \alpha_1 x_2 + \beta_1$$

This expression reduces to the first Eq. (8.82) if α_1 and β_1 are chosen as follows:

$$\alpha_1 = -\frac{c_1}{b_1} \quad \text{and} \quad \beta_1 = \frac{d_1}{b_1}$$

Recurrence formulas (8.85) can then be used to compute all α's and β's. Eq. (8.83) written for $i = I - 1$ yields:

$$x_{I-1} = \alpha_{I-1}x_I + \beta_{I-1} \tag{8.86}$$

Eliminating x_{I-1} between this relation and the third Eq. (8.82), we obtain:

$$x_I = \frac{d_I - a_I\beta_{I-1}}{b_I + a_I\alpha_{I-1}} \tag{8.87}$$

Recurrence formula (8.83) is then used to compute all x_i successively for $i = I - 1,\ldots,1$.

It is possible to prove (Ref. [13]) that the different recurrencies will not lead to any stability problem or to an unacceptable accumulation of round-off errors, if the tridiagonal matrix is diagonally dominant, i.e. if:

$$|b_1| \geq |c_1|$$
$$|b_i| \geq |c_i| + |a_i| \quad \text{for all } i = 2,\ldots,I-1 \tag{8.88}$$
$$|b_I| \geq |a_I|$$

These sufficient conditions which are, however, far from being necessary in practice, should be satisfied in boundary layer calculations.

This short description of the Thomas algorithm is taken from Arina and Essers [14] and can be found described in more detail including a Fortran program in Roache [15].

8.6 Sample Calculations

8.6.1 Three Dimensional Boundary Layer Calculation Along Lines of Symmetry

Boundary layer calculations on bodies along lines of symmetry (Ref. [16]) have to be performed in three dimensions (Fig. 8.6). The flow around bodies at high incidence moves from the windward to the leeward side. This obviously causes a thickening of the leeward boundary layer for continuity reasons. If the influence of the cross flow is neglected the three dimensionality of the flow cannot be taken into account.

The numerical calculation of such boundary layer flows is eased if the outer boundary conditions, i.e. the inviscid flow, are known analytically. The ellipsoid is one of the very few bodies for which this is the case concerning all angles of attack and ratios of the half-axes (Lamb [17]; Maruhn [18]).

The first results ever to be presented for three-dimensional laminar and incompressible boundary layer flow over an ellipsoid were performed by Eichelbrenner and Oudart [19] using integral methods. They used a streamline coordinate system that needs a lot of precalculations to be done as an additional numerical effort. Later on, results were presented by Geissler [20] and Schönauer et al. [21] who also used the streamline coordinate system but the calculation was carried out by using finite difference methods.

Other authors such as Wang [22] and Hirsch and Cebeci [23] introduced a surface-oriented elliptical, orthogonal coordinate system, which has the disadvantage of creating geometrical singularities at the poles of the ellipsoid. By this means it becomes difficult to calculate boundary layer flows at high incidence without us-

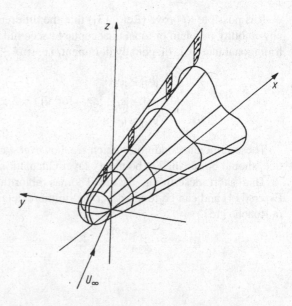

Fig. 8.6 Sample of a leeside boundary layer along the line of symmetry of an inclined body

ing additional transformations. Blottner & Ellis [24] as well as Stock [25] developed coordinate systems which are similar to the streamline coordinate system but easier to apply.

Now it will be shown that a surface oriented curved and non-orthogonal system eliminates the singularities at the poles, no transformations are needed and the numerical effort can be kept small.

8.6.2 Geometrical Conditions

At first the ellipsoid has to be split into a nose region where the stagnation point is situated and into an afterbody region (Fig. 8.7). In the front portion the new coordinate system will be applied while in the rear part the elliptical coordinate system may be used.

The coordinate system in the nose region is a modified spherical system for which the radius depends on the geometry of the ellipsoid. Using the definitions of Fig. 8.8 the geometrical conditions between the new and the cartesian coordinate system are:

$$x = r\cos\theta\cos\varphi,$$
$$y = r\cos\theta\sin\varphi,$$
$$z = r\sin\theta. \tag{8.89}$$

The radius r is a function of the new independent variables θ and φ and of the half-axes, a and b, of the ellipsoid:

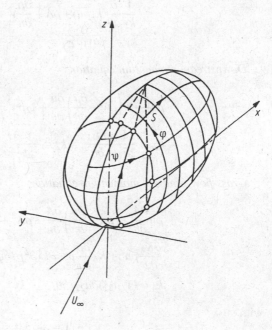

Fig. 8.7 Coordinate system in the nose region of an ellipsoid

Fig. 8.8 Specific coordinate
system on an ellipsoid

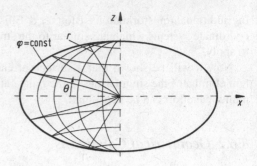

$$r = ab[b^2 \cos^2 \theta \cos^2 \varphi + a^2(\cos^2 \theta \sin^2 \varphi + \sin^2 \theta)]^{-1/2} \qquad (8.90)$$

8.6.3 Fluid Mechanical Equations

The system of differential equations describing the quasi three-dimensional, laminar, compressible flow along the symmetry line of an inclined ellipsoid with the coordinate system demonstrated in Fig. 8.8 was developed following the method of Robert [26]. In the case of incompressible flow and considering the different coordinate systems and their geometrical conditions it corresponds to those of Wang [22] as well as Cebeci, Khattab and Stewartson [27].

Continuity equations:

$$\frac{1}{k_1} \frac{\partial}{\partial \theta}(k_1 \rho u) + \rho A + \frac{\partial \rho}{\partial n} w + \rho \frac{\partial w}{\partial n} = 0$$
$$k_1 = \sqrt{a_{11} a_{22}} \qquad (8.91)$$

Downstream momentum equation:

$$\rho u \frac{\partial u}{\partial \theta} + \left(\rho w - \frac{\partial \mu}{\partial n}\right) \frac{\partial u}{\partial n} + k_2 \rho u^2 = k_3 \frac{\partial p}{\partial \theta} + \mu \frac{\partial^2 u}{\partial n^2}$$
$$k_2 = \frac{1}{2 a_{11}} \frac{\partial a_{11}}{\partial \theta}, \quad k_3 = -\frac{a_{22}}{k_1^2} \qquad (8.92)$$

Cross-flow gradient momentum equation:

$$\rho u \frac{\partial A}{\partial \theta} + \left(\rho w - \frac{\partial \mu}{\partial n}\right) \frac{\partial A}{\partial n} + \rho A^2 + k_4 \rho u A +$$
$$\frac{\partial K_4}{\partial \varphi} \rho u^2 + K_4 \frac{\partial \rho}{\partial \varphi} u^2 = k_5 \partial^2 p / \partial \varphi^2 + \mu \partial^2 A / \partial n^2, \qquad (8.93)$$
$$k_4 = (1/a_{22}) \partial a_{22} / \partial \theta, \qquad k_5 = -a_{11}/k_1^2$$

with

$$A = \partial v / \partial \varphi.$$

Energy equation:

$$c_{\mathrm{p}}\rho u\frac{\partial T}{\partial \theta} + \left(c_{\mathrm{p}}\rho w - \frac{1}{\mathrm{Pr}}\frac{\partial k}{\partial n}\right)\frac{\partial T}{\partial n} = \mathrm{Ec}\, u\frac{\partial p}{\partial \theta} + \frac{k}{\mathrm{Pr}}\frac{\partial^2 T}{\partial n^2} + \mathit{Ec}\,\mu a_{11}\left(\frac{\partial u}{\partial n}\right)^2 \quad (8.94)$$

Herein, u, v and w are the non-dimensional velocity components in the non-dimensional coordinate directions, θ, φ and n, respectively. T, ρ and p are the temperature, the density and the pressure. The quantities μ, k and c_{p} are the dynamic viscosity, the heat conductivity and the specific heat coefficient at constant pressure. Pr and Ec denote the Prandtl and the Eckert number.

Equation (8.93) describes the derivative of the momentum equation in the cross flow direction with the velocity component v with respect to the φ-direction. Since along the line of symmetry the cross flow component v vanishes, but not its gradient, the three-dimensional status of the flow still can be considered by the use of this cross-flow gradient equation.

The geometrical coefficients k_{i} appearing in the set of equations combine themselves from the elements of the metric tensor of the surface a_{ij} and their derivatives. Along the line of symmetry and for the chosen coordinate system, there exist only two elements of the metric tensor:

$$a_{11} = r^2 + (\partial r/\partial \theta)^2, \qquad a_{12} = \frac{\partial r}{\partial \theta}\frac{\partial r}{\partial \varphi}, \qquad a_{22} = r^2\cos^2\theta \quad (8.95)$$

Besides the stretching of the normal coordinate no transformation will be applied. The choice of this coordinate system excludes geometrical singularities at the poles of the ellipsoid.

8.6.4 Boundary Conditions

The ellipsoid is one of the very few bodies for which the three-dimensional, inviscid, incompressible velocity distribution is given analytically. The potential ϕ for the surface can be written as follows (Refs. [17]– [22]):

$$\begin{aligned}
\phi &= Bx + Cz \\
B &= (1+k_{\mathrm{a}})\cos\alpha, \\
C &= (1+k_{\mathrm{c}})\sin\alpha, \\
k_{\mathrm{a}} &= \frac{(1/2e)\ln[(1+e)/(1-e)]-1}{[1/(1-e^2)]-(1/2e)\ln[(1+e)/(1-e)]} \\
k_{\mathrm{c}} &= 1/(1+2k_{\mathrm{a}}) \\
e &= (1-b^2/a^2)^{1/2}
\end{aligned} \quad (8.96)$$

with the half-axes a and b

where α is the angle of attack. Further details formulating the potential are described in Ref. [22]. This potential is given in cartesian coordinates and can be transformed through Eq. (8.89) into the new surface-oriented coordinate system

$$\phi = Br \cos \theta \cos \varphi + Cr \sin \theta. \tag{8.97}$$

The derivatives of the potential with respect to the coordinate axes give the inviscid velocity component u_e in downstream direction and the crossflow gradient A_e along the line of symmetry

$$u_e = (1/a_{11})\partial\phi/\partial\theta, \tag{8.98}$$

$$A_e = \left.\frac{\partial v}{\partial \varphi}\right|_e = (1/a_{22})\left(\frac{\partial^2 \phi}{\partial \varphi^2} - \frac{\partial a_{12}}{\partial \varphi}u_e\right) \tag{8.99}$$

Equations (8.98) and (8.99) characterize the boundary conditions for the outer edge of the boundary layer. At the wall, the no-slip condition is

$$u_w = 0 \tag{8.100}$$

$$A_w = \left.\frac{\partial v}{\partial \varphi}\right|_w = 0. \tag{8.101}$$

Since the system of Eqs. (8.91), (8.92), (8.93) and (8.94) describes compressible flow, boundary conditions for the temperature have to be established. Requiring that the total enthalpy should stay constant, the relation for the temperature T_e at the outer edge of the boundary layer becomes:

$$T_e = 1 + \left(u_\infty^2/2c_p T_\infty\right)(1 - u_e^2). \tag{8.102}$$

At the wall, the temperature distribution or the establishment of the adiabatic wall temperature can be prescribed.

$$T_w = T_w(\theta) \qquad \left.\frac{\partial T}{\partial n}\right|_w = 0. \tag{8.103}$$

8.6.5 Solution Scheme

The differential equations are discretized in the sense of an implicit finite difference method. The difference molecule is chosen in the way to calculate three unknown values with three known values coming from the last computational step. The centre of discretization is placed in the middle of the molecule, so that the truncation error is not greater than the square of the step sizes.

The solution scheme follows the Richtmyer algorithm (Ref. [13]). The solution vector contains three components, which are the downstream velocity component, the cross-flow gradient and the temperature. The normal velocity component is calculated by the use of the Thomas algorithm, separately. This solution scheme is described in detail in Refs. [14] and [15].

8.6.6 Numerical Results

Measurements on an ellipsoid with the ratio of the half-axes $b/a = 1/6$ were performed by Meier & Kreplin [28]. An example is given in Fig. 8.9 showing results along the windward side at an inclination angle of $\alpha = 10°$. The dimensional wall shear stress τ_w along the non dimensional x-axis is compared with other theoretical results given by Geissler [20] and Stock [25]. The agreement of all these results is very good.

In Fig. 8.10 the result of a symmetry line boundary layer calculation on an ellipsoid with the ratio of its half-axis $b/a = 1/4$ at zero angle of attack is shown. The non-dimensional skin friction $c_f Re^{1/2}$ is plotted against the non-dimensional x-axis, where the Reynolds number $Re = u_\infty a/\nu$ is formed with the free stream velocity u_∞, the half-axis a and the kinematic viscosity ν. The comparison with other theoretical results by Wang [22] and Hirsch & Cebeci [23] gives a good agreement with the present results.

Figure 8.11 presents comparable results performed with the above described method for an ellipsoid with $b/a = 1/6$ at high angles of attack. The non-dimensional skin friction $c_f Re^{1/2}$ is plotted here against the angle θ, starting at the stagnation point and following along the leeward side up to a prescribed point. These results are qualitatively the same as those given by Cebeci, Khattab and Stewartson [27]. In general it can be stated if the angle of attack exceeds $\alpha = 41°$ separation occurs immediately downstream of the nose of the ellipsoid.

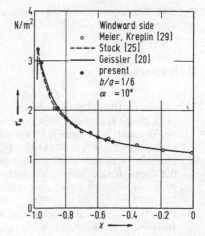

Fig. 8.9 Dimensional wall shear stress along the windward symmetry line of an ellipsoid at angle of attack $\alpha = 10°$, $b/a = 1/6$

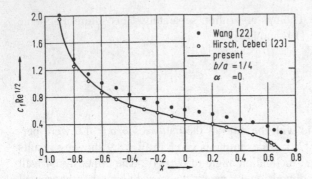

Fig. 8.10 Dimensionless wall shear stress along the symmetry line of an ellipsoid for zero angle of attack, $b/a = 1/4$

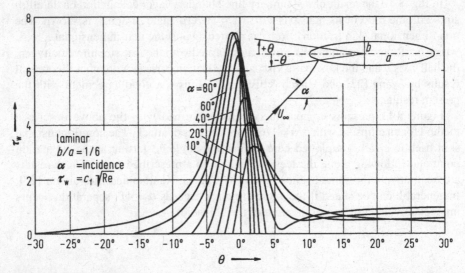

Fig. 8.11 Dimensionless wall shear stress along the leeside symmetry line of an ellipsoid at different angles of attack, $b/a = 1/6$

References

1. Prandtl, L., Über Flüssigkeitsbewegung bei sehr kleiner Reibung. Verhandlung d. III Intern. Math. Kongr. Heidelberg, 1904, pp. 484–491. Wieder abgedruckt in 'Vier Abhdl. zur Hydro- u. Aerodynamik', Göttingen 1927; vgl. auch. Gesammelte Abhandg., Bd. II, 1961, pp. 575–584; English translation NACA TM 452, 1928.
2. Proceedings of the 1980 Conference on Complex Turbulent Flows (Eds. S.J. Kline, B.J. Cantwell, G.M. Lilley), Stanford University, September 3–6, 1980, Vols. I, II, III.
3. Bradshaw, P., *The Effect of Streamline Curvature on Turbulent Flow*. AGARDograph 169, 1973.
4. Rodi, W., 'Turbulence models for practical applications', in '*Introduction to the Modeling of Turbulence*', von Karman Institute Short Course 1985–01, June 10–11, 1985.

5. Arina, R. and Benocci, C., 'Turbulent boundary layers: examples of turbulence models', in *Introduction to Computational Fluid Dynamics*, von Karman Institute LS 1985–01, January 21–25, 1985, also von Karman Institute Preprint 1985–05.
6. Schlichting, H., *Boundary Layer Theory*, (Seventh edn), McGraw-Hill Book Company, 1979.
7. Lees, L., 'Laminar heat transfer over blunt-nosed bodies at hypersonic flight speeds', *Jet Propulsion*, April 1956.
8. Probstein, R.F. and Elliot, D., 'The transverse curvature effect in compressible axially symmetric laminar boundary layer flow', *Journal of the Aeronautical Sciences*, Vol. 23, No. 3, March 1956, pp. 208–244, 236.
9. Falkner, V.M. and Skan, S.W., 'Some approximate solutions of the boundary layer equations', *ARC* R&M 1314, 1930.
10. Hartree, D.R., 'On an equation occurring in Falkner and Skan's approximate treatment of the equations of the boundary layer', *Proc. Cambridge Phil. Soc.*, Vol. 33, Part II, 1937.
11. Illingworth, C.R., 'The laminar boundary layer associated with the retarded flow of a compressible fluid', *ARC* R&M 2590, 1953.
12. Levy, S., 'Effect of large temperature changes (including viscous heating) upon laminar boundary layers with variable free stream velocity', *Journal of the Aeronautical Sciences*, Vol. 21, No. 7, July 1954, pp. 459–474.
13. Richtmyer, R.D. and Morton, K.W., *Difference Methods for Initial Value Problems*, New York, Interscience, 1967.
14. Arina, R. and Essers, J.A., *A Finite Difference Technique for Laminar and Turbulent Compressible Boundary Layers*, von Karman Institute LS 1983–01, January 24–28, 1983; also von Karman Institute Preprint 1983–03.
15. Roache, P.J., *Computational Fluid Dynamics*, Albuquerque, Hermosa Publ., 1976.
16. Grundmann, R., 'Dreidimensionale Grenzschichtrechnungen entlang Symmetrielinien auf Körpern', in *Zeitschrift für Flugwissenschaften und Weltraumforschung*, Vol. 5, No. 6, 1981, pp. 389–395.
17. Lamb, H., *Hydrodynamics*, London, Cambridge University Press, 1932.
18. Maruhn, K., 'Drukverteilungsrechnungen an elliptischen Rümpfen und in ihrem Aussenraum', *Jahrb. dtsch. Luftfahrtforschung*, 1941, pp. 135–147.
19. Eichelbrenner, E.A. and Oudart, A., *Méthode de calcul de la couche limite tridimensionalle. Application à un corps fuselé incliné sur le vent.* ONERA Publ. 76, 1955.
20. Geissler, W., *Berechnung der laminaren, dreidimensionalen Grenzschicht an unsymmetrisch umströmten Rotationskörpern mittels Differenzenverfahren*, Institut für Strömungsmechanik der DFLVR, DFVLR-Interner Bericht, 1973.
21. Schönauer, W. et al., 'Selfadaptive solution of the incompressible 3-D boundary layer equations with controlled error', *Proc. Second GAMM-Conference on Numerical Methods in Fluid Mechanics*, DFVLR, Köln 1977, pp. 184–191.
22. Wang, K.C., 'Three-dimensional boundary layer near the plane of symmetry of a spheroid at incidence', *Journal of Fluid Mechanics*, Vol. 43, Part I, August 1970, pp. 187–209.
23. Hirsch, R.S. and Cebeci, T., 'Calculation of three dimensional boundary layers with negative cross flow on bodies of revolution', *AIAA Paper* 77–683, 1977.
24. Blottner, F.G. and Ellis, M.A., 'Finite difference solution of the incompressible three-dimensional boundary layer equations for a blunt body', *Computers and Fluids*, Vol. 1, No. 2, June 1973, pp. 133–158.
25. Stock, H.W., 'Laminar boundary layers on inclined ellipsoids', *Zeitschrift für Flugwiss. Weltraumforsch.*, Vol. 4, No. 2, 1980, pp. 217–224.
26. Robert, K., 'Higher-order boundary layer equations for three-dimensional, compressible flow', in *Beiträge zur Gasdynamik und Aerodynamik*, DFVLR FB 77-36, 1977, pp. 273–288.
27. Cebeci, T. et al., 'Studies on three dimensional laminar and turbulent boundary layers on bodies revolution at incidence. I. Nose separation', *AIAA Paper* 79–138, 1979.
28. Meier, H.U. and Kreplin, H.P., 'Experimental investigation of the boundary layer transition and separation phenomena on a body of revolution', *Zeitschrift für Flugwiss. Weltraumforsch.*, Vol. 4, Heft 2, 1980, pp. 65–71.

Chapter 9
Implicit Time-Dependent Methods for Inviscid and Viscous Compressible Flows, with a Discussion of the Concept of Numerical Dissipation

G. Degrez

9.1 Introduction

The compressible Euler and Navier-Stokes equations represent the most sophisticated models of single-phase flows of single-component Newtonian fluids. As such, they allow the analysis of complex inviscid and viscous flow phenomena including rotational flows caused by curved shock waves or viscous/inviscid interactions leading to flow separation. As a counterpart, the numerical techniques required to solve these equations are also the most sophisticated and the numerical effort needed to obtain them is also the greatest. This is schematically represented in Fig. 9.1 taken from Green's [18] review of the state- of-the-art in numerical methods in aeronautical fluid dynamics.

The difficulties of solving complex steady compressible flows were already pointed out in the first part of this volume, in which the blunt body problem was taken as an illustrative example. It was shown that the crux of the difficulty lies in the mixed character of the flow, involving regions governed by "elliptic" equations and others governed by "hyperbolic" equations. Finally, the solution to the problem was found by solving the time dependent equations using a time marching method, taking advantage of the uniform nature of the unsteady equations with respect to time, independently of the subsonic or supersonic character of the flow.[1]

Following that breakthrough, many methods were developed to integrate the unsteady Euler or Navier-Stokes equations. These methods can be classified in two main categories: explicit and implicit methods (Part I, Sect. 5.3).

Historically, explicit methods were developed earlier, because of their greater simplicity. Several examples were given in Part I, Chap. 7. The major limitation of these methods is their stability characteristics which impose an upper bound on the usable integration time step. In recent years, implicit methods have been developed to overcome this limitation and have proved more efficient than the former explicit methods, which justifies their study.

G. Degrez
Université Libre de Bruxelles, Brussels, Belgium, e-mail: gdegrez@ulb.ac.be

[1] Alternatively, it is possible to address the problem from a completely different standpoint, i.e. to look for more sophisticated iterative methods for solving the (non-linear) algebraic system of equations resulting from the space discretization, as briefly discussed below.

J.F. Wendt (ed.), *Computational Fluid Dynamics*, 3rd ed.,
© Springer-Verlag Berlin Heidelberg 2009

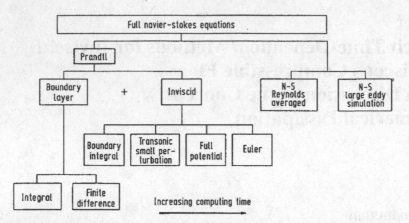

Fig. 9.1 Hierarchy of computational fluid dynamics after [18]

In Sect. 9.2, we shall examine solution techniques for simpler flows and explain why these techniques fail for the solution of the steady compressible Euler/Navier-Stokes equations. In Sect. 9.3, stability properties of numerical integration techniques will be studied in detail first for ordinary differential equations, then for partial differential equations. In Sect. 9.4, it will be shown how an implicit method can be constructed to solve partial differential equations such as the Euler or Navier-Stokes equations. It will be seen that this can be subdivided into three steps, the choice of an explicit discretization scheme, the choice of an implicit operator and finally the choice of a solution strategy, which will be discussed in turn. For the first step, the issue of numerical dissipation will turn out to be crucial, and this concept will be discussed in detail. As in Part I, only the finite difference method is considered as the space discretization technique, but, as will be mentioned in the lecture, most of the concepts discussed and of the basic methods described apply equally to finite volume discretizations (especially on structured meshes) and some to finite element discretizations.

The content of these notes will remain rather basic except in a few instances, in accordance with the objectives of this book. In particular, no individual scheme will be examined in great detail. For additional information, the reader is referred to the very comprehensive survey of CFD methods by C. Hirsch [22, 23] and, finally, to the original literature.

9.2 Solution Techniques for Simpler Flows and Reason of Their Failure for Euler/Navier-Stokes Equations

9.2.1 Solution Strategies for Simple Flow problems

In Part I, Sect. 1.1, the blunt body problem was considered and it was stated that, because of the non-uniform nature of the governing equations, there existed up to the mid sixties no technique to solve the (steady flow) problem, in particular to determine the shock shape, and that finally a numerical solution was made possible by employing the time-dependent approach. Let us now return to this question and wonder why the classical solution techniques used for solving simpler flow problems do not work for solving the Euler or Navier-Stokes equations. For this purpose, let us consider a small disturbance problem. The subsonic/supersonic small disturbance equation in 2D is

$$(1 - M_\infty^2)\frac{\partial^2 \varphi}{\partial x^2} + \frac{\partial^2 \varphi}{\partial y^2} = 0 \tag{9.1}$$

According to the classification of partial differential equations (Part I, Sect. 4.2), this equation is elliptic for $M_\infty < 1$ and hyperbolic for $M_\infty > 1$. Now, suppose that this equation is to be solved on a uniform cartesian mesh and that central differences are used to discretize it, as central differences were shown to be more accurate for a given stencil span (Part I, Sect. 5.2). For simplicity, let us consider a very coarse mesh of 4×4 points with $\Delta x = \Delta y$.

9.2.1.1 Subsonic Flow ($M_\infty = 0$)

In subsonic flow (elliptic problem), one boundary condition is required at each boundary. For simplicity, let us assume φ is given on the boundaries (Dirichlet boundary conditions). Then, the set of finite difference equations can be written in matrix form as

$$A\Phi = b$$

with

$$A = \begin{pmatrix}
1 & & & & & & & & & & & \\
& 1 & & & & & & & & & & \\
& & 1 & & & & & & & & & \\
-1 & & & -1 & 4 & -1 & & & -1 & & & \\
& -1 & & & -1 & 4 & -1 & & & -1 & & \\
& & & & & & 1 & & & & & \\
& & & & & & & 1 & & & & \\
& & & & -1 & & & -1 & 4 & -1 & & -1 \\
& & & & & -1 & & & -1 & 4 & -1 & & -1 \\
& & & & & & & & & 1 & & \\
& & & & & & & & & & 1 & \\
& & & & & & & & & & & 1
\end{pmatrix}$$

Notice that A is a non trivial matrix (it is not lower or upper triangular). For the present model problem, it is easy to find the solution using a direct method such as the Gaussian elimination method but, when the problem becomes large (10^6 unknowns are often encountered nowadays), direct methods become prohibitively expensive both in storage and CPU time, and one uses rather an iterative (also called relaxation) technique. The principle of such techniques is to replace the original system by a simplified one and thus construct a sequence of approximations of the solution

$$B\Phi^{(k+1)} = b + (B-A)\Phi^{(k)} \tag{9.2}$$

Simple (and yet efficient) choices for the (preconditioning) matrix B are

$$\begin{cases} B = D & \text{Jacobi method} \\ B = L+D \text{ or } U+D & \text{Gauss-Seidel method} \\ B = L+\dfrac{D}{\omega} \text{ or } U+\dfrac{D}{\omega} & \text{Successive overrelaxation (SOR) method} \\ B = (L+D)D^{-1}(U+D) & \text{Symmetric Gauss-Seidel method} \\ B = -\dfrac{\omega}{2}\left(\dfrac{I}{\omega}-A_x\right)\left(\dfrac{I}{\omega}-A_y\right) & \text{Alternating direction implicit (ADI) method} \end{cases}$$

where D, L and U denote the diagonal, lower and upper parts of A respectively. Now, it can be shown that these iterative methods *converge* if the original system is *diagonally dominant*. It turns out that, for central discretizations of the small disturbance potential equation considered here, this remains the case as long as the original equation is elliptic, i.e. as long as the incoming flow is *subsonic*. This in fact remains true for central discretizations of the full potential equation provided the flow is subsonic at all points of the domain. For central space discretizations, the previous discussion can thus be summarized as follows:

SUBSONIC $\rightarrow$ ELLIPTIC $\rightarrow$ DIAGONALLY DOMINANT $\rightarrow$ ITERATIVE METH-
FLOW   EQUATION   ALGEBRAIC SYSTEM   ODS WORK

9.2.1.2 Supersonic Flow ($M_\infty = \sqrt{2}$)

In supersonic flow, two boundary conditions are to be specified at the inlet boundary, i.e. φ (which allows the calculation of $\partial\varphi/\partial y$, the vertical perturbation velocity) and $\partial\varphi/\partial x$ (the horizontal perturbation velocity) as discussed previously Part I, Sect. 4.3. Using this latter boundary condition, it is possible to discretize the flow equation (9.1) at points 1 and 2 of the boundary as well. Indeed, introducing a fictitious point a outside of the domain, the centrally discretized equation at point 1 reads

$$-\underbrace{\frac{\varphi_4+\varphi_a-2\varphi_1}{\Delta x^2}}_{\partial^2\varphi/\partial x^2}+\underbrace{\frac{\varphi_2+\varphi_c-2\varphi_1}{\Delta y^2}}_{\partial^2\varphi/\partial y^2}=0$$

Now, φ_c is given since c is on the boundary and φ_a may be evaluated by the boundary condition $\partial\varphi/\partial x = u_1$, whose central space discretization reads

$$\frac{\varphi_4 - \varphi_a}{2\Delta x} = u_1$$

It is then possible to eliminate the fictitious φ_a yielding the following discretized flow equation at point 1.

$$-\frac{2\varphi_4 - 2\varphi_1}{\Delta x^2} + \frac{\varphi_2 - 2\varphi_1}{\Delta y^2} = -\frac{2u_1}{\Delta x} + \frac{\varphi_c}{\Delta y^2}$$

or, with $\Delta x = \Delta y = h$,

$$-2\varphi_4 + \varphi_2 = \varphi_c - 2hu_1 \tag{9.3}$$

Likewise, the discretized flow equation at point 2 reads

$$-2\varphi_5 + \varphi_1 = \varphi_d - 2hu_2 \tag{9.4}$$

where $u_2 = \partial\varphi/\partial x)_2$ is specified.

Using a central discretization at interior points as well and gathering all finite difference expressions in matrix format, the system matrix A is

$$A = \begin{pmatrix}
1 & & & & & & & & & & \\
& 1 & & & & & & & & & \\
& & 1 & & & & & & & & \\
& & & 1 & -2 & & & & & & \\
& 1 & & & & -2 & & & & & \\
& & & & & & 1 & & & & \\
& & & & & & & 1 & & & \\
-1 & 1 & & 1 & & & & -1 & & & \\
& -1 & 1 & & 1 & & & & -1 & & \\
& & & & & & & & & 1 & \\
& & & -1 & & 1 & 1 & & & & -1 & \\
& & & & -1 & & 1 & 1 & & & & -1 \\
\end{pmatrix}$$

and it appears that the matrix is *lower-triangular*. This implies that the solution can be obtained directly by *forward substitution*. It is quite interesting to observe that the physical nature of the problem (finite region of influence, allowing the solution to be computed by forward space marching) is mimicked by the structure of the discretized algebraic system (lower triangular system, allowing the solution to be computed by forward substitution). Now, for such triangular systems, the only concern is the possible amplification of round-off errors, which can be evaluated by a von Neumann stability analysis (Part I, Sect. 5.4). Let us consider the general finite difference discretization of the flow equation

$$-\frac{\varphi_{i+1\,j} - 2\varphi_{ij} + \varphi_{i-1\,j}}{\Delta x^2} + \frac{\varphi_{i\,j+1} - 2\varphi_{ij} + \varphi_{i\,j-1}}{\Delta y^2} = 0$$

and suppose that the solution is computed by forward substitution, i.e. the previous equation is solved for $\varphi_{i+1\,j}$ with all other quantities known. Then, assuming as in Part I, Sect. 5.4 (Eq. 5.34)

$$\varepsilon(x,y) = e^{ax}e^{ik_m y}$$

and posing $g = e^{a\Delta x}$ and $\eta = k_m \Delta y$, we find that the amplification factor g satisfies

$$-\frac{g^2 - 2g + 1}{\Delta x^2} - \frac{4g \sin^2 \eta/2}{\Delta y^2} = 0$$

or

$$g^2 - 2g\left(1 - 2\frac{\Delta x^2}{\Delta y^2}\sin^2 \eta/2\right) + 1 = 0$$

Now, because the product of roots equals 1, stability is achieved only if the equation has complex conjugate roots, i.e.

$$\left(1 - 2\frac{\Delta x^2}{\Delta y^2}\sin^2 \eta/2\right)^2 \le 1$$

and thus

$$\frac{\Delta x^2}{\Delta y^2} \le 1 \tag{9.5}$$

which is just satisfied in the present case $\Delta x = \Delta y = 1$. Again, the conclusion applies equally to the more general problem governed by the full potential equation in supersonic flow (except that the stability condition depends on the local flow Mach number rather than on the incoming flow Mach number in the present example).

In conclusion, for supersonic flow, we can summarize the results as follows

SUPERSONIC $\rightarrow$ HYPERBOLIC $\rightarrow$ INITIAL VALUE PROB- $\rightarrow$ SPACE MARCH-
FLOW   EQUATION   LEM WELL POSED   ING WORKS

provided stability of the marching is ensured.

9.2.1.3 Ill-posed Problems

Suppose now that we would like to solve the following (ill-posed) problems: a supersonic flow treated as a boundary value problem and a subsonic flow treated as an initial-value problem.

In the first case, because of the prescription of boundary conditions everywhere on the boundary, the matrix of the resulting discretized problem will be non-triangular. But on the other hand, because of the hyperbolic nature of the equation, the linear system to be solved will not be diagonally dominant, so that iterative (relaxation) methods will not work.

Conversely, if a subsonic flow is treated as an initial value problem, the resulting matrix of the discretized flow problem will be lower triangular, but the marching (forward substitution) solution procedure will be unstable as shown below. Indeed, for Laplace's equation, the amplification factor g of the marching scheme must satisfy

$$\frac{g^2 - 2g + 1}{\Delta x^2} - \frac{4g \sin^2 \eta/2}{\Delta y^2} = 0$$

or

$$g^2 - 2g(1 + 2\frac{\Delta x^2}{\Delta y^2} \sin^2 \eta/2) + 1 = 0$$

Since the discriminant is positive, this equation has real roots, and since the product of roots equals one, one of them must have a modulus larger than one and hence the marching scheme is unstable.

In conclusion, if one attempts to solve an ill-posed problem, the classical solution techniques (relaxation methods, marching schemes) will protest by becoming unstable. This is in one sense fortunate since it issues a warning when trying to address a wrong problem, but it is also a source of difficulty for more complex problems.

9.2.2 More Complex Problems

Now that we have understood why relaxation and marching methods work respectively for wholly subsonic or supersonic potential flows, we can explain why they break down in more complicated flow situations.

As a first example of a more complicated flow field, let us consider a potential flow in which there exist both regions of subsonic and supersonic flow. Such a situation occurs for flows over aerofoils at supercritical Mach numbers for which the maximum flow Mach number remains moderate (≈ 1.3) (see Fig. 9.2 below). Since the maximum Mach number remains moderate, shock waves are weak so that the potential flow assumption remains valid but, similar to the blunt body problem, there exists one region where the flow equation is elliptic and another where it is hyperbolic. Therefore, for a central space discretization, we are faced with the following problem. If we select a relaxation method, the solution will be unstable in the supersonic region. We could think of switching methods in the supersonic region, but this is impossible as we don't know a priori the location of the subsonic/supersonic boundary.

As a second example, we shall consider a wholly subsonic (even incompressible) flow but which is no longer irrotational. In two dimensions, the flow equations are

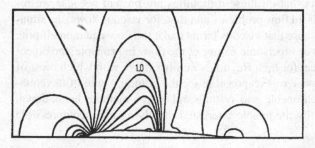

Fig. 9.2 Mach number contours in potential flow over NACA 0012 ($M_\infty = 0.8$)

$$\frac{\partial u}{\partial x} + \frac{\partial v}{\partial y} = 0$$

$$u\frac{\partial u}{\partial x} + v\frac{\partial u}{\partial y} = -\frac{1}{\rho}\frac{\partial p}{\partial x} \tag{9.6}$$

$$u\frac{\partial v}{\partial x} + v\frac{\partial v}{\partial y} = -\frac{1}{\rho}\frac{\partial p}{\partial y}$$

or

$$A\frac{\partial U}{\partial x} + B\frac{\partial U}{\partial y} = 0$$

with

$$A = \begin{pmatrix} 1 & 0 & 0 \\ u & 0 & \frac{1}{\rho} \\ 0 & u & 0 \end{pmatrix} \qquad B = \begin{pmatrix} 0 & 1 & 0 \\ v & 0 & 0 \\ 0 & v & \frac{1}{\rho} \end{pmatrix} \qquad U = \begin{pmatrix} u \\ v \\ p \end{pmatrix}$$

Multiplying by A^{-1}, we have

$$\frac{\partial U}{\partial x} + \underbrace{\begin{pmatrix} 0 & 1 & 0 \\ 0 & \frac{v}{u} & \frac{1}{\rho u} \\ \rho v & -\rho u & 0 \end{pmatrix}}_{C}\frac{\partial U}{\partial y} = 0 \tag{9.7}$$

The characteristic slopes are the eigenvalues of C, given by the characteristic equation

$$\left(\frac{v}{u} - \lambda\right)(\lambda^2 + 1) = 0 \tag{9.8}$$

and we see that one eigenvalue is real $\lambda = \frac{v}{u}$ and the remaining two are imaginary. The first one is associated with the "hyperbolicity" of the convection of vorticity and the latter two with the "elliptic" character of the pressure field. Consequently, the overall flow equations are neither hyperbolic nor elliptic; they are of hybrid type and neither relaxation nor marching methods are suitable to solve this flow problem. As a matter of fact, although this may appear paradoxical, incompressible inviscid rotational flows are among the most difficult to compute, more so than compressible rotational flows.

Finally, one might object that all these difficulties are present because we restricted our attention to inviscid flow problems and that, for viscous flows, the situation may be simpler in the sense that viscous terms make the flow equation elliptic, irrespective of the subsonic or supersonic nature of the flow. In principle, the objection is correct but in practice, for high Reynolds number flows with which most of us are concerned, the viscous terms responsible for the elliptic nature of the equation are so small that, for reasonable grid systems, relaxation methods break down. This has long been known by the people interested in solving the incompressible Navier-Stokes equations.

9.2.3 A Solution: The Time-Dependent Approach

From the previous discussion, it appears that no method exists that is suitable to solve steady flow problems in the more complicated flow situations considered (at least for central space discretizations). To overcome this problem, one may think of various strategies.

First, for boundary value problems, one may think of investigating more sophisticated solution techniques than the classical relaxation techniques described above. When the problem is not too large, direct methods, in particular adapted to the sparse nature of the linear systems resulting from the discretization of PDEs may be considered. Otherwise, more sophisticated iterative techniques may be considered. Two examples are the multigrid method and Newton conjugate gradient-like methods. Both approaches have been used successfully in CFD (e.g. [20,28,31] for the multigrid method and [24, 27, 47] for Newton conjugate gradient-like methods). These methods are more involved and fall clearly beyond the scope of an introductory text such as this one. In addition, as will be discussed later, this approach does not address a second problem of equal importance, namely that of the accuracy of the final solution.

A second strategy relates to the space discretization. Indeed, as was emphasized several times, only central discretizations were considered up to now. One could then wonder whether the problems described earlier are associated with the choice of the discretization technique and whether the use of biased difference formulas could help solving the problem. This avenue was investigated very early for the transonic potential flow problem and led to the important breakthrough of Murman and Cole [32] in which a solution-adaptive discretization technique was used. The application of this idea to the Euler and Navier-Stokes equations proved more difficult. Research in this area has been motivated mainly by accuracy rather than by solvability (stability) considerations, but when solution-adaptive biased discretizations started appearing in the early eighties [34, 45], their beneficial effect on the solvability of the resulting discretized equations was soon realized.

Finally, the third, and certainly the most well known approach is the time-dependent approach, already introduced in Part I, Chap. 7. In this approach, the steady solution is found as the asymptotic result of the unsteady process governed by the time-dependent equations. As mentioned already in the introduction, the advantage of using the time dependent equations lies in the fact that they provide well-posed initial value problems, irrespective of the particular type of flow considered, be it supersonic or subsonic, inviscid or viscous. Indeed, the inviscid equations are always hyperbolic in time and the viscous equations are hyperbolic/parabolic in time. As a consequence, marching methods can be used to integrate the flow equations in time. Actually, this approach may be considered as asking nature to provide a converging iterative technique for the mixed/hybrid problems under consideration. In itself, it does not address the accuracy question either.

Starting from the partial differential equations governing the flow, let us first discretize them in space by some suitable method as for instance central space differencing. The partial differential equation, together with the boundary conditions is then transformed into a set of first-order non-linear ordinary differential equations which can be written

$$\frac{dU}{dt} = -R(U) \tag{9.9}$$

where U is the vector of unknowns and R is some (residual) vector which depends on the vector U. The system can be very large as the number of unknowns equals the number of grid points times the number of unknowns per point (5 for 3-D compressible Euler or Navier-Stokes equations). Now, this system can be integrated by any method suited for the integration of Ordinary Differential Equations (ODE) [Runge-Kutta, Euler, Adams]. These methods can be divided into two large classes: explicit and implicit methods. Let us illustrate this by considering the most simple one-step time discretization.

$$\frac{U^{n+1} - U^n}{\Delta t} = -\left[\theta\, R(U^{n+1}) + (1-\theta)R(U^n)\right] \tag{9.10}$$

where the superscript denotes the time level. When $\theta = 0$, the value of U^{n+1} can be calculated directly from the value of U^n by simple mathematical operations like function evaluations, multiplications and additions. The method is said to be *explicit*. On the other hand, when $\theta \neq 0$, a system of equations (in general non linear) must be solved in order to determine the value of U^{n+1}. The method is called *implicit*. Implicit methods clearly appear more complicated to use because of this need to solve a system of equations at each time step. On the other hand, it was shown previously in Part I, Sect. 5.4 that marching methods exhibit a stability limit, i.e. there exists a limiting value of the time step Δt above which the error will grow without bound. It turns out that, in general, implicit methods have significantly larger stability bounds than explicit methods. This is particularly advantageous when one is looking for a steady solution. Using a bigger Δt will mean reaching faster the limiting state ($t \rightarrow \infty$). In conclusion, implicit methods will be advantageous with respect to explicit methods in those cases when the increase in usable time step will more than overbalance the increase in computational effort per time step.

To end this section, it is important to mention that the three strategies identified to overcome the problems associated with complex mixed/hybrid problems are not exclusive but rather complementary. As a matter of fact, many recent Euler/Navier-Stokes solvers combine solution-adaptive differencing with a sophisticated (accelerated) iterative strategy whose preconditioner is derived from an implicit time-stepping scheme. The objective of the present lecture is to discuss the way in which efficient implicit schemes can be constructed. The issue of accuracy will introduce itself naturally and will lead to the description of solution-adaptive biased discretizations (upwind schemes).

9.3 Stability Properties of Explicit and Implicit Methods

Since the outcome of the competition between explicit and implicit methods is governed by their respective stability properties, a closer look must be given to this issue. First, we observe that the space-discretization of a time-dependent partial differential equation produces a system of ordinary differential equations. Consider for example the diffusion equation

$$\frac{\partial u}{\partial t} = \frac{\partial^2 u}{\partial x^2} \tag{9.11}$$

After discretization in space using central finite differences, we obtain the following system of ordinary differential equations

$$\frac{du_i}{dt} = \frac{1}{\Delta x^2}(u_{i-1} - 2u_i + u_{i+1}) \qquad \text{or} \qquad \frac{dU}{dt} = AU + F(t) \tag{9.12}$$

with

$$U = \begin{pmatrix} u_1 \\ u_2 \\ \vdots \\ u_{n-1} \\ u_n \end{pmatrix} \qquad A = \frac{1}{\Delta x^2} \begin{pmatrix} \text{B.C.} \\ 1 & -2 & 1 \\ & \ddots & \ddots & \ddots \\ & & 1 & -2 & 1 \\ & & & & \text{B.C.} \end{pmatrix} \tag{9.13}$$

Therefore, the analysis of the stability of a time stepping scheme for solving the PDE reduces to the analysis of the stability of a time stepping scheme for solving a system of ODEs.

Furthermore, when we consider a periodic solution in space, i.e. $u = u(t)e^{ik_m x}$, the system of ODEs reduces to a single ODE. Indeed, inserting the periodic solution hypothesis in (9.12) and realizing that

$$u_{i+1} = u(t)e^{ik_m x_{i+1}} = u(t)e^{ik_m(x_i + \Delta x)} = u(t)e^{ik_m x_i}e^{ik_m \Delta x} = u_i e^{ik_m \Delta x}$$

and similarly

$$u_{i-1} = u_i e^{-ik_m \Delta x}$$

we obtain

$$\frac{du_i}{dt} = \frac{e^{ik_m \Delta x} - 2 + e^{-ik_m \Delta x}}{\Delta x^2} u_i = \underbrace{-\frac{4}{\Delta x^2}\sin^2\left(\frac{k_m \Delta x}{2}\right)}_{q} u_i$$

i.e. an ODE whose coefficient q depends on the reduced wavenumber $k_m \Delta x$, the locus of q (in the complex plane) being called the Fourier footprint of the discretized equation. The stability analysis can then be reduced to the stability analysis for a single ordinary differential equations $du/dt = qu$, where q is a complex coefficient.[2]

---

[2] We obtain such a simple linear ODE because the original PDE and the discretization scheme are both linear. Despite these restrictions, this linear equation remains of great practical significance.

9.3.1 Definition–Examples

Stability of the numerical integration of an ordinary differential equation is usually defined by the following statement. A method is said to be stable (weakly-stable) if the numerical solution remains bounded when the number of steps n goes to infinity and the time step size Δt goes to zero with the product $n\Delta t$ remaining constant.

Mathematically, the sequence u_n is to be bounded

with
$$\Delta t \to 0$$
$$n \to \infty$$
$$n\Delta t = \text{constant}$$

When the method is stable, according to that definition, we have the important consequence that, provided the discretization of the equation is consistent, the method is convergent, which means that the numerical solution tends towards the analytical solution for Δt going to zero. This important theorem stating that stability + consistency imply convergence is due to P. Lax [23].

Stability properties of methods for integrating ODEs are generally studied by considering the linear test equation

$$\frac{du}{dt} = qu \qquad u(0) = 1 \tag{9.14}$$

The analytical solution is $u = e^{qt}$. When a numerical method is applied to that problem, the solution can always be written

$$u_{n+1} = g(q, \Delta t)u_n$$

where g is called an amplification factor

Consequently,
$$u_{n+1} = [g(q, \Delta t)]^n \; u_1$$

The stability condition then requires that the numbers $[g(q, \Delta t)]^n$ must be uniformly bounded for $0 < \Delta t < t$, $0 \le n\Delta t < T$. A necessary condition for this is that $|g(q, \Delta t)| \le 1 + 0(\Delta t)$, in particular $|g(q, 0)| \le 1$

Let us consider a couple of examples. The 4th-order Runge-Kutta method reads for the test equation.

$$k_1 = \Delta t q u_n \qquad\qquad = q\Delta t(1)u_n$$
$$k_2 = \Delta t q\left(u_n + \tfrac{1}{2}k_1\right) = q\Delta t\left(1 + q\tfrac{\Delta t}{2}\right)u_n$$
$$k_3 = \Delta t q\left(u_n + \tfrac{1}{2}k_2\right) = q\Delta t\left(1 + q\tfrac{\Delta t}{2} + q^2\tfrac{\Delta t^2}{4}\right)u_n$$
$$k_4 = \Delta t q\left(u_n + \tfrac{1}{2}k_3\right) = q\Delta t\left(1 + q\tfrac{\Delta t}{4} + q^2\tfrac{\Delta t^2}{4} + q^3\tfrac{\Delta t^3}{4}\right)u_n$$

$$u_{n+1} = u_n + \frac{1}{6}(k_1 + 2k_2 + 2k_3 + k_4)$$

$$= \left(1 + q\Delta t + \frac{(q\Delta t)^2}{2} + \frac{(q\Delta t)^3}{6} + \frac{(q\Delta t)^4}{24}\right)u_n$$

$$\rightarrow \quad g = 1 + q\Delta t + \frac{(q\Delta t)^2}{2} + \frac{(q\Delta t)^3}{6} + \frac{(q\Delta t)^4}{24}$$

Thus $|g| \leq 1 + 0(\Delta t)$, $|g(q, 0)| = 1 \leq 1$ and the 4th-order Runge-Kutta method is stable.

Another example is the two-step explicit mid-point method which reads for the test equation

$$u_{n+1} = u_{n-1} + 2q\Delta t\, u_n \tag{9.15}$$

Let us look for g such that

$$u_{n+1} = g\, u_n = g^2\, u_{n-1}$$

Then,

$$g^2 u_{n-1} = u_{n-1} + 2q\Delta t g\, u_{n-1} \quad \text{or} \quad (g^2 - 2q\Delta t g - 1)u_{n-1} = 0$$

Consequently, g must be a root of

$$g^2 - 2q\Delta t g - 1 = 0 \quad \rightarrow \quad g = q\Delta t \pm \sqrt{1 + (q\Delta t)^2}$$

This expression is such that $|g(q, \Delta t)| \leq 1 + 0(\Delta t)$ and the two-step explicit mid-point method is stable.

9.3.2 Weak Instability

Since it is consistent and stable, let us apply the 2-step explicit mid-point method to the test equation defined above with $q = -1$ and $\Delta t = 0.1$. Since the method is a 2-step method, two initial values are required. One is given by the initial condition of the problem : $u_0 = 1$ but a value of u_1 is also required. The calculation has been performed with two approximations of the exact solution $e^{-0.1} = 0.90484$, i.e. $u_1 = 0.85$ and $u_1 = 0.9$.

The results of the calculation are displayed in Fig. 9.3. One notices that the perturbation on u_1 gives rise to amplifying oscillations. In fact, as small as the initial perturbation may be - and there will always be one because of round off errors - it will eventually lead to an explosion of the numerical solution. This phenomenon is clearly inacceptable. It is named *weak instability*.

Before advancing further, let us look for the cause of this phenomenon. It was seen in the previous section that the 2-step explicit mid-point method allowed solutions of the form $u_{n+1} = g u_n = g^n u_1$ with two possible values for g, say g_1 and g_2. An expression of the form

Fig. 9.3 Numerical solution of $\frac{du}{dt} = -u$; $u(o) = 1$ with the 2-step explicit mid point method

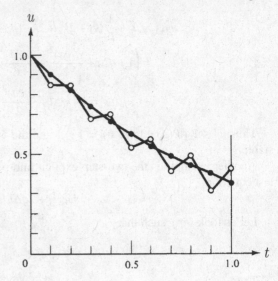

$$u_n = c_1 g_1^n + c_2 g_2^n$$

also satisfies the difference equation

$$u_{n+1} = u_{n-1} + 2q\Delta t u_n \qquad (q = -1)$$

and the constants c_1 and c_2 an determined by the initial values

$$u_o = c_1 + c_2 \qquad u_1 = c_1 g_1 + c_2 g_2$$

But

$$g_1 = q\Delta t + \sqrt{1 + (q\Delta t)^2} = -\Delta t + \sqrt{1 + \Delta t^2}$$
$$g_2 = q\Delta t - \sqrt{1 + (q\Delta t)^2} = -\Delta t - \sqrt{1 + \Delta t^2} \qquad (q = -1)$$

so that $|g_1| < 1$ and $|g_2| > 1$ with $g_2 < 0$ Proceeding with the analysis, one obtains

$$g_1 = 1 - \Delta t + \frac{\Delta t^2}{2} + O(\Delta t^3) = e^{\Delta t} + O(\Delta t^3) = e^{-\Delta t + O(\Delta t^3)}$$
$$g_2 = -1 - \Delta t - \frac{\Delta t^2}{2} + O(\Delta t^3) = (-1)(e^{\Delta t} + O(\Delta t^3)) = (-1)e^{\Delta t + O(\Delta t^3)}$$
$$g_1^n = e^{-n\Delta t} e^{O(n\Delta t^3)} = e^{-t} e^{o(\Delta t^2)}$$
$$g_2^n = (-1)^n e^{n\Delta t} e^{O(n\Delta t^3)} = (-1)^n e^t e^{O(\Delta t^2)} \qquad \text{since } n\Delta t = t$$

The oscillations observed are therefore explained by the g_2^n term which oscillates because of the $(-1)^n$ factor and is amplified like e^t. This term has no relation with the exact solution and is a purely numerical artefact. The closer u_1 gets to $g_1 u_o$, the smaller the c_2 coefficient will be, which delays the observation of the amplifying oscillations, but it will always be impossible to have $c_2 = 0$, which would avoid the instability.

The previous example shows that the definition of stability stated in Sect. 9.2.1 is insufficient as it does not warn us of the possibility of the unacceptable weak instability phenomenon. The reason for this is that the definition of stability in Sect. 9.2.1 gives only information in the (rather useless) limiting case $\Delta t \to 0$, not for actual computations with finite Δt. This result leads to the introduction of a new concept, called region of (absolute) stability.

9.3.3 Region of (absolute) stability

The concept of region of (absolute) stability was introduced by Dahlquist [10]. The region of (absolute) stability of a numerical algorithm for integrating an O.D.E. is the set of values of $z = q\Delta t$ (q = complex parameter of the test equation $\frac{du}{dt} = qu$) such that the sequence u_n of numerical values remains bounded as $n \to \infty$. As the previous definition of stability (Sect. 9.3.1) required that the sequence u_n remain bounded for $n \to \infty$, $\Delta t \to 0$, this is equivalent to stating that the origin lies in the region of (absolute) stability [$\Delta t \to 0$ implies $z = q\Delta t \to 0$].

Let us illustrate this concept by considering a couple of examples. First, let us consider the forward time difference (forward Euler) method.

$$\frac{u^{n+1} - u^n}{\Delta t} = qu^n$$

The amplification factor is readily computed

$$\frac{g-1}{\Delta t} = q \to g = 1 + q\Delta t = 1 + z$$

Hence, we deduce that the region of stability is the region of the complex plane $|1 + z| \le 1$, which is represented in Fig. 9.4. With $q = -1$ (test problem), this means

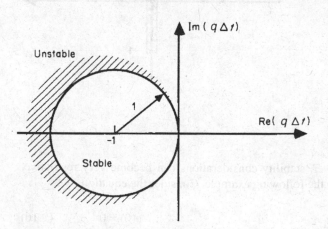

Fig. 9.4 Region of stability of the forward Euler method

that we need $\Delta t \le 2$ to obtain a bounded numerical solution. This is *not* a severe restriction since, to follow the exact solution $u = e^{-t}$, one would rather choose values of Δt of the order of 0.1 to obtain a satisfactory accuracy. In other words, for this test problem, Δt is limited by accuracy rather than by stability considerations.

Let us consider next the central finite time difference (explicit mid-point) method.

$$\frac{u_{n+1} - u_{n-1}}{2\Delta t} = q u_n$$

The amplification factor was already computed to be

$$g^2 - 2q\Delta t g - 1 = 0 \quad \rightarrow \quad g = q\Delta t \pm \sqrt{1 + (q\Delta t)^2}$$

The two roots g_1 and g_2 are such that $g_1 g_2 = 1$. The only possibility for having $|g_{1,2}| \le 1$ is therefore that $|g_1| = |g_2| = 1 \rightarrow g = e^{i\alpha}$, which corresponds to

$$z = q\Delta t = \frac{g - 1/g}{2} = \frac{e^{i\alpha} - e^{-i\alpha}}{2} = 2i\sin\alpha$$

from which we deduce that the region of stability is the segment of the imaginary axis between $-i$ and $+i$, as shown in Fig. 9.5. It is therefore not surprising that the calculation performed with $q = -1$, $\Delta t = 0.1 \rightarrow z = -0.1$ led to an explosion of the calculation, since $z = -0.1$ lies outside of the region of (absolute) stability for that scheme.

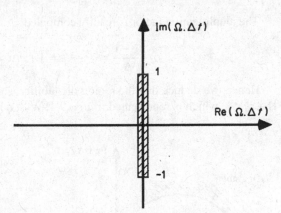

Fig. 9.5 Region of stability of the mid-point method

9.3.4 Stiff Problems

In some instances, however, stability considerations can become very restrictive. This will be illustrated by the following example. Consider the equation

$$\frac{du}{dt} = 100(\underbrace{\sin t}_{\text{Forcing term}} - \underbrace{u}_{\text{Homogeneous term}}) \qquad u(0) = 0 \qquad (9.16)$$

The exact solution is

$$u(t) = \frac{\sin t - 0.01\cos t + 0.01 e^{-100t}}{1.0001} \qquad (9.17)$$

The exponential term (natural response) dies out very quickly and the solution becomes a periodic function of period 2π (forced reponse). If one is interested in that periodic behaviour and not in the exponential transient, one would like to choose a time step of the order of, say, a twentieth of the period, or $\Delta t = \frac{2\pi}{20} \approx 0.3$. However, using the 4th-order Runge-Kutta method, the following results are obtained:

Δt	0.015	0.020	0.025	0.030
Number of steps	200	150	120	100
$u(3)$	0.151004	0.150996	0.150943	$6.7\ 10^{11}$

i.e. with Δt as small as 0.03, the calculation blows up. The explanation is provided by the presence of the homogeneous term. If one ignores the forcing term, the equation is of the same type as the test equation $\frac{du}{dt} = qu$ with $q = -100$. Now, the region of stability of the Runge-Kutta method is shown in the following figure. The stability bound for the Runge-Kutta method is seen to be $q\Delta t \leq -2.8$ or, in this case, $\Delta t \leq 0.028$. It thus appears that it is the presence of the homogeneous term which is responsible for the severe time step limitation, although it produces a fast decaying (very stable) natural response. The fundamental cause of the difficulty lies in the coexistence of two phenomena with very different time scales (the periodic behaviour and the exponential transient) and in the fact that it is the shortest time scale which determines the maximum allowable time step [even though the corresponding phenomenon is disappearing very fast].

Problems where there is such a coexistence of phenomena with very disparate time scales are called *stiff problems*. Unfortunately they are not uncommon in many fields of engineering and in particular in fluid mechanics. For those problems, it would be desirable to have at our disposal schemes such that a physically stable problem would lead to a bounded solution irrespective of the value of the time step Δt. That property is called absolute stability or A - Stability and will be discussed in the next section.

9.3.5 Absolute Stability

Absolute stability was defined in the previous section as a property by which the numerical solution of a physically stable problem would be bounded, irrespective of the time step. Let us translate this in mathematical terms. Test problems of the type $\frac{du}{dt} = qu$ are stable if $Re(q) \leq 0$. Therefore, the set of values of $q\Delta t$, corresponding to stable problems is the left half plane. Absolute stability is thus equivalent to requiring that the region of stability include the left-half plane.

Fig. 9.6 Region of stability of
the Runge-Kutta method

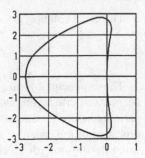

Absolute stability ≡ the region of stability includes the left-half plane.

Now that we have a working definition of absolute stability, we may look for Absolutely stable (or A-stable) schemes. We first observe that the 4th-order Runge-Kutta method is not A-stable (Fig. 9.6). Let us next examine the backward Euler method. Applied to the test equation, the backward Euler method reads

$$\frac{u_n - u_{n-1}}{\Delta t} = qu_n$$

Therefore,

$$u_n(1 - q\Delta t) = u_{n-1} \quad or \quad u_n = \frac{1}{1 - q\Delta t}u_{n-1} \tag{9.18}$$

The sequence u_n will be bounded if $|\frac{1}{1-q\Delta t}| < 1$ or $|1 - q\Delta t| > 1$, which in the $z(= q\Delta t)$ plane is the region outside of a circle of radius one centred at 1, as shown in Fig. 9.7. That region includes the whole left-half plane and the method is thus A-stable. Now, one may observe that the backward Euler method is an *implicit method* and requires the solution of an equation to obtain u_n. In the test example, of course, the equation is a linear scalar equation which is solved immediately, but in the case of systems of non-linear equations, the problem is not so simple.

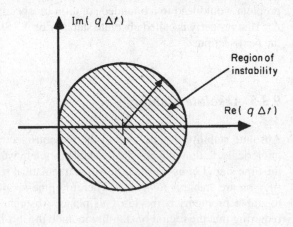

Fig. 9.7 Region of stability of
the backward Euler method

Since explicit methods are easier to set up than implicit methods, one might wonder whether there exist A-stable explicit methods. Unfortunately, the answer to this is no, after a theorem by Dahlquist [9] stating that

- An A-stable method must be IMPLICIT
- An A-stable method has an accuracy of order $p \leq 2$.

Consequently for stiff problems, one will frequently choose among the simplest 2nd order A-stable methods, in particular the trapezoidal method, which for the test equation reads

$$\frac{u_{n+1} - u_n}{\Delta t} = \frac{q}{2}(u_n + u_{n+1}) \tag{9.19}$$

It can easily be shown that the stability region of the trapezoidal method is the left-half plane, so that the method is indeed A-stable.

To summarize, implicit methods are more costly per time step than explicit methods but, because they may be A-stable, the allowed increase in integration step size Δt more than compensates the extra cost per time step. That is particularly true for problems in which one is only interested in the steady-state solution but it is also true for unsteady problems in which the period of interest is much larger than the limit imposed by stability properties. This is the reason why implicit methods have gained in popularity and are worth studying.

9.4 Construction of Implicit Methods for Time-Dependent Problems

Having established under which circumstances implicit methods are competitive, it remains to discuss how they can be constructed in such a way as to be efficient. This will be the subject of the present chapter.

9.4.1 An Essential Ingredient: Linearization

Let us consider the system of ODEs obtained by the space discretization of a PDE

$$\frac{dU}{dt} = -R(U)$$

As mentioned above, this system is in general nonlinear. Suppose that, knowing that the system is stiff, one wishes to integrate it with an implicit one-step method such as the trapezoidal ($\theta = \frac{1}{2}$) or backward Euler ($\theta = 1$) methods. The time-discretization of the system of ODE's thus reads

$$U^{n+1} = U^n - \Delta t \left[\theta R(U^{n+1}) + (1 - \theta)R(U^n) \right] \tag{9.20}$$

Now, if the original system is nonlinear, this equation represents a nonlinear system of equations, which is as difficult to solve as the steady problem $R(U) = 0$. In this case, the increase in computational work per time step of implicit methods with respect to explicit methods would be so big that they would be of no interest. To simplify the system produced by the implicit method, the following linearization is performed

$$R(U^{n+1}) \text{ is replaced by } R(U^n) + \frac{\partial R^n}{\partial U}(U^{n+1} - U^n)$$

where $\frac{\partial R^n}{\partial U}$ is the Jacobian matrix evaluated at time level n. Then, the scheme becomes

$$U^{n+1} = U^n - \Delta t R(U^n) - \theta \Delta t \frac{\partial R^n}{\partial U}(U^{n+1} - U^n) \qquad (9.21)$$

Introducing $\delta U = U^{n+1} - U^n$, this equation can be written

$$\left[I + \theta \Delta t \frac{\partial R}{\partial U}\right] \delta U = -\Delta t R(U^n) \qquad (9.22)$$

which is a linear system for the vector of unknowns δU. The (linear) stability properties of this scheme are of course the same as those of the original scheme since both schemes are identical for linear problems. The linearization process outlined above is not limited to one-step methods. The extension to general multistep methods is presented in [4, 9]. In particular, 2-step methods are of interest. For the one step method, the linearized implicit scheme (9.22) is in fact equivalent to performing one step of a Newton iteration on the non-linear system (9.20). Indeed, posing

$$F(U^{n+1}) = \theta R(U^{n+1}) + (1 - \theta)R(U^n) + \frac{U^{n+1} - U^n}{\Delta t}$$

Newton's method reads

$$J_F^k \tilde{\delta} U = -F(U^{n+1,k})$$

$$U^{n+1,k+1} = U^{n+1,k} + \tilde{\delta} U$$

where J_F is the Jacobian of F with respect to U^{n+1}, i.e.

$$J_F^k = \theta \frac{\partial R}{\partial U}(U^{n+1,k}) + \frac{I}{\Delta t}$$

Taking $U^{n+1,0} = U^n$ as the first guess for Newton's iteration, the first step reads

$$\left[\frac{I}{\Delta t} + \theta \frac{\partial R}{\partial U}(U^n)\right] \tilde{\delta} U = -R(U^n)$$

$$U^{n+1,1} = U^n + \tilde{\delta} U$$

which is exactly the linearized implicit scheme (9.22).

At this stage, it is important to observe that the steady solution is entirely controlled by the right hand side. The scheme is in the so-called δ-form, so that, upon convergence ($\delta U \to 0$), the solution is independent of the transient process, in particular of the time step. One can take advantage of this property to modify (simplify) the matrix of the linear system, as the modifications will only affect the convergence behaviour and not the solution accuracy. The simplifications may be conceptually divided up in two steps:

1. The residual Jacobian $\partial R/\partial U$ is substituted by some approximation $L(U)$ called the *implicit operator*.

$$\frac{I}{\Delta t} + \theta \frac{\partial R}{\partial U} \qquad \to \qquad \frac{I}{\Delta t} + \theta L(U)$$

2. The simplified matrix $(I/\Delta t) + \theta L(U)$ is approximately factorized.

$$\frac{I}{\Delta t} + \theta L(U) \qquad \to \qquad B(U)$$

so that the general form of an implicit time-stepping scheme is

$$B(U^n)\delta U = -R(U^n) \tag{9.23}$$

In summary, the construction of an implicit time-stepping scheme appears as a three-stage process:

Choice of $R(U)$ For a given PDE, $R(U)$ may differ by the choice of the space discretization formula. As this choice has to be made also for explicit time-stepping schemes, this is called the choice of the *explicit operator*. It controls the accuracy of the steady state solution.

Choice of $L(U)$ Upon the selection of the explicit operator, one must select the implicit operator $L(U)$. As it should be in some sense as close as possible to the residual Jacobian $\partial R/\partial U$, the choice of $L(U)$ is not independent of the choice of $R(U)$. The choice of the implicit operator influences the scheme convergence [speed of residual drop as a function of the number of iterations] and efficiency [computational work per iteration].

Choice of the factorization scheme Finally, the simplified linear system is approximately factorized to minimize the computational work per time step.[3] The choice of a factorization scheme influences primarily the scheme efficiency but also the convergence because of the factorization error.

All three stages will now be examined in turn. As it appears clearly from the discussion, the first stage (choice of the explicit operator) applies equally to explicit and implicit methods. It really is a (somewhat more advanced) sequel to Part I, Chap. 7.

---

[3] Alternatively, one can solve the unfactored linear system by some linear iterative technique described in Sect. 9.2.1.1. In this case, a preconditioner is needed, which in most cases is precisely some approximate factorization of the system matrix. The factorized implicit scheme is in fact equivalent to performing a single linear iteration on the system corresponding to the unfactored scheme (see Sect. 9.4.4).

9.4.2 Choice of the Explicit Operator: Central Versus Upwind

For a given difference stencil span, central discretizations have the highest formal
accuracy, as determined by truncation error analysis (Part I, Sect. 5.2 and also [23]).
However, for advection dominated problems as encountered in fluid dynamics, central space discretizations lead to severe problems of spurious oscillations due to a
lack of *numerical dissipation*. First, the origin of the problems will be determined
by introducing the concept of numerical dissipation and explaining why some numerical dissipation is required for advection dominated problems. Then, the various
ways by which numerical dissipation can be generated will be reviewed and, finally
it will be shown how conservative upwind discretizations of hyperbolic systems
such as the Euler equations can be constructed.

9.4.2.1 Numerical Dissipation, What it is and Why it is Needed

Definition of Numerical Dissipation

As mentioned above, the question of numerical dissipation arises for advection dominated problems. Numerical dissipation is therefore defined in reference to the advection (wave) equation:

$$\frac{\partial u}{\partial t} + c \frac{\partial u}{\partial x} = 0 \tag{9.24}$$

This equation describes the transport of the quantity u with speed c. Its general
solution is $u = f(x - ct)$. A particular solution is the periodical solution

$$u = e^{ik(x-ct)} = e^{ikx} e^{-i\omega t} \qquad \text{with } \omega = kc \tag{9.25}$$

which represents the unattenuated propagation of a wave of length $\frac{2\pi}{k}$ with speed c.

Let us compute the amplification factor $u(x, t + \Delta t)/u(x, t)$ for the exact solution.
We find

$$\frac{u(x, t + \Delta t)}{u(x, t)} = e^{-i\omega \Delta t} = e^{-ikc\Delta t} = e^{-i\eta v} \tag{9.26}$$

with

$$v = \frac{c\Delta t}{\Delta x} \qquad \text{CFL number}$$
$$\eta = k\Delta x \qquad \text{dimensionless wave number}$$

A numerical solution will yield

$$\frac{u_i^{n+1}}{u_i^n} = g(\eta, v) \tag{9.27}$$

When one wishes to accurately follow a true unsteady phenomenon, one obvi-
ously desires to have $g(\eta, v)$ as close as possible to $e^{-i\eta v}$. For stability, one must

have $|g(\eta, v)| \leq 1$ for all η. The difference between $|g(\eta, v)|$ and 1 is called *dissipation* or else *dissipative error*, and the difference between $\arg(g(\eta, v))$ and $-\eta v$ is called *dispersion* or *dispersive error*.

Physical interpretation Let us attempt now to give an interpretation of numerical dissipation, to understand what it relates to. For this purpose, let us consider the advection-diffusion equation.

$$\frac{\partial u}{\partial t} + c\frac{\partial u}{\partial x} = \epsilon\frac{\partial^2 u}{\partial x^2}$$

A particular solution of the same type as above, i.e. a periodic solution is

$$u = e^{ik(x-ct)}e^{-\epsilon k^2 t}$$

The amplification factor is now

$$\frac{u(x, t+\Delta t)}{u(x, t)} = e^{ikc\Delta t}e^{-\epsilon k^2\Delta t} = e^{-i\eta v}e^{-\epsilon k^2\Delta t} = e^{-i\eta v}e^{\frac{-\epsilon}{c\Delta x}v\eta^2} \qquad (9.28)$$

i.e. the same expression as before multiplied by the damping factor $e^{-\epsilon k^2\Delta t}$. In this case, a pure wave propagates with speed c but is attenuated with time, as illustrated in Fig. 9.8. When, in a numerical solution, $|g(\eta, v)| < 1$ for a particular dimensionless wave number η, this produces an attenuation of the wave with time analogous to that observed in the case of the advection - diffusion equation. That is the reason why dissipation, defined by the difference between $|g(\eta, v)|$ and 1, is associated with the phenomenon of viscous diffusion.

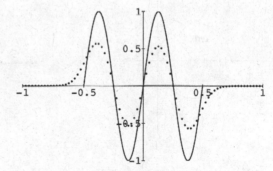

Fig. 9.8 Advection-diffusion of a wave packet of wavenumber $k = 4\pi$ with $\epsilon = 2.5\,10^{-3}$. Solution at $t = 1.6$

Not only diffusion terms of second order such as considered above can produce attenuation. Fourth order terms and, in general, all even derivatives in the space variable produce attenuation. The dissipation of a numerical scheme is generally a blend of dampings produced by some or all of those terms. The nature of the blending depends of the particular scheme considered and can be determined by the so - called modified equation technique [For a thorough treatment of that technique, see the textbook by Anderson et al. [1]].

Example 1. Let us illustrate the effects of numerical dissipation by considering a couple of examples. Let us suppose we are looking for a numerical solution of the advection equation (9.24). Let us first select the space discretization scheme. We consider two possibilities: the central finite difference and the backward (upwind) finite difference formulas.

BACKWARD (UPWIND)   CENTRAL

$$\frac{du_i}{dt} + c\frac{u_i - u_{i-1}}{\Delta x} = 0 \qquad\qquad \frac{du_i}{dt} + c\frac{u_{i+1} - u_{i-1}}{2\Delta x} = 0$$

In order to select the time-integration scheme, let us first calculate the Fourier footprints of the discretized equations. Introducing the periodic solution hypothesis $(u = U(t)e^{Ikx} \rightarrow u_{i\pm1}^n = u_i^n e^{\pm I\eta})$, we get

$$\frac{du_i}{dt} = -c\underbrace{\frac{1 - e^{-I\eta}}{\Delta x}}_{q} u_i \qquad\qquad \frac{du_i}{dt} = -c\frac{e^{I\eta} - e^{-I\eta}}{2\Delta x} u_i$$

$$= -I\underbrace{\frac{c\sin\eta}{\Delta x}}_{q} u_i$$

i.e. both discretized equations reduce to the model equation (9.14) where the q coefficient depends on the reduced wave number η. The locus of q is the Fourier footprint of the discretized equation. They are shown below for the two discretization schemes

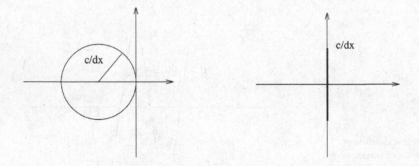

Now, the time-integration scheme should be selected so that $q\Delta t$ can lie within the region of stability. By comparing the respective loci of q with the region of stabilities of some of the schemes examined previously (see Figs. 9.4 and 9.5), it appears quite clearly that the forward Euler scheme cannot be used together with the central space discretization, but the mid-point method, on the contrary, can, and the opposite conclusion applies to the backward finite difference discretization. Applying the mid-point method and the forward Euler method to the central and backward space discretizations respectively, the fully discrete schemes and their truncation errors are respectively

FIRST-ORDER UPWIND-FORWARD EULER

$$\frac{u_i^{n+1} - u_i^n}{\Delta t} + c\frac{u_i - u_{i-1}}{\Delta x} = 0$$

$$TE = O(\Delta x, \Delta t)$$

LEAPFROG (CENTRAL SPACE-MID POINT)

$$\frac{u_i^{n+1} - u_i^{n-1}}{2\Delta t} + c\frac{u_{i+1} - u_{i-1}}{2\Delta x} = 0$$

$$TE = O(\Delta x^2, \Delta t^2)$$

and, being central in both space and time, the leapfrog method is seen to be of superior accuracy. Let us now compute the amplification factors of both schemes.

$$g = 1 + q\Delta t \quad \rightarrow$$

$$g = 1 - \frac{c\Delta t}{\Delta x}(1 - e^{-I\eta})$$

$$= 1 - \nu(1 - e^{-I\eta})$$

$$= 1 - \nu(1 - \cos\eta) - I\nu\sin\eta$$

$$|g|^2 = 1 - 2\nu(1 - \cos\eta) +$$

$$\nu^2[(1 - \cos\eta)^2 + \sin^2\eta]$$

$$= 1 - 2\nu(1 - \cos\eta) + 2\nu^2(1 - \cos\eta)$$

$$= 1 - 2\nu(1 - \nu)(1 - \cos\eta) < 1 \text{ for } 0 < \nu < 1$$

$$g = q\Delta t \pm \sqrt{1 + (q\Delta t)^2} \quad \rightarrow$$

$$g = -I\frac{c\Delta t}{\Delta x}\sin\eta \pm \sqrt{1 - (\frac{c\Delta t}{\Delta x}\sin\eta)^2}$$

$$= -I\nu\sin\eta \pm \sqrt{1 - \nu^2\sin^2\eta}$$

$$= Ie^{\pm I(\frac{\pi}{2}+\alpha)} \text{ with } \sin\alpha = \nu\sin\eta$$

$$|g|^2 = 1$$

and we observe that the first order upwind-forward Euler method is dissipative whereas the leapfrog method is not. It thus seems that the leapfrog method is in all ways (truncation error, dissipative properties) superior to the first order upwind-forward Euler method. Let us check this conclusion by looking at numerical examples. We first consider the advection of a wave packet of period 0.5 ($k = 4\pi$) on a mesh of size $\Delta x = 1/40$ (hence $\eta = k\Delta x = \pi/10$), using a time step such that the CFL number $\nu = 0.8$. The numerical results obtained by both methods shown below confirm the conclusions of the analysis: the dissipative properties of the first order upwind-forward Euler method result in a serious reduction of the wave amplitude, whereas in contrast, the leapfrog solution almost perfectly agrees with the exact solution. We can however observe some trailing oscillations in the leapfrog solution.

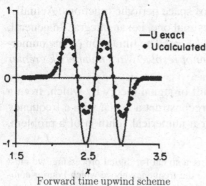

Forward time upwind scheme

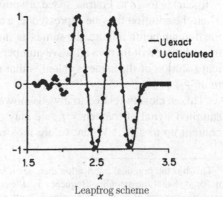

Leapfrog scheme

Let us now consider a second example, namely the advection of a front. The mesh size and the time step are the same as in the previous calculation.

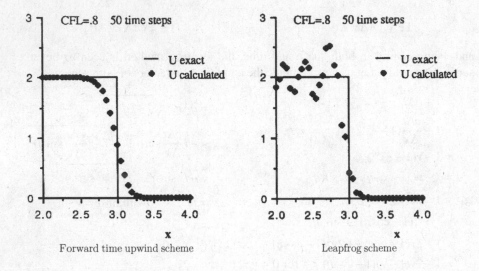

Forward time upwind scheme   Leapfrog scheme

In this case, the leapfrog solution exhibits large spurious oscillations whereas the first order upwind-forward Euler method solution is smooth and monotone. For this particular example, the first order upwind-forward Euler method is thus clearly preferable. It remains to analyse why this is the case, and whether it is possible to construct numerical methods that are simultaneously accurate for smooth problems and provide smooth solutions for rough problems such as the advection of a front.

The Need for Dissipation

The inherent limitation of numerical methods. Let us consider a linear problem governed by a PDE on a 1-D infinite domain. A way of solving the problem (since it is linear) is to go to Fourier space, assuming that conditions to do this are fulfilled. Then the solution may be expressed as a sum of space periodic functions. Actually, the domain being infinite, the sum is an infinitesimal one, i.e. an integral. In general, the solution will involve all wave numbers $0 \le k \le \infty$. The limitation of any numerical solution of the same problem is that *it cannot resolve wave numbers k greater than $\frac{\pi}{\Delta x}$*.

This is closely related to a well-known result of signal theory by which, from a sampled signal of frequency f_s, we may only reconstruct a signal with a frequency content up to $f_s/2$.[4] When we are looking for a numerical solution of a problem,

---

[4] This has the practical implication that, when sampling a signal for digital processing, we must make sure that the sampling frequency is at least twice the frequency above which Fourier components become of negligible amplitude (cutting frequency). For example, for audio signals the

we are actually trying to calculate space samples of the exact solution. The sampling length is Δx. Therefore, the sampling wavenumber is $\frac{2\pi}{\Delta x}$ and we can only resolve wavenumbers up to $\frac{1}{2}\frac{2\pi}{\Delta x} = \frac{\pi}{\Delta x}$. The wave with wavenumber $\frac{\pi}{\Delta x}$ is generally called the extreme numerical mode and exhibits an oscillating pattern of the form $+1, -1, +1 \ldots$ at mesh points.

Instances in which dissipation is needed: inherent inadequacy of the mesh. Now, as in signal sampling, the mesh is generally chosen such that the wavenumber content of the solution does not extend up to the extreme numerical mode, i.e. the wavenumbers of interest are such that $k\Delta x$ is small compared to 1. But there are instances in which high wavenumbers are inherently present. One of those instances is when the solution exhibits discontinuities like fronts, shock waves or contact discontinuities. Starting from an initially smooth solution, discontinuities are produced by the non-linear interaction of waves. Numerically, the higher wavenumbers generated by non-linear wave interactions eventually reach the resolution limit of the mesh. Then, two possibilities exist:

1. the high wavenumbers bounce back into low wavenumbers and alter the accuracy of the numerical solution.
2. they pile up at the high frequency end.

In addition, both effects may possibly result in numerical instability. Let us illustrate this problem by an example. Suppose we want to compute the solution of the non-linear advection equation $\frac{\partial u}{\partial t} + u\frac{\partial u}{\partial x} = 0$ or $\frac{\partial u}{\partial t} + \frac{\partial}{\partial x}\frac{u^2}{2} = 0$ in conservation form, using the explicit mid-point method and central differencing (leapfrog method). As shown in the previous section, this method is non-dissipative. Applying the leapfrog method to non-linear advection of a discontinuity (shock wave), the result shown in Fig. 9.9 is obtained after 80 time steps (the parameters of the calculation were $\Delta x = .025$, $\Delta t = .0125$), where the piling up of waves at the high frequency end of the spectrum is clearly visible. Obviously, such a result is unsatisfactory.

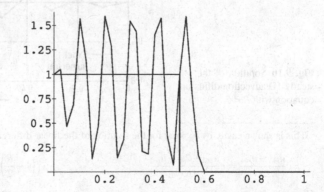

Fig. 9.9 Numerical solution of non-linear advection of a discontinuity using the leapfrog method. Solution at $t = 1$

---

cutting frequency is 20 kHz since we can't hear above that frequency. This explains why CD systems generally operate with a sampling frequency of 44 kHz.

There is another important source of high wavenumbers, which is illustrated by the following steady 1D advection-diffusion problem.

$$c\frac{du}{dx} = \epsilon\frac{d^2x}{dx^2} \qquad\qquad u(0) = 0; u(1) = 1 \qquad\qquad (9.29)$$

The analytical solution of this equation is

$$u = \frac{e^{\frac{c}{\epsilon}x} - 1}{e^{\frac{c}{\epsilon}} - 1} \qquad\qquad (9.30)$$

When $\epsilon \to 0$, this solution is of boundary-layer type where most of the variation of the solution takes place in a narrow region of size proportional to $\frac{\epsilon}{c}$ [with $\frac{\epsilon}{c} \ll 1$, the problem is said of singular perturbation nature]. Consequently, as ϵ tends to zero [which is analogous to the Reynolds number Re tending to infinity in fluid mechanics], the wavenumber content of the solution shifts towards the higher end of the wavenumber spectrum. When this problem is solved numerically using central differences (as would be obtained in the limit $n \to \infty$ applying any time integration scheme to the central space discretization of the unsteady equation $\frac{\partial u}{\partial t} + c\frac{\partial u}{\partial x} = \epsilon\frac{\partial^2 u}{\partial x^2}$), undesirable oscillations are produced when $\frac{c\Delta x}{\epsilon} = R$ the so-called mesh Reynolds number is greater than 2^5 as seen in Fig. 9.10.

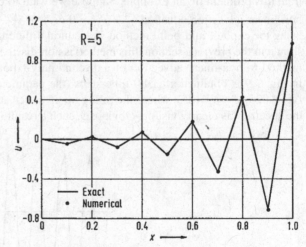

Fig. 9.10 Solution of the steady 1D advection diffusion equation with $R = 5$

---

[5] This is shown easily by looking for the solution of the linear difference equation

$$c\frac{u_{i+1} - u_{i-1}}{2\Delta x} = \epsilon\frac{u_{i+1} - 2u_i + u_{i-1}}{\Delta x^2} \qquad \to \qquad u_{i+1}\left(1 - \frac{R}{2}\right) - 2u_i + u_{i-1}\left(1 + \frac{R}{2}\right) = 0$$

General solutions of this equation are $u = g^n$. Inserting in the difference equation, one finds

$$g^2\left(1 - \frac{R}{2}\right) - 2g + \left(1 + \frac{R}{2}\right) = 0 \qquad \to \qquad g = 1 \quad \text{or} \quad g = \frac{1 + R/2}{1 - R/2}$$

so that g becomes negative when $R > 2$ and the solution is oscillatory.

Now, one may object that in the latter case, oscillations develop because the mesh is not appropriately chosen, that is, if the length scale $\frac{\epsilon}{c}$ of the boundary layer near $x = 1$ gets small, then the mesh must be refined in order to capture that boundary layer. For the present advection diffusion problem, the objection is valid since, after all, if we retain the term $\epsilon \frac{\partial^2 u}{\partial x^2}$ in the equation, that means that we are interested in the structure of the boundary layer at $x = 1$. But in fluid dynamics calculations, even though we may want to retain the viscous terms, we may not be interested in all viscous effects. Let us illustrate this point.

The direct analogy in fluid dynamics of the advection/diffusion model problem discussed above is the problem of the structure of a shock wave. It is well known from Gas Dynamics that the shock wave relative thickness $\frac{t}{L} = 0\left(\frac{1}{Re}\right) = 0\left(\frac{\epsilon}{u_\infty L}\right)$ or $t = 0\left(\frac{\epsilon}{u_\infty}\right)$ which is an expression very similar to that found for the model problem. In order to appropriately capture the shock wave structure, one would therefore need to use a Δx that would be a fraction of $\frac{\epsilon}{u_\infty}$. Practically this is impossible since $\frac{\epsilon}{u_\infty}$ is of the order of a few mean free paths.[6] In addition, in most instances, we are not interested in the detailed shock wave structure.

On the contrary, the development of boundary layers along bodies in 2D or 3D flows yields viscous layers of relative thickness $\frac{t}{L} = 0\left(\frac{1}{\sqrt{Re}}\right)$ [for laminar flows]. Those boundary layers are generally of interest and need to be captured by the mesh. In high Reynolds number viscous flows, we have a coexistence of viscous phenomena, some of which are of interest (wall boundary layers, shear layers: transverse diffusion phenomena) and some of which are of no interest (shock wave structure: streamwise diffusion phenomena) in most instances. If an algorithm without dissipation is used, since the mesh will be inappropriate to resolve streamwise diffusion phenomena, this will result in unacceptable oscillations but on the other hand, if a dissipative scheme is used, the artificial dissipation may, especially for high Reynolds number flows, completely overshadow the physical dissipation, resulting in unrealistic results. This suggests that for high Reynolds number viscous flows much care will be needed to produce a suitable dissipation and that the suitable dissipation will necessarily be anisotropic, i.e. stronger in the streamwise direction than in the transverse direction where it should ideally vanish.

9.4.2.2 Dissipation for Steady State Problems — Control of Steady Oscillations

Dissipation was defined above in reference to unsteady problems and it was shown that the essential effect of a lack of dissipation was the generation of spurious oscillations near discontinuities or quasi-discontinuities. It was also shown that such oscillations may also appear in steady state solutions. Now, the unsteady definition of dissipation is clearly inadequate to study the question of steady state oscillations.

---

[6] Since the mean free path is of order M/Re_u where Re_u is the unit Reynolds number, it is clear that for high Reynolds numbers, the mean free path is very small.

Indeed, when space and time differencing are performed independently (as in the class of implicit time -stepping methods discussed in this text — see Sect. 9.4.1), the steady state solution is independent of the time-stepping scheme and thus of its dissipative properties. One has therefore to develop some alternative concept to characterize the (non)oscillatory properties of space discretization formulas.

For this purpose, let us consider the semi-discretization of the linear advection equation — actually, the theory can easily be extended to non-linear conservation laws and to several dimensions, see e.g [25]. A general semi-discretization is

$$\frac{du_i}{dt} = \sum_k c_{ik} u_k \tag{9.31}$$

where the only non-zero coefficients c_{ik} correspond to the points of the computational stencil. For example, for central space discretization, one has

$$c_{ii+1} = -\frac{c}{2\Delta x} \qquad\qquad c_{ii-1} = \frac{c}{2\Delta x} \tag{9.32}$$

For consistency, one must have $\sum_k c_{ik} = 0$. Therefore, the semi-discretization (9.31) can be rewritten

$$\frac{du_i}{dt} = \sum_k c_{ik} u_k - \left(\sum_k c_{ik}\right) u_i = \sum_k c_{ik}(u_k - u_i)$$

Now, scalar conservation laws (and the linear advection equation in particular) have the following properties [29]:

- No new local extremum can appear in the solution as time increases;
- The value of a local maximum cannot increase and the value of a local minimum cannot decrease.

To avoid the appearance of spurious oscillations, a numerical approximation should have both properties as well. Now, if u_i is a local maximum, then all $(u_k - u_i)$ are negative and consequently du_i/dt will be negative provided all coefficients $c_{ik}(k \neq i)$ are *positive*. The same condition ensures that $du_i/dt > 0$ if u_i is a local minimum. By the condition $\sum_k c_{ik} = 0$, it results that $c_{ii} = -\sum_{k \neq i} c_{ik} < 0$. Consequently, at steady state,

$$u_i = -\frac{\sum_{k \neq i} c_{ik} u_k}{c_{ii}}$$

i.e. u_i is a convex average of the nodal values at the other points of the stencil. Hence, no local maximum can exist, which guarantees the absence of spurious oscillations. The use of the positivity constraint was proposed by Spekreijse [39] as a natural extension to several dimensions of the TVD concept by Harten [19]. A similar approach has been followed by Jameson [25] to develop the concept of local extremum diminishing schemes. The central space discretization (9.32) is not positive, which explains its oscillatory behaviour for the advection-diffusion problem.

In contrast, the (first-order) backward discretization

$$\frac{du_i}{dt} + c\frac{u_i - u_{i-1}}{\Delta x} = 0$$

is positive since $c_{i\,i-1} = c/\Delta x$ (assuming $c > 0$). It should be mentioned at this point that the use of a backward discretization has a sound justification in this case. Indeed, as was mentioned earlier, the advection equation represents the transport of quantity u with speed c, i.e. from left to right. Mathematically, this corresponds to the fact that the characteristics of the advection equation have a positive slope in the $x - t$ plane. It is thus quite logical to bias the discretization in the upwind direction (i.e. backward for $c > 0$) to account for the propagation of information in a preferential direction.

The control of oscillations can also be analyzed from a different viewpoint. The idea here is to append to the baseline central difference discretization some artificial diffusion terms. For a scalar conservation law,

$$\frac{\partial u}{\partial t} + \frac{\partial f(u)}{\partial x} = 0$$

[note that the linear advection equation corresponds to $f = cu$], a conservative (finite-volume like) semi-discretization is

$$\frac{du_i}{dt} + \frac{h_{i+1/2} - h_{i-1/2}}{\Delta x} = 0 \tag{9.33}$$

where $h_{i+1/2}$ is a so-called numerical flux function. Taking

$$h_{i+1/2} = \frac{f_i + f_{i+1}}{2}$$

results in the classical central difference discretization. The introduction of artificial diffusion results in modifying the flux formula as follows:

$$h_{i+1/2} = \frac{f_i + f_{i+1}}{2} - \underbrace{d_{i+1/2}(u_{i+1} - u_i)}_{\text{artificial diffusion term}} \tag{9.34}$$

That the additional term is diffusive is easily realized by observing that $d_{i+1/2}$ $(u_{i+1} - u_i) \approx d_{i+1/2}\Delta x \partial u/\partial x$, $d_{i+1/2}$ being the artificial viscosity coefficient. One can then wonder what is the minimum diffusion needed to produce a positive discretization. For the linear advection equation, it can easily be shown that $(d_{i+1/2})_{\min} = |c|/2$, so that the resulting flux formula becomes (for $c > 0$)

$$h_{i+1/2} = c\frac{u_i + u_{i+1}}{2} - \frac{c}{2}(u_{i+1} - u_i) = cu_i \tag{9.35}$$

i.e. nothing else than the (first-order) upwind discretization, which leads Jameson to state that [25] "in this sense, upwinding is a natural approach to the construction of non-oscillatory schemes".

Fig. 9.11 Advection of a
front, first order upwind
scheme, $\nu = .8$, solution after
50 time steps (from [21])

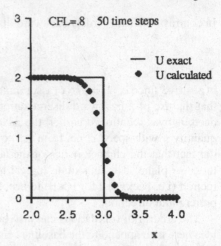

Now, the big drawback of the (first order) upwind scheme is precisely that it
is only first order accurate. As a consequence, there results an important smearing of discontinuities and a mediocre accuracy in smooth regions. The former
is illustrated in Fig. 9.11 which displays the numerical solution of a front advection problem, obtained using the first order upwind scheme and the explicit
1-step (forward Euler) time integration scheme. One can wonder whether higher
order accuracy could be obtained by enlarging the stencil — i.e. by considering
higher order upwind discretizations or higher order artificial diffusion terms. It turns
out that for linear schemes,[7] a theorem by Godunov states that

(a) a positive scheme can only be first order accurate,
(b) among first order schemes, the first order upwind scheme is the least diffusive.

It is however possible to evade Godunov's theorem by considering non-linear
schemes. The basic idea of such schemes is to append to the first order upwind
scheme anti-diffusive terms controlled by some non-linear limiter function in order
to ensure positivity. It was first proposed by Boris and Book [5] and in a different
context by van Leer [44]. The discussion of such schemes falls beyond the scope
of the present lecture. For more information, the reader is referred to [22] and the
recent tutorial presentations by Deconinck [11] and Jameson [25].

Before closing this section, a word should be said about the Lax-Wendroff family
of schemes which was discussed in detail in Part I, Sects. 7.2 and 7.3. The first
thing to observe is that in the Lax-Wendroff scheme and its variants, space and time
differencing *are not* performed independently. It results that the steady state solution
depends on the time step used in the time-stepping process. This is clearly shown
by considering the linear advection equation. For this equation, the Lax-Wendroff
scheme reads

---

[7] A scheme is linear if, for a linear equation such as the linear advection equation, the coefficients
c_{ik} in the discretization (9.31) do not depend on the numerical solution.

$$u_i^{n+1} = u_i^n - \frac{c\Delta t}{2\Delta x}(u_{i+1}^n - u_{i-1}^n) + \frac{c^2\Delta t^2}{2\Delta x^2}(u_{i+1}^n - 2u_i^n + u_{i-1}^n)$$

so that, at steady state, the solution satisfies

$$0 = -\frac{c}{2\Delta x}(u_{i+1} - u_{i-1}) + \frac{c^2\Delta t}{2\Delta x^2}(u_{i+1} - 2u_i + u_{i-1})$$

which corresponds to the numerical flux

$$h_{i+1/2} = c\frac{u_i + u_{i+1}}{2} - \frac{c^2\Delta t}{2\Delta x}(u_{i+1} - u_{i-1}) = c\frac{u_i + u_{i+1}}{2} - \frac{cv}{2}(u_{i+1} - u_{i-1})$$

It results that the Lax-Wendroff scheme involves some artificial diffusion with $d_{i+1/2} = \frac{cv}{2}$ (where v is the CFL number as usual). This explains why the numerical solution of the front advection problem considered above obtained with the Lax-Wendroff scheme is much less oscillatory than that obtained with the non-dissipative leapfrog method (see Fig. 9.12). Now, the amount of diffusion is seen to be a function of the time step (CFL number) used in the time-stepping process. In addition, since the stability limit of the Lax-Wendroff method is $|v| \le 1$ as can easily be shown

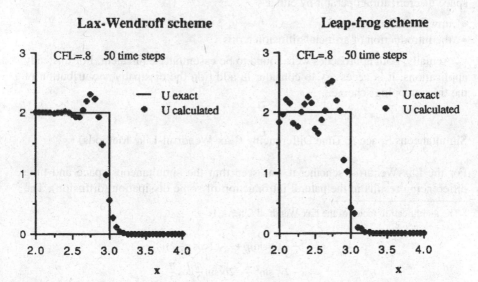

Fig. 9.12 Advection of a front, leapfrog and Lax-Wendroff schemes, $v = .8$, solution after 50 time steps (from [21])

by Fourier analysis,[8] it results that the Lax-Wendroff method is not positive except in the limiting case $v = 1$. It is however possible to construct positive variants of the Lax-Wendroff scheme by the introduction of non-linear limiting functions just as for upwind or artificial viscosity methods (see e.g. [35, 41]).

9.4.2.3 Dissipation for Scalar Problems—Summary

Up to now, the discussion has dealt only with scalar problems, with emphasis on the linear advection equation. Before examining the extension to systems, let us briefly summarize the results obtained so far:

Independent Space and Time Differencing (Method of Lines)

The use of independent space and time differencing is particularly suited for *steady problems*, insofar as the steady state solution is then entirely controlled by the space differencing operator, independently of the time-stepping scheme. To prevent (or limit) oscillations in the steady solution, some diffusion must be introduced in the space discretization operator by either

- upwind differencing, or
- the introduction of artificial diffusion terms.

Actually, both approaches were found to be essentially equivalent. For unsteady applications, it is necessary to consider in addition the dissipative contribution of the time-stepping scheme.

Simultaneous Space & Time Differencing (Lax-Wendroff-Like Methods)

For the Lax-Wendroff scheme, it was seen that the simultaneous space and time differencing results in the natural introduction of some dissipation (diffusion). The

---

[8] The amplification factor of the Lax-Wendroff scheme is

$$g = 1 - \frac{v}{2}(2i\sin\eta) + \frac{v^2}{2}(2\cos\eta - 2)$$
$$= 1 - 2v^2\sin^2\frac{\eta}{2} - 2iv\sin\frac{\eta}{2}\cos\frac{\eta}{2}$$

Therefore,

$$|g|^2 = 1 - 4v^2\sin^2\frac{\eta}{2} + 4v^4\sin^4\frac{\eta}{2} + 4v^2\sin^2\frac{\eta}{2}\cos^2\frac{\eta}{2}$$
$$= 1 - 4v^2\sin^4\frac{\eta}{2} + 4v^4\sin^4\frac{\eta}{2} = 1 - 4v^2(1 - v^2)\sin^4\frac{\eta}{2}$$

The stability condition is thus

$$1 - v^2 \geq 0 \qquad \rightarrow \qquad |v| \leq 1$$

amount of dissipation depends on the time step though, so that steady solutions depend on the time step used in the time-stepping process. It is for unsteady problems that schemes of the Lax-Wendroff family (including the finite element Taylor-Galerkin schemes [14]) prove particularly useful, especially when some oscillation control mechanism is included.

As mentioned previously, since our primary interest concerns steady problems, only schemes of the first family are considered in this lecture.

9.4.2.4 Extension to Systems of Conservation Laws

When systems of conservation laws like the Euler equations are considered, the extension of upwind schemes poses a problem, in the sense that wave speeds of both signs can be simultaneously present. Indeed, the characteristic speeds associated with the unsteady 1D Euler equations

$$\frac{\partial \mathbf{U}}{\partial t} + \frac{\partial \mathbf{F}}{\partial x} = \frac{\partial \mathbf{U}}{\partial t} + \mathbf{A}\frac{\partial \mathbf{U}}{\partial x} = 0 \qquad (9.36)$$

are u, $u + a$ and $u - a$, so that speeds of both signs exist when the flow is subsonic. It is then impossible to use a biased discretization of the whole flux vector $\mathbf{F}$ since this would lead to a downwind discretization for one of the waves. If one considers the quasi-linear form of the equations, then one can decompose the original system in characteristic equations and upwind each equation according to the corresponding wave speed sign (Courant-Isaacson-Rees scheme [8]) but this approach does not satisfy the conservation property which is crucial for the correct treatment of discontinuities (Part I, Sect. 2.9). This is the main reason why schemes based on a central space discretization such as the Lax-Wendroff scheme and schemes involving artificial diffusion have been so popular in the sixties and seventies. Indeed, these schemes are indifferent to wave speed sign and therefore extend readily to systems:

$$\text{artificial diffusion} \qquad h_{i+1/2} = \frac{\mathbf{F}_i + \mathbf{F}_{i+1}}{2} - d(\mathbf{U}_{i+1} - \mathbf{U}_i) \qquad (9.37)$$

$$\text{Lax-Wendroff} \qquad h_{i+1/2} = \frac{\mathbf{F}_i + \mathbf{F}_{i+1}}{2} - \frac{\mathbf{A}_{i+1/2}\Delta t}{2\Delta x}(\mathbf{F}_{i+1} - \mathbf{F}_i) \qquad (9.38)$$

The early eighties have seen the development of conservative upwind schemes, which have since become extremely popular, because of their crisp resolution of discontinuities and their superior ability in following moving shock waves. The remainder of this section will therefore be devoted to a brief presentation of the two major families of conservative upwind schemes.

Flux Difference Splitting (FDS) Schemes — Approximate Riemann Solvers

The starting point of Flux Difference Splitting schemes is the scheme developed in the late fifties by the Russian mathematician Godunov [17] for the unsteady 1D

Euler equations. This scheme is based on the integral form of the equations.[9] The integral form of the unsteady 1D Euler equations (9.36) is

$$\frac{d}{dt} \int_a^b \mathbf{U} dx + \mathbf{F}(\mathbf{U}_b) - \mathbf{F}(\mathbf{U}_a) = 0 \qquad (9.39)$$

For the numerical solution of the problem, the domain of interest is divided up into intervals (cells in the finite volume terminology) and the unknowns of the numerical solution $\mathbf{U}_i$ are the *average* flow quantities over the corresponding interval (see Fig. 9.13) rather than point values as in the finite difference method. The boundaries of interval i are noted $i \pm 1/2$. As illustrated in the figure, the intervals need not be of constant length ($h_{i-1} \neq h_i \neq h_{i+1}$). The first step in Godunov's method consists in *reconstructing* a piecewise continuous distribution of the flow variables from the cell averages. The simplest choice is a piecewise constant reconstruction as illustrated in the figure.[10] At the interval interfaces, the flow variable distributions are thus discontinuous. Now, there exists an *exact* solution of the 1D Euler equations for initial data consisting of two constant states separated by a discontinuity—this problem is known in the literature as the Riemann problem, and applies in particular to the flow in a shock tube. The solution consists of elementary waves (shock wave, contact discontinuity, expansion wave) originating from the interface, as illustrated in Fig. 9.14 for the shock tube problem. An interesting property of the solution is that flow variables are constant along straight lines in $x-t$ space (which implies that the solution is self-similar). In particular, it is constant in time at the location of the interface. As long as the two solutions at each interface of an interval do not interact (which imposes an upper bound on the time step), it is thus possible to compute the *exact* solution at the new time level from the initial piecewise constant data. This constitutes the second step in Godunov's method, called the *evolution* step. From

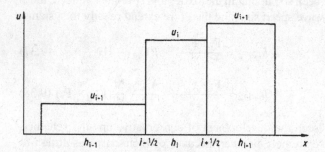

Fig. 9.13 Finite volume representation

---

[9] The integral form of the equations is the basis of the finite volume method — see Chap. 11. The introduction of the finite volume method as such is however posterior to the development of Godunov's scheme.

[10] Within this framework, higher accuracy is then achieved by using piecewise higher order polynomial reconstructions. This approach, called the variable extrapolation approach [22], is widely used with piecewise linear reconstructions and known in the literature as the MUSCL (Monotonic Upstream centred Scheme for Conservation Laws) approach.

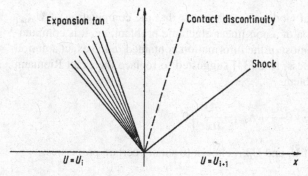

Fig. 9.14 Schematic representation of the solution of the Riemann problem

the exact solution at the new time level, it is then possible to compute the new cell averages in order to restart the process. This constitutes the third step of the method, called *projection* step.

Actually, it is possible to compute directly the cell averages at the new time level without computing the details of the solution. Indeed, integrating in time between t^n and $t^{n+1} = t^n + \Delta t$ the integral form of the equations applied to interval i, one finds

$$\int_{i-1/2}^{i+1/2} \mathbf{U}^{n+1} dx - \int_{i-1/2}^{i+1/2} \mathbf{U}^n dx + \int_{t^n}^{t^{n+1}} \mathbf{F}_{i+1/2} dt - \int_{t^n}^{t^{n+1}} \mathbf{F}_{i-1/2} dt = 0 \qquad (9.40)$$

This expression simplifies greatly since $\mathbf{U}^n = \mathbf{U}_i^n = \text{const}$ over the interval and $\mathbf{F}_{i\pm1/2}$ are constant over the time step. In addition, $\mathbf{U}_i^{n+1}$ being the average over the interval of the solution at the new time level, $\int_{i-1/2}^{i+1/2} \mathbf{U}^{n+1} dx = \mathbf{U}_i^{n+1} h_i$. The final result is thus

$$(\mathbf{U}_i^{n+1} - \mathbf{U}_i^n) h_i + (\mathbf{F}_{i+1/2} - \mathbf{F}_{i-1/2}) \Delta t = 0$$

or dividing through by $h_i \Delta t$,

$$\frac{\mathbf{U}_i^{n+1} - \mathbf{U}_i^n}{\Delta t} + \frac{\mathbf{F}_{i+1/2} - \mathbf{F}_{i-1/2}}{h_i} = 0 \qquad (9.41)$$

from which we deduce that Godunov's scheme is a conservative discretization of the 1D Euler equations with the numerical flux function

$$h_{i+1/2} = \mathbf{F}(\mathbf{U}_{\text{exact}}(x_{i+1/2}, t)) \qquad (9.42)$$

combined with forward Euler time stepping. That this is an upwind discretization clearly shows up by applying it to the linear advection equation. Since the exact solution of the linear advection equation is the initial solution moving with speed c, it results that (for $c > 0$)

$$h_{i+1/2} = c\, u_i \qquad \text{and} \qquad h_{i-1/2} = c\, u_{i-1}$$

and one recovers the first-order upwind discretization.

The essential drawback of Godunov's scheme is that the computation of $\mathbf{U}_{exact}$ $(x_{i+1/2}, t)$ requires the solution of a non-linear algebraic problem, i.e. it is computationally expensive. Now, as most of the information contained in the exact solution is lost by the averaging process, Roe [34] suggested to replace the exact Riemann problem by a linearized problem

$$\frac{\partial \mathbf{U}}{\partial t} + \tilde{\mathbf{A}}_{i+1/2} \frac{\partial \mathbf{U}}{\partial x} = 0$$

where $\tilde{\mathbf{A}}_{i+1/2}$ is a function of $\mathbf{U}_i$ and $\mathbf{U}_{i+1}$, chosen to satisfy certain properties:

1. $\tilde{\mathbf{A}}(\mathbf{U}, \mathbf{U}) = \mathbf{A}(\mathbf{U})$;
2. $\tilde{\mathbf{A}}_{i+1/2}$ has a complete set of real eigenvalues for any pair of $\mathbf{U}_i$, $\mathbf{U}_{i+1}$;
3. $\tilde{\mathbf{A}}_{i+1/2}(\mathbf{U}_{i+1} - \mathbf{U}_i) = \mathbf{F}_{i+1} - \mathbf{F}_i$.

The first condition is required for consistency, the second ensures that the linearized problem has a solution, and the third condition is a sufficient condition for the scheme to be conservative. It also has the nice additional property that the solution of the linearized problem is identical to the solution of the exact problem when a single wave is present.

Now, the solution of the linearized problem is found quite easily by the theory of characteristics. Multiplying the linearized equation by the matrix L of left eigenvectors of $\tilde{\mathbf{A}}_{i+1/2}$, one obtains

$$L\frac{\partial \mathbf{U}}{\partial t} + L\tilde{\mathbf{A}}_{i+1/2} \frac{\partial \mathbf{U}}{\partial x} = L\frac{\partial \mathbf{U}}{\partial t} + \Lambda L\frac{\partial \mathbf{U}}{\partial x} = 0$$

where Λ is the (diagonal) matrix of eigenvalues of $\tilde{\mathbf{A}}_{i+1/2}$. These are decoupled linear advection equations for the characteristic variables, components of the vector $L\mathbf{U}$. For the 1D Euler equations, there are three components. Noting

$$L\mathbf{U}_i = (\alpha_1, \alpha_2, \alpha_3)^t; \qquad L\mathbf{U}_{i+1} = (\beta_1, \beta_2, \beta_3)^t$$

and arranging the eigenvalues in increasing order, the solution of the linear problem is schematically shown in Fig. 9.15 (in terms of characteristic variables) and for the case of the figure ($\lambda_1 < 0$, λ_2, $\lambda_3 > 0$),

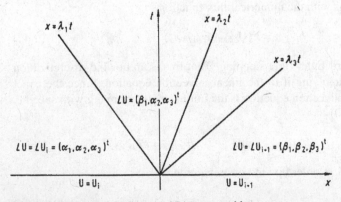

Fig. 9.15 Solution of the linearized Riemann problem

$$LU_{i+1/2} = (\beta_1, \alpha_2, \alpha_3)^t \qquad \rightarrow U_{i+1/2} = R(\beta_1, \alpha_2, \alpha_3)^t$$

where R is the matrix of right eigenvectors of $\tilde{A}_{i+1/2}$ (inverse of L). The corresponding flux is thus (the flux being linear)

$$
\begin{aligned}
\mathbf{F}_{i+1/2} &= \mathbf{F}_i + \tilde{A}_{i+1/2}(\mathbf{U}_{i+1/2} - \mathbf{U}_i) = \mathbf{F}_i + \tilde{A}_{i+1/2}R(\beta_1 - \alpha_1, 0, 0)^t \\
&= \mathbf{F}_i + R\Lambda(\beta_1 - \alpha_1, 0, 0)^t = \mathbf{F}_i + R\Lambda^- L(\mathbf{U}_{i+1} - \mathbf{U}_i) \\
&= \mathbf{F}_i + \tilde{A}^-_{i+1/2}(\mathbf{U}_{i+1} - \mathbf{U}_i)
\end{aligned}
$$

or

$$
\begin{aligned}
\mathbf{F}_{i+1/2} &= \mathbf{F}_{i+1} - \tilde{A}_{i+1/2}(\mathbf{U}_{i+1} - \mathbf{U}_{i+1/2}) = \mathbf{F}_{i+1} - \tilde{A}_{i+1/2}R(0, \beta_2 - \alpha_2, \beta_3 - \alpha_3)^t \\
&= \mathbf{F}_{i+1} - R\Lambda(0, \beta_2 - \alpha_2, \beta_3 - \alpha_3)^t = \mathbf{F}_{i+1} - R\Lambda^+ L(\mathbf{U}_{i+1} - \mathbf{U}_i) \\
&= \mathbf{F}_{i+1} - \tilde{A}^+_{i+1/2}(\mathbf{U}_{i+1} - \mathbf{U}_i)
\end{aligned}
$$

relations from which it appears that the flux difference $\mathbf{F}_{i+1} - \mathbf{F}_i$ has been split into a positive and a negative part to calculate $\mathbf{F}_{i+1/2}$, whence the name Flux Difference Splitting. By this splitting of the flux difference, the scheme automatically adapts the difference scheme to the local flow quantities. It is thus a solution-adaptive differencing scheme as alluded to in the introduction.

Averaging the two previous expressions, the following (third) form of Roe's scheme is obtained:

$$\mathbf{F}_{i+1/2} = \frac{\mathbf{F}_i + \mathbf{F}_{i+1}}{2} - \frac{1}{2}|\tilde{A}_{i+1/2}|(\mathbf{U}_{i+1} - \mathbf{U}_i) \tag{9.43}$$

where $|\tilde{A}_{i+1/2}| = \tilde{A}^+_{i+1/2} - \tilde{A}^-_{i+1/2}$. Now, this has exactly the same form as the artificial diffusion flux formula (9.37) except that the diffusion coefficient is replaced here by a diffusion matrix.

The Flux Difference Splitting approach pioneered by Roe has met with a considerable success. Several schemes of this type, also called Approximate Riemann solvers, were developed since the beginning of the eighties [13, 15, 33, 38], among which the most popular is certainly Osher's scheme [33].

Flux Vector Splitting (FVS) Schemes

The idea of flux vector splitting was introduced in computational fluid dynamics by Steger and Warming [40]. The idea had been previously proposed in astrophysics by Sanders and Prendergast [36] but was rediscovered independently by Steger and Warming. The starting point of Steger and Warming's scheme is the observation that the compressible inviscid fluxes are homogeneous functions of degree 1 in the conservative variables. Consequently, by a theorem due to Euler,

$$\mathbf{F}(\mathbf{U}) = \frac{\partial \mathbf{F}(\mathbf{U})}{\partial \mathbf{U}}\mathbf{U} = \mathbf{A}(\mathbf{U})\mathbf{U} \tag{9.44}$$

Now, the flux Jacobian matrix $\mathbf{A}$ is fully diagonalizable and it is possible to split it between its positive and negative parts (see previous paragraph)

$$\mathbf{A} = R\Lambda L = R(\Lambda^+ + \Lambda^-)L = \underbrace{R\Lambda^+ L}_{\mathbf{A}^+} + \underbrace{R\Lambda^- L}_{\mathbf{A}^-} \tag{9.45}$$

to which correspond the split fluxes

$$\mathbf{F}^+ = \mathbf{A}^+\mathbf{U} \qquad\qquad \mathbf{F}^- = \mathbf{A}^-\mathbf{U} \tag{9.46}$$

Now, the split fluxes $\mathbf{F}^\pm$ being associated with positive (respectively negative) eigenvalues only, it is possible to use upwind difference formulas to discretize the corresponding flux derivatives.

The Steger and Warming flux vector splitting suffers from a lack of continuity at those points where an eigenvalue of $\mathbf{A}$ vanishes (stagnation and sonic points). To remedy this problem, van Leer developed an alternative, continuous, flux vector splitting [45], which is no longer based on the homogeneity property of the inviscid flux vectors. The basic requirements are

- the split fluxes sum up to the whole flux: $\mathbf{F}^+ + \mathbf{F}^- = \mathbf{F}$;
- the split fluxes Jacobians have positive (respectively negative) eigenvalues only;
- $\mathbf{F}^- = 0$ for supersonic flow (respectively $\mathbf{F}^+ = 0$ for supersonic flow with negative velocity).

Van Leer imposed a few additional requirements in particular to ensure the crisp resolution of discontinuities.

The flux vector splitting approach and van Leer's scheme in particular have become extremely popular in the CFD community [37, 43], but it was soon realized that flux vector splitting schemes are excessively dissipative at contact discontinuities (boundary and shear layers) [46]. To avoid this while keeping the robustness of flux vector splitting schemes, an improved flux vector splitting scheme was recently developed by Liou and Steffen [30](AUSM scheme). Jameson's CUSP scheme [25], although formulated in the artificial diffusion formalism, appears essentially equivalent to this latter scheme. Finally, Coquel and Liou [6] have proposed a procedure to construct hybrid flux vector/flux difference splitting schemes to combine the robustness of the flux vector splitting schemes with respect to strong shock and expansion waves and the accuracy of flux difference splitting schemes with respect to contact discontinuities. They examine in particular the van Leer/Osher hybrid, which provides results of comparable accuracy as Osher's scheme for viscous flow calculations at a cost only slightly superior to van Leer's FVS scheme.

9.4.2.5 Concluding Remarks

For systems of conservation laws such as the Euler equations (and for advection dominated advection diffusion problems), it has been shown that some dissipation must be introduced in the explicit discretization operator to control (and even prevent) the appearance of spurious oscillations. This can be achieved in basically two ways, i.e. by the introduction of artificial diffusion or by the use of FDS/FVS upwind schemes. Actually, it was shown (see also [6, 46]) that the FDS/FVS upwind schemes are in fact equivalent to an artificial diffusion scheme with a diffusion *matrix*.

The only pending question at this stage is to select between the two approaches. Some general guidelines, which probably reflect the author's own bias and can thus be argued, are now presented.

- For transonic steady flows, for which the shock waves are not too strong, scalar artificial diffusion is probably the best approach. It is computationally inexpensive, simple to program and, according to Jameson [25], "combined with a high resolution switching scheme, [it] captures shock waves in about three interior points".
- On the contrary, for high Mach number flows associated with strong shock waves and for unsteady flows with moving shock waves, matrix diffusion, i.e. upwind FDS/FVS schemes are required. They are more costly but they provide the best possible resolution of shock waves and, more important for viscous applications, of contact discontinuities.

9.4.3 Choice of the Implicit Operator

The basic linearized implicit scheme (9.22) can be rewritten as

$$\left[\frac{I}{\Delta t} + \theta \frac{\partial R}{\partial U} \right] \delta U = -R(U^n)$$

In this section, we examine briefly how this can be effectively simplified. For simplicity, let us consider the 2D Euler equations discretized on a cartesian mesh. Then the most general expression for $R_{i,j}$ is

$$R_{i,j} = \frac{\mathbf{F}_{i+1/2,j} - \mathbf{F}_{i-1/2,j}}{\Delta x} + \frac{\mathbf{G}_{i,j+1/2} - \mathbf{G}_{i,j-1/2}}{\Delta y} \tag{9.47}$$

where $\mathbf{F}_{i\pm1/2,j}$ and $\mathbf{G}_{i,j\pm1/2}$ are artificial diffusion based or upwind FDS/FVS numerical flux functions. Now, as mentioned in Sect. 9.4.2.2, flux functions involving only the neighbouring points (i, j and $i + 1, j$ for the flux $\mathbf{F}_{i+1/2,j}$) are only first order accurate, and for higher accuracy, one needs to enlarge the computational stencil. As a result, the Jacobian matrix $\frac{\partial R}{\partial U}$ has more entries (and thus a larger bandwidth) than for a first order discretization. The first simplification generally made is to use the Jacobian matrix of the *first-order* flux formula even when a high order scheme is used in the explicit operator:

First simplification

$$\frac{I}{\Delta t} + \theta \frac{\partial R}{\partial U} \qquad \rightarrow \qquad \frac{I}{\Delta t} + \theta \frac{\partial R^{1st}}{\partial U}$$

This simplification reduces the linear system bandwidth and (because the first order flux formula is more dissipative) improves its diagonal dominance as will be shown shortly. The net result is that the solution of the linear system is greatly simplified, reducing the computational cost per iteration. On the other hand, this simplification has an (adverse) effect on the iterative scheme convergence rate. This effect can be studied by a linear fixed point Fourier or full matrix analysis. Several studies [2, 16, 26] of 1D inviscid problems have shown that as a result of the simplification, there appears an optimum time step, corresponding to a CFL number between 10 and 100, at which convergence is fastest, whereas in contrast convergence continuously improves as Δt increases if no simplification is made, corresponding to the fact that the linearized time-stepping scheme tends towards Newton's method in the limit $\Delta t \rightarrow \infty$. Nevertheless, for a time step chosen around the optimum (in a rather wide range), the decrease in CPU time/iteration outweighs the convergence deterioration, especially for 2D and 3D problems.

At this stage, let us write down the full expression of the simplified scheme for the 2D Euler equations. Since

$$R_{i,j} = \frac{\mathbf{F}_{i+1/2,j} - \mathbf{F}_{i-1/2,j}}{\Delta x} + \frac{\mathbf{G}_{i,j+1/2} - \mathbf{G}_{i,j-1/2}}{\Delta y}$$

the simplified implicit operator reads

$$\left(\frac{\partial R^{1st}}{\partial U} \delta U \right)_{i,j} = \underbrace{\frac{1}{\Delta x} \frac{\partial \mathbf{F}_{i+1/2,j}}{\partial \mathbf{U}_{i+1,j}} \delta \mathbf{U}_{i+1,j}}_{C_i} + \underbrace{\frac{-1}{\Delta x} \frac{\partial \mathbf{F}_{i-1/2,j}}{\partial \mathbf{U}_{i-1,j}} \delta \mathbf{U}_{i-1,j}}_{B_i}$$

$$+ \underbrace{\frac{1}{\Delta y} \frac{\partial \mathbf{G}_{i,j+1/2}}{\partial \mathbf{U}_{i,j+1}} \delta \mathbf{U}_{i,j+1}}_{C_j} + \underbrace{\frac{-1}{\Delta y} \frac{\partial \mathbf{G}_{i,j-1/2}}{\partial \mathbf{U}_{i,j-1}} \delta \mathbf{U}_{i,j-1}}_{B_j}$$

$$+ \underbrace{\left[\frac{1}{\Delta x} \left(\frac{\partial \mathbf{F}_{i+1/2,j}}{\partial \mathbf{U}_{i,j}} - \frac{\partial \mathbf{F}_{i-1/2,j}}{\partial \mathbf{U}_{i,j}} \right) + \frac{1}{\Delta y} \left(\frac{\partial \mathbf{G}_{i,j+1/2}}{\partial \mathbf{U}_{i,j}} - \frac{\partial \mathbf{G}_{i,j-1/2}}{\partial \mathbf{U}_{i,j}} \right) \right]}_{\mathcal{A}} \delta \mathbf{U}_{i,j} \quad (9.48)$$

where the flux formulas are to be taken at first order. Now, for some schemes, especially the flux difference schemes in general and Roe's FDS scheme in particular, the exact expression of the numerical flux Jacobians is prohibitively complex. The second simplification then consists of substituting the exact numerical flux Jacobian by some approximation thereof.

Second simplification Substitute the exact numerical flux Jacobians by some suitable approximation.

Now, this second simplification may also have adverse effects on the convergence of the iterative process and in particular certain approximations may be more effective than others, depending on the numerical flux formula used in the explicit operator. This problem has also been studied using fixed point analysis in [2, 16, 26], from which guidelines can be obtained for the choice of a suitable flux Jacobian approximation. The main results are summarized below:

For scalar artificial diffusion The first order numerical flux function is then

$$\mathbf{F}_{i+1/2,j} = \frac{\mathbf{F}_{i,j} + \mathbf{F}_{i+1,j}}{2} - d_{i+1/2,j}(\mathbf{U}_{i+1,j} - \mathbf{U}_{i,j})$$

Good approximations of the flux Jacobians are

$$\frac{\partial \mathbf{F}_{i+1/2,j}}{\partial \mathbf{U}_{i+1,j}} \approx \frac{\mathbf{A}_{i+1,j}}{2} - d_{i+1/2,j}I \approx \frac{\mathbf{A}_{i+1,j} - \gamma_{A_{i+1,j}}I}{2}$$

$$\frac{\partial \mathbf{F}_{i+1/2,j}}{\partial \mathbf{U}_{i,j}} \approx \frac{\mathbf{A}_{i,j}}{2} + d_{i+1/2,j}I \approx \frac{\mathbf{A}_{i,j} + \gamma_{A_{i,j}}I}{2} \tag{9.49}$$

where γ_A is a real larger than the spectral radius of A. This approximation was proposed by Yoon and Jameson [48]. It ensures that the linear system to be solved is (block) diagonally dominant. If the artificial diffusion contribution is omitted, which corresponds to pure central differencing, the diagonal dominance is lost for large time steps, which makes the linear system harder to solve. Indeed, for pure central differencing, the diagonal term vanishes altogether in the expression of the implicit operator (9.48).

For FVS schemes For FVS schemes, the first order numerical flux function is

$$\mathbf{F}_{i+1/2,j} = \mathbf{F}_{i,j}^+ + \mathbf{F}_{i+1,j}^-$$

In general, and for van Leer's FVS scheme in particular, the exact numerical flux Jacobian is relatively simple to evaluate and no simplification is needed.

For Roe's FDS scheme In this case, the numerical flux function is

$$\mathbf{F}_{i+1/2,j} = \frac{\mathbf{F}_{i,j} + \mathbf{F}_{i+1,j}}{2} - \frac{1}{2}|\tilde{\mathbf{A}}_{i+1/2,j}|(\mathbf{U}_{i+1,j} - \mathbf{U}_{i,j})$$

Good approximations of the flux Jacobians are

$$\frac{\partial \mathbf{F}_{i+1/2,j}}{\partial \mathbf{U}_{i+1,j}} \approx \frac{\mathbf{A}_{i+1,j}}{2} - \frac{1}{2}|\tilde{\mathbf{A}}_{i+1/2,j}| \approx \tilde{\mathbf{A}}_{i+1/2,j}^- \quad \text{or} \quad \mathbf{A}_{i+1,j}^-$$

$$\frac{\partial \mathbf{F}_{i+1/2,j}}{\partial \mathbf{U}_{i,j}} \approx \frac{\mathbf{A}_{i,j}}{2} + \frac{1}{2}|\tilde{\mathbf{A}}_{i+1/2,j}| \approx \tilde{\mathbf{A}}_{i+1/2,j}^+ \quad \text{or} \quad \mathbf{A}_{i,j}^+ \tag{9.50}$$

The approximation proposed by Yoon and Jameson (9.49) was also found to be effective when used in combination with Roe's scheme.

For the two latter cases, diagonal dominance of the linear system is a direct consequence of the use of first-order upwind differencing.

In summary, all simplifications made, the linear equation corresponding to point i, j is

$$\left(\underbrace{\frac{I}{\Delta t} + \theta \mathcal{A}}_{\tilde{\mathcal{A}}}\right) \delta \mathbf{U}_{i,j} + \theta B_i \delta \mathbf{U}_{i-1,j} + \theta C_i \delta \mathbf{U}_{i+1,j} + \theta B_j \delta \mathbf{U}_{i,j-1} + \theta C_j \delta \mathbf{U}_{i,j+1} = -R_{i,j} \quad (9.51)$$

Introducing the shift operators $\varepsilon_i^{\pm 1}$, $\varepsilon_j^{\pm 1}$ in the i and j indices respectively, this can be rewritten in compact form (for $\theta = 1$ or including the factor θ in the B and C matrices)

$$\left[\tilde{\mathcal{A}} + B_i \varepsilon_i^{-1} + C_i \varepsilon_i^{+1} + B_j \varepsilon_j^{-1} + C_j \varepsilon_j^{+1}\right] \delta \mathbf{U}_{i,j} = -R_{i,j} \quad (9.52)$$

9.4.4 Choice of the Linear System Solution Strategy

9.4.4.1 1D Problems

Let us first consider a 1D problem. In that case, the linear system to solve reduces to

$$\left[\tilde{\mathcal{A}} + B_i \varepsilon_i^{-1} + C_i \varepsilon_i^{+1}\right] \delta \mathbf{U}_i = -R_i$$

This would be the case for the analysis of a quasi one-dimensional flow in a nozzle (problem considered in Part I, Sects. 7.2 and 7.3) for example. Now, this system is a block-tridiagonal system which can be solved very efficiently by direct elimination (LU decomposition or Thomas algorithm, see Sect. 8.5.4). Therefore, the additional computational workload per time step required by the implicit method with respect to an explicit method is relatively small and the implicit method is computationally much more efficient. Actually, when no approximation is introduced in the implicit operator, the best (asymptotic) convergence and thus the best efficiency is obtained in the limit $\Delta t \to \infty$ in which the implicit time-stepping method in fact becomes Newton's method known for its quadratic convergence property.

9.4.4.2 2D Problems

For 2D problems, the linear system of four equations associated to point i, j (9.52) couples the following five flow variables updates: $\delta \mathbf{U}_{i,j}$, $\delta \mathbf{U}_{i+1,j}$, $\delta \mathbf{U}_{i-1,j}$, $\delta \mathbf{U}_{i,j+1}$ and $\delta \mathbf{U}_{i,j-1}$. This coupling can be schematically represented by the following cross. In order to determine the nature of the linear system to solve, the unknowns have

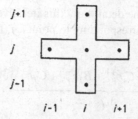

to be ordered. Let us assume they are ordered by rows (lexicographic ordering): $\delta \mathbf{U}_{11}, \ldots, \delta \mathbf{U}_{m1}, \delta \mathbf{U}_{12}, \ldots, \delta \mathbf{U}_{m2}, \ldots$, if there are m points per row. Then, the equation for point i, j is the $(i + (j-1)m)$th equation and couples the unknowns having the numbers $i + (j-2)m$, $i - 1 + (j-1)m$, $i + (j-1)m$, $i + 1 + (j-1)m$, $i + jm$. i.e. the system has the structure shown below. Actually, since there are four unknowns per point, the structure shown is the block structure of the linear system, each of the

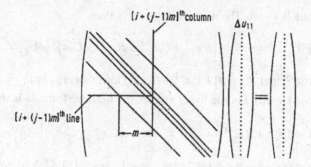

entries being a 4×4 block. One observes that the system is still sparse but no longer block-tridiagonal. Under these circumstances, direct elimination becomes prohibitively expensive both in terms of CPU time and storage, and this becomes even worse for 3D problems. Alternatively, the iterative techniques mentioned in Sect. 9.2.1.1 may be used. At this stage, two comments must be made:

- For upwind biased (or diffusive) implicit operators such as those mentioned in the previous section, the linear system (9.52) is block-diagonally dominant. This ensures the convergence of the classical iterative techniques and constitutes a side benefit of using upwind-biased schemes.
- Because of the various approximations made in the second stage (choice of the implicit operator), it is not generally worthwhile to solve the linear system (9.52) with great accuracy, so that an approximate but fast solver is generally preferable. This is why only a few, and in most cases only one, linear iterations are performed.[11]

---

[11] As mentioned in the previous section, when no approximation is introduced in the implicit operator, the best convergence is obtained for $\Delta t \to \infty$ (Newton method), in which case it becomes worthwhile to solve the linear system accurately. This is the approach followed in Newton iterative methods. One key issue in these methods is to find an economical way of using the unapproximated implicit operator. More information about these methods which fall out of the scope of this lecture can be found in [12].

Let us now describe in some detail how this iterative strategy is set up. First, let us recall the linear equation corresponding to point i, j (for simplicity, let us consider $\theta = 1$, fully implicit scheme).

$$\underbrace{\left[\tilde{\mathcal{A}} + B_i \varepsilon_i^{-1} + C_i \varepsilon_i^{+1} + B_j \varepsilon_j^{-1} + C_j \varepsilon_j^{+1}\right]}_{A_{i,j}} \delta \mathbf{U}_{i,j} = -R_{i,j}$$

where $\tilde{\mathcal{A}} = (I/\Delta t) + \mathcal{A}$. The general form of an iterative method for solving the corresponding linear system is (Sect. 9.2.1.1)

$$B\delta \mathbf{U}^{(k+1)} = -R + (B - A)\delta \mathbf{U}^{(k)}$$

The various iterative methods correspond to different choices for B. Let us now list a few popular methods for implicit time-stepping schemes.

Point Jacobi $B_{i,j} = \tilde{\mathcal{A}}$. The iterative formula is then

$$\tilde{\mathcal{A}}\delta \mathbf{U}_{i,j}^{(k+1)} = -R_{i,j} - \left[B_i \varepsilon_i^{-1} + C_i \varepsilon_i^{+1} + B_j \varepsilon_j^{-1} + C_j \varepsilon_j^{+1}\right]\delta \mathbf{U}_{i,j}^{(k)} \tag{9.53}$$

Point Gauss-Seidel $B_{i,j} = \tilde{\mathcal{A}} + B_i \varepsilon_i^{-1} + B_j \varepsilon_j^{-1}$ (lower sweep) or $B_{i,j} = \tilde{\mathcal{A}} + C_i \varepsilon_i^{+1} + C_j \varepsilon_j^{+1}$ (upper sweep). For the lower sweep, the iterative formula is thus

$$\left[\tilde{\mathcal{A}} + B_i \varepsilon_i^{-1} + B_j \varepsilon_j^{-1}\right]\delta \mathbf{U}_{i,j}^{(k+1)} = -R_{i,j} - \left[C_i \varepsilon_i^{+1} + C_j \varepsilon_j^{+1}\right]\delta \mathbf{U}_{i,j}^{(k)} \tag{9.54}$$

Symmetric point Gauss-Seidel $B_{i,j} = \left[\tilde{\mathcal{A}} + B_i \varepsilon_i^{-1} + B_j \varepsilon_j^{-1}\right]\tilde{\mathcal{A}}^{-1}\left[\tilde{\mathcal{A}} + C_i \varepsilon_i^{+1} + C_j \varepsilon_j^{+1}\right]$. This method is equivalent to a lower Gauss-Seidel sweep followed by an upper Gauss-Seidel sweep. This shows up nicely when considering the first iteration starting from the initial guess $\delta \mathbf{U}_{i,j}^{(0)} = 0$,

$$\left[\tilde{\mathcal{A}} + B_i \varepsilon_i^{-1} + B_j \varepsilon_j^{-1}\right]\underbrace{\tilde{\mathcal{A}}^{-1}\left[\tilde{\mathcal{A}} + C_i \varepsilon_i^{+1} + C_j \varepsilon_j^{+1}\right]\delta \mathbf{U}_{i,j}^{(1)}}_{\delta \mathbf{U}^*} = -R_{i,j} \tag{9.55}$$

which can be decomposed into two steps, i.e.

$$\begin{cases} \left[\tilde{\mathcal{A}} + B_i \varepsilon_i^{-1} + B_j \varepsilon_j^{-1}\right]\delta \mathbf{U}^* = -R_{i,j} \\ \tilde{\mathcal{A}}^{-1}\left[\tilde{\mathcal{A}} + C_i \varepsilon_i^{+1} + C_j \varepsilon_j^{+1}\right]\delta \mathbf{U}_{i,j}^{(1)} = \delta \mathbf{U}^* \end{cases}$$

Now, the first step is clearly a lower Gauss-Seidel sweep. As far as the second step is concerned, multiplying it through with $\tilde{\mathcal{A}}$, it comes

$$\left[\tilde{\mathcal{A}} + C_i \varepsilon_i^{+1} + C_j \varepsilon_j^{+1}\right]\delta \mathbf{U}_{i,j}^{(1)} = \tilde{\mathcal{A}}\delta \mathbf{U}^* = -R_{i,j} - \left[B_i \varepsilon_i^{-1} + B_j \varepsilon_j^{-1}\right]\delta \mathbf{U}^*$$

which is indeed an upper Gauss-Seidel sweep. In practise, the most efficient implementation is the following two step procedure

$$\begin{cases} \left[\tilde{\mathcal{A}} + B_i\varepsilon_i^{-1} + B_j\varepsilon_j^{-1}\right]\delta\mathbf{U}^* = -R_{i,j} \\ \left[\tilde{\mathcal{A}} + C_i\varepsilon_i^{+1} + C_j\varepsilon_j^{+1}\right]\delta\mathbf{U}_{i,j}^{(1)} = \tilde{\mathcal{A}}\delta\mathbf{U}^* \end{cases} \tag{9.56}$$

Notice that for this method, the iteration (preconditioning) matrix B appears as a product of terms which represent an *approximate factorization* of the system matrix A. The nature of this approximate factorization is schematically illustrated in

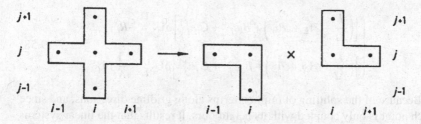

the above figure. The symmetric point Gauss-Seidel method is particularly effective when used in combination with Yoon & Jameson's implicit operator as outlined in their original paper [48].

Approximate LU factorization The approximate LU factorization (ALU) method is similar to the symmetric point Gauss-Seidel factorization in the sense that it involves a lower and an upper sweep, but the factorization involves a decomposition of $\tilde{\mathcal{A}}$. Let us first recall the expression of $\mathcal{A}$ (see definition in (9.48)).

$$\mathcal{A} = \left[\underbrace{\frac{1}{\Delta x}\frac{\partial\mathbf{F}_{i+1/2,j}}{\partial\mathbf{U}_{i,j}}}_{\mathcal{A}_E} + \underbrace{\frac{-1}{\Delta x}\frac{\partial\mathbf{F}_{i-1/2,j}}{\partial\mathbf{U}_{i,j}}}_{\mathcal{A}_W} + \underbrace{\frac{1}{\Delta y}\frac{\partial\mathbf{G}_{i,j+1/2}}{\partial\mathbf{U}_{i,j}}}_{\mathcal{A}_N} + \underbrace{\frac{-1}{\Delta y}\frac{\partial\mathbf{G}_{i,j-1/2}}{\partial\mathbf{U}_{i,j}}}_{\mathcal{A}_S} \right]$$

The ALU factorization method is defined by $B_{i,j} = [((I/\Delta t) + \mathcal{A}_E + \mathcal{A}_N) + B_i\varepsilon_i^{-1} + B_j\varepsilon_j^{-1}]\Delta t[((I/\Delta t) + \mathcal{A}_W + \mathcal{A}_S) + C_i\varepsilon_i^{+1} + C_j\varepsilon_j^{+1}]$, which, for FVS discretizations, corresponds to grouping all positive subflux (F^+, G^+) contributions in the lower sweep and all negative subflux contributions in the upper sweep. Just as the symmetric Gauss-Seidel method, it is implemented in two steps, which read for the first iteration

$$\begin{cases} \left[\left(\frac{I}{\Delta t} + \mathcal{A}_E + \mathcal{A}_N\right) + B_i\varepsilon_i^{-1} + B_j\varepsilon_j^{-1}\right]\delta\mathbf{U}^* = -R_{i,j} \\ \left[\left(\frac{I}{\Delta t} + \mathcal{A}_W + \mathcal{A}_S\right) + C_i\varepsilon_i^{+1} + C_j\varepsilon_j^{+1}\right]\delta\mathbf{U}_{i,j}^{(1)} = \frac{1}{\Delta t}\delta\mathbf{U}^* \end{cases} \tag{9.57}$$

and the nature of the factorization (schematic representation) is the same as for the symmetric Gauss-Seidel method.

Approximate directional factorization This method, which is closely related to the ADI method was introduced by Beam and Warming [3]. The approximate directional factorization (ADF) method corresponds to grouping all terms associated with the i-coordinate line in a first term and all terms associated with the j-coordinate line in a second term, i.e. $B_{i,j} = \left[((I/\Delta t) + \mathcal{A}_E + \mathcal{A}_W) + B_i \varepsilon_i^{-1} + C_i \varepsilon_i^{+1} \right] \Delta t \left[((I/\Delta t) + \mathcal{A}_N + \mathcal{A}_S) + B_j \varepsilon_j^{-1} + C_j \varepsilon_j^{+1} \right]$. It is implemented in two steps, which read for the first iteration

$$\begin{cases} \left[\left(\dfrac{I}{\Delta t} + \mathcal{A}_E + \mathcal{A}_W \right) + B_i \varepsilon_i^{-1} + C_i \varepsilon_i^{+1} \right] \delta \mathbf{U}^* = -R_{i,j} \\[2mm] \left[\left(\dfrac{I}{\Delta t} + \mathcal{A}_N + \mathcal{A}_S \right) + B_j \varepsilon_j^{-1} + C_j \varepsilon_j^{+1} \right] \delta \mathbf{U}_{i,j}^{(1)} = \dfrac{1}{\Delta t} \delta \mathbf{U}^* \end{cases} \tag{9.58}$$

Because of the splitting of implicit terms along gridline directions, and since each point is only coupled with its neighbours, it results that the linear systems corresponding to each step of the ADF method are block tridiagonal systems which can be solved very efficiently as mentioned previously. The schematic illustration of the Approximate Directional Factorization is shown in the following figure.

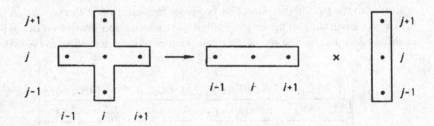

j-**line (vertical line) Jacobi** $B_{i,j} = \tilde{\mathcal{A}} + B_j \varepsilon_j^{-1} + C_j \varepsilon_j^{+1}$. The corresponding iterative formula is

$$\left[\tilde{\mathcal{A}} + B_j \varepsilon_j^{-1} + C_j \varepsilon_j^{+1} \right] \delta \mathbf{U}_{i,j}^{(k+1)} = -R_{i,j} - \left[B_i \varepsilon_i^{-1} + C_i \varepsilon_i^{+1} \right] \delta \mathbf{U}_{i,j}^{(k)} \tag{9.59}$$

Similar to the ADF method, the line Jacobi method requires the solution of block tridiagonal systems.

j-**line (vertical line) Gauss-Seidel** $B_{i,j} = \tilde{\mathcal{A}} + B_j \varepsilon_j^{-1} + C_j \varepsilon_j^{+1} + B_i \varepsilon_i^{-1}$ (lower sweep) or $B_{i,j} = \tilde{\mathcal{A}} + B_j \varepsilon_j^{-1} + C_j \varepsilon_j^{+1} + C_i \varepsilon_i^{+1}$ (upper sweep). The iterative formula for the lower sweep is then

$$\left[\tilde{\mathcal{A}} + B_j \varepsilon_j^{-1} + C_j \varepsilon_j^{+1} + B_i \varepsilon_i^{-1} \right] \delta \mathbf{U}_{i,j}^{(k+1)} = -R_{i,j} - C_i \varepsilon_i^{+1} \delta \mathbf{U}_{i,j}^{(k)} \tag{9.60}$$

Symmetric *j*-line (vertical line) Gauss-Seidel Just as for the point scheme, combining a lower and an upper line Gauss-Seidel sweeps yields the following algorithm: $B_{i,j} = \left[\tilde{\mathcal{A}} + B_j \varepsilon_j^{-1} + C_j \varepsilon_j^{+1} + B_i \varepsilon_i^{-1} \right] \left[\tilde{\mathcal{A}} + B_j \varepsilon_j^{-1} + C_j \varepsilon_j^{+1} \right]^{-1} \left[\tilde{\mathcal{A}} + B_j \varepsilon_j^{-1} \right.$

$+C_j\varepsilon_j^{+1}+C_i\varepsilon_i^{+1}\big]$, which is most efficiently implemented in two steps as follows (1 iteration)

$$\begin{cases} \big[\tilde{\mathcal{A}}+B_j\varepsilon_j^{-1}+C_j\varepsilon_j^{+1}+B_i\varepsilon_i^{-1}\big]\delta\mathbf{U}^* = -R_{i,j} \\ \big[\tilde{\mathcal{A}}+B_j\varepsilon_j^{-1}+C_j\varepsilon_j^{+1}+C_i\varepsilon_i^{+1}\big]\delta\mathbf{U}_{i,j}^{(1)} = -R_{i,j}-B_i\varepsilon_i^{-1}\delta\mathbf{U}^* \end{cases} \tag{9.61}$$

In practise, in most cases, only one iteration of the linear iterative method is performed. This is in fact equivalent to solving exactly a simplified linear system as was presented in Sect. 9.4.1. Notice that the simplification is in many cases (symmetric point and line Gauss-Seidel, ALU and ADF methods) an approximate factorization of the original matrix. Because of the simplification of the linear system (factorization error), there generally appears an optimum time step at which convergence is fastest (even when the implicit operator suffers no simplification as for a central space discretization [3] or for first-order FVS computations), as shown in particular by the analysis of Thomas et al. [42]. When the implicit operator is simplified as for second order computations using upwind schemes, the effect of this simplification is generally dominant and the optimum time step is little affected by the factorization error.

A rather extensive investigation of the efficiencies of the various implicit schemes and solution strategies for upwind biased discretizations of the 2D Euler equations was performed by P. Corbett [7]. The ADF, ALU and line Gauss-Seidel schemes were considered, as well as the point symmetric Gauss-Seidel scheme together with the Yoon and Jameson implicit operator. In all cases, only one iteration of the linear iterative method was performed. The most effective schemes were found to be the ADF and line Gauss-Seidel schemes. When Roe's scheme is used in the explicit operator, Yoon & Jameson's scheme [48] (Yoon & Jameson's implicit operator + point symmetric Gauss-Seidel iterative method) constitutes a competitive alternative, especially for second order computations.

9.5 Conclusions

For complex fluid dynamics problems such as those governed by the compressible Euler and Navier-Stokes equations, the classical methods used for simpler problems were shown to fail because of the mixed/hybrid type of the steady equations. One strategy to circumvent this difficulty is to use time-dependent techniques, thanks to the fact that the time-dependent equations provide a well-posed initial value problem for all flow situations.

Now, the systems of ordinary differential equations resulting from the space discretization of the time-dependent equations are often very stiff, in particular for viscous flow problems, and for those types of problems, it was shown that it is generally more efficient to use implicit time-stepping schemes.

To construct efficient implicit time-stepping schemes, it is essential to perform a linearization. The resulting linear implicit time-stepping scheme then appears to be made of three buiding blocks, i.e. an explicit discretization scheme which controls the accuracy of the final steady solution, an implicit operator defining the linear system to be solved and a strategy to solve the linear system, the latter two controlling the scheme convergence and efficiency.

As far as the explicit discretization scheme is concerned, it was shown that for advection dominated problems such as those governed by the Euler and Navier-Stokes equations, which involve regions with large gradients like shocks and shear layers, it is necessary to introduce some amount of dissipation in order to control (prevent) the appearance of spurious oscillations. For scalar problems, it was shown that this could be achieved equivalently by using upwind biased discretization schemes or by appending artificial diffusion terms to a baseline central difference scheme. For systems of equations like the Euler equations, the extension of upwind biased discretization schemes was made possible by the development of conservative upwind schemes. These schemes, which are more complex and thus more costly than artificial diffusion based schemes, allow a better capture of discontinuities and a better resolution of moving discontinuities.

For implicit operators, first-order schemes are generally preferred, first because they keep the linear system narrow-banded, which reduces the solution cost, and also because they ensure its diagonal dominance. As a result, many efficient iterative solution strategies are available, including classical relaxation (Jacobi,Gauss-Seidel and their symmetric and line versions) and factorization schemes. In some cases, it is even possible to let the time step tend to infinity and the time-stepping scheme in fact becomes a Newton or quasi-Newton iterative scheme for solving directly the steady equations. That time-stepping schemes which were introduced in the first place to circumvent the difficulty of solving directly the steady state problem eventually led to an iterative process for solving this problem constitutes a rather amusing conclusion.

References

1. D. A. Anderson, J. C. Tannehill, and R. H. Pletcher. *Computational fluid mechanics and heat transfer*. Hemisphere Publishing Company, Washington, 1984.
2. T. J. Barth. Analysis of implicit local linearisation techniques for upwind and TVD algorithms. AIAA Paper 87-0595.
3. R. M. Beam and R. F. Warming. An implicit factored scheme for the compressible Navier-Stokes equations. *AIAA Journal*, 16(4):393–402, 1978.
4. R. M. Beam and R. F. Warming. Implicit numerical methods for compressible Navier-Stokes and Euler equations. VKI LS 1982–04, 1982.
5. J. P. Boris and D. L. Book. Flux-corrected transport, I. SHASTA, a fluid transport algorithm that works. *Journal of computational physics*, 43:357–352, 1973.
6. F. Coquel and M.-S. Liou. Field by field hybrid upwind splitting methods. AIAA Paper 93-3302 CP.
7. P. Corbett. A comparison of various flux splitting operators and implicit solution techniques for the Euler equations. Project Report 1992–31, VKI, 1992.

8. R. Courant, E. Isaacson, and M. Rees. On the solution of nonlinear hyperbolic differential equations by finite differences. *Communications in pure and applied mathematics*, 5: 243–255, 1952.
9. G. Dahlquist. The theory of linear multistep methods and related mathematical topics. Lecture notes (microfilm), Department of Numerical Analysis, Royal Institute of Technology, Stockholm, 1976.
10. G. Dahlquist and Å. Björk. *Numerical methods*. Prentice Hall, Englewood Cliffs, 1974.
11. H. Deconinck. Introduction to artificial dissipation and shock capturing high resolution upwind schemes for the Euler equations. V.K.I. LS 1994–06, 1994.
12. G. Degrez and E. Issman. Acceleration of compressible flow solvers by Krylov subspace methods. V.K.I. LS 1994–05, 1994.
13. E. Dick. A flux-difference splitting method for steady Euler equations. *Journal of computational physics*, 76:19–32, 1988.
14. J. Donea. A Taylor-Galerkin method for convective transport problems. *International Journal for Numerical Methods Engineering*, 20:101–119, 1984.
15. B. Einfeldt, C. D. Munz, P. L. Roe, and B. Sjögreen. On Godunov-type methods near low densities. *Journal of computational physics*, 92:273–295, 1991.
16. F. Fortin. *Simulation d'écoulements compressibles non-visqueux et visqueux par les méthodes de fractionnement des flux*. PhD thesis, Université de Sherbrooke, 1991.
17. S. K. Godunov. A finite difference method for numerical computation of discontinuous solutions of the equations of fluid dynamics. *Mat.Sb.*, 47:271–306, 1959.
18. J. E. Green. Numerical methods in aeronautical fluid dynamics—An introduction. In P. L. Roe, editor, *Numerical methods in aeronautical fluid dynamics*, pp. 1–32, 1982.
19. A. Harten. High resolution schemes for hyperbolic conservation laws. *Journal of computational physics*, 49:357–393, 1983.
20. P. W. Hemker and S. P. Spekreijse. *Multigrid solution of the steady Euler equations*, Vol. 11 of *Notes on numerical fluid mechanics*, pp. 33–44. Vieweg Verlag, 1985.
21. C. Hirsch. *Numerical computation of internal and external flows*, Vol. 1, Fundamentals of numerical discretization. Wiley & Sons, Chichester, 1988.
22. C. Hirsch. *Numerical computation of internal and external flows*, Vol. 2, Computational methods for inviscid and viscous flows. Wiley, Chichester, 1990.
23. C. Hirsch. *Numerical computation of internal and external flows*, Vol. 1, The fundamentals of Computational Fluid Dynamics. Butterworth-Heinemann, Oxford, second edition, 2007.
24. E. Issman and G. Degrez. Convergence acceleration of a 2D Euler/Navier-Stokes solver by Krylov subspace methods. Second European Computational Fluid Dynamics Conference, Stuttgart, Sep. 1994.
25. A. Jameson. Artificial diffusion, upwind biasing, limiters and their effect on accuracy and multigrid convergence in transonic and hypersonic flows. AIAA Paper 93–3359.
26. D. C. Jespersen and T. H. Pulliam. Flux vector splitting and approximate Newton methods. AIAA Paper 83–1899.
27. Z. Johan, T. J. R. Hughes, and F. Shakib. A globally convergent matrix-free algorithm for implicit time-marchingschemes arising in finite element analysis in fluids. *Computer Methods in Applied Mechanics and Engineering*, 87:281–304, 1991.
28. B. Koren. *Multigrid and defect correction for the Navier-Stokes equations*. PhD thesis, T. U. Delft, 1989.
29. P. D. Lax. Hyperbolic systems of conservation laws and the mathematical theory of shock waves. In *Regional conference series in Applied Mathematics 11*. SIAM, 1973.
30. M.-S. Liou and C. J. Steffen. A new flux splitting scheme. *Journal of computational physics*, 107:23–39, 1993.
31. D. J. Mavriplis. Three-dimensional unstructured multigrid for the Euler equations. AIAA Paper 91-1549-CP.
32. E. M. Murman and J. D. Cole. Calculation of plane steady transonic flows. *AIAA Journal*, 9:114–121, 1971.
33. S. Osher and F. Solomon. Upwind difference schemes for hyperbolic systems of conservation laws. *Mathematics of Computation*, 38:339–374, 1982.

34. P. L. Roe. Approximate Riemann solvers, parameter vectors and difference schemes. *Journal of computational physics*, 43:357–352, 1981.
35. P. L. Roe. Generalized formulation of TVD Lax-Wendroff schemes. Report 84–53, ICASE, 1984.
36. R. H. Sanders and K. H. Prendergast. The possible relation of the 3-kiloparsec arm to explosions in the galactic nucleus. *Astrophysical Journal*, 188:489–500, 1974.
37. G. Simeonides, W. Haase, and M. Manna. Experimental, analytical and computational methods applied to hypersonic compression ramp flows. In *Theoretical and experimental methods in hypersonic flows*. AGARD CP 514, May 1992.
38. L. B. Simpson and D. L. Whitfield. A flux-difference split algorithm for unsteady thin layer Navier-Stokes solutions. AIAA Paper 89–1995.
39. S. P. Spekreijse. *Multigrid solution of the steady Euler equations*. PhD thesis, T. U. Delft, 1987.
40. J. L. Steger and R. F. Warming. Flux vector splitting of the inviscid gasdynamic equations with application to finite difference methods. *Journal of computational physics*, 40:263–293, 1981.
41. P. K. Sweby. High resolution schemes using flux limiters for hyperbolic conservation laws. *SIAM Journal on Numerical Analysis*, 21:995–1011, 1984.
42. J. L. Thomas, B. van Leer, and R. W. Walters. Implicit flux-split schemes for the Euler equations. AIAA Paper 85–1680.
43. J. L. Thomas and R. W. Walters. Upwind relaxation algorithms for the Navier-Stokes equations. AIAA Paper 85–1501.
44. B. van Leer. Towards the ultimate conservative difference scheme, II. Monotonicity and conservation combined in a second-order scheme. *Journal of computational physics*, 14: 361–376, 1974.
45. B. van Leer. Flux-vector splitting for the Euler equations. *Lecture Notes in Physics*, 170: 507–512, 1982.
46. B. van Leer, J. L. Thomas, P. L. Roe, and R. W. Newsome. A comparison of numerical flux formulas for the Euler and Navier-Stokes equations. AIAA Paper 87–1104.
47. V. Venkatakrishnan. Preconditioned conjugate gradient methods for the compressible Navier-Stokes equations. *AIAA Journal*, 29(7):1092–1110, 1991.
48. S. Yoon and A. Jameson. An LU-SSOR scheme for the Euler and Navier-Stokes equations. AIAA Paper 87–0600.

Chapter 10
Introduction to Finite Element Methods in Computational Fluid Dynamics

E. Dick

10.1 Introduction

The finite element method (FEM) is a numerical technique for solving partial differential equations (PDE's). Its first essential characteristic is that the continuum field, or *domain*, is subdivided into cells, called *elements*, which form a grid. The elements (in 2D) have a triangular or a quadrilateral form and can be rectilinear or curved. The grid itself need not be structured. With *unstructured* grids and *curved cells*, complex geometries can be handled with ease. This important advantage of the method is not shared by the finite difference method (FDM) which needs a structured grid, which, however, can be curved. The finite volume method (FVM), on the other hand, has the same geometric flexibility as the FEM.

The second essential characteristic of the FEM is that the solution of the discrete problem is assumed a priori to have a prescribed form. The solution has to belong to a *function space*, which is built by varying function values in a given way, for instance linearly or quadratically, between values in nodal points. The nodal points, or *nodes*, are typical points of the elements such as vertices, mid-side points, mid-element points, etc. Due to this choice, the representation of the solution is strongly linked to the geometric representation of the domain. This link is, for instance, not as strong in the FVM.

The third essential characteristic is that a FEM does not look for the solution of the PDE itself, but looks for a solution of an integral form of the PDE. The most general integral form is obtained from a *weighted residual formulation*. By this formulation the method acquires the ability to naturally incorporate *differential type boundary conditions* and allows easily the construction of higher order accurate methods. The ease in obtaining higher order accuracy and the ease of implementation of boundary conditions form a second important advantage of the FEM. With respect to accuracy, the FEM is superior to the FVM, where higher order accurate formulations are quite complicated.

E. Dick
Department of Flow, Heat and Combustion Mechanics, Ghent University,
Sint-Pietersnieuwstraat 41, 9000 Gent, Belgium
e-mail: Erik.Dick@ugent.be

J.F. Wendt (ed.), *Computational Fluid Dynamics*, 3rd ed.,
© Springer-Verlag Berlin Heidelberg 2009

The combination of the representation of the solution in a given function space, with the integral formulation treating rigorously the boundary conditions, gives to the method an extremely *strong and rigorous mathematical foundation*.

A final essential characteristic of the FEM is the modular way in which the discretization is obtained. The discrete equations are constructed from contributions on the element level which afterwards are *assembled*.

Historically, the finite element method originates from the field of structural mechanics. This has some remnants in the terminology. In structural mechanics, the partial differential formulation of a problem can be replaced by an equivalent *variational formulation*, i.e. the minimization of an energy integral over the domain. The variational formulation is a natural integral formulation for the FEM. In fluid dynamics, in general, a variational formulation is not possible. This makes it less obvious how to formulate a finite element method. The history of computational fluid dynamics (CFD) shows that every essential break-through has first been made in the context of the finite difference method or the finite volume method and that it always has taken considerable time, often much more than a decade, to incorporate the same idea into the finite element method. The history of CFD, on the other hand, also shows that, once a suitable FEM-formulation has been found, the FEM is almost exclusively used. This is due to the advantages with respect to the treatment of complex geometries and obtaining higher order accuracy.

The development of the finite element method in fluid dynamics is at present still far from ended. For the simplest problems such as potential flows, both compressible and incompressible, and incompressible Navier-Stokes flows at low Reynolds numbers, the finite element method is more or less full-grown, although new evolutions, certainly for Navier-Stokes problems, are still continuing. More complex problems like compressible flows governed by Euler- or Navier-Stokes equations or incompressible viscous flows at high Reynolds numbers still form an area of active research.

In this introductory text, the option is taken to explain the basic ingredients of the finite element method on a very simple, purely mathematical, problem and to give fluid dynamics illustrations in detail only for simple problems. For more complex problems, only a basic description is given with reference to further literature. Also in the explanation of the method, all mathematical aspects are systematically avoided. For the mathematical aspects, reference is made to further literature. This makes the text accessible for a reader with almost no knowledge of functional analysis and numerical analysis. For the fluid dynamics illustrations, the option has been taken to use only simple techniques, so that the detailed examples can be reproduced by the reader not really familiar with general computational fluid dynamics or even general fluid dynamics. This text therefore is to be seen as the absolute minimum introduction to the subject. The text is in no way complete and the author deliberately has taken the risk to be seen as naive by a more informed reader. A reference list is given for a deeper introduction. A reader beginning with computational fluid dynamics should be aware that a complete study of the finite element method may take considerable time and may necessitate, depending on background, a considerable effort. The method is much less intuitive than the finite difference method and the finite volume method and requires a more fundamental attitude for mathematical formulations. This introductory text therefore is also meant to create some enthusiasm for the method by showing its power with simple examples and to justify in this way the need for further study. It is the conviction of the author that a practitioner of CFD, even if it is not his or her intention to use the FEM, should have at least a basic knowledge of the method. This is in particular useful with respect to the treatment of boundary conditions. Also one should consider that the impact of the FEM in CFD is already extremely important and that it probably will grow in the future.

10.2 Strong and Weak Formulations of a Boundary Value Problem

10.2.1 Strong Formulation

Consider as an example, the following simple one-dimensional boundary value problem, consisting of the *differential equation*

$$\frac{d}{dx}\left(\lambda\frac{du}{dx}\right) = f \quad \text{on} \quad 0 \leq x \leq X \tag{10.1}$$

and the *boundary conditions*

$$u(0) = u_0 \tag{10.2}$$

and

$$\lambda\frac{du}{dx}(X) = q \tag{10.3}$$

More generally, the differential equation is denoted by

$$a(u) = f \tag{10.4}$$

The *domain* to which it applies is denoted by Ω. The boundary condition of type (10.2) is called a *Dirichlet boundary condition*. More generally, it is denoted by

$$b_0(u) = g_0 \tag{10.5}$$

The boundary condition of type (10.3), which is formulated on the *flux* of the variable, is called a *Neumann boundary condition*. More generally, it is denoted by

$$b_1(u) = g_1 \tag{10.6}$$

The boundary of the domain Ω is denoted by Γ. The part to which the Dirichlet boundary condition applies is Γ_0 and the part to which the Neumann boundary condition applies is Γ_1.

The *boundary value problem* (10.1, 10.2 and 10.3) is said to be in its *strong form*, requiring the satisfaction of the differential equation (10.1) in all points of the domain Ω, the satisfaction of the Dirichlet boundary condition (10.2) in all points (here one) of the part of the boundary Γ_0 and the satisfaction of the Neumann boundary condition (10.3) in all points (here one) of the part of the boundary Γ_1.

One way of relaxing the requirements of the boundary value problem, notably the *finite difference way*, consists in requiring the approximate satisfaction of the differential equation and the boundary conditions in a finite number of points in the domain and at the boundary. These points usually are chosen to belong to a *mesh* with some form of regularity. For the one-dimensional domain, a typical mesh or grid is obtained by choosing equally spaced *grid points*, as shown on Fig. 10.1.

The *grid spacing* is denoted by Δx. Following standard finite difference methodology, du/dx is approximated in the mid-point of the interval (x_i, x_{i+1}) by

$$\left(\frac{du}{dx}\right)_{i+1/2} \approx \frac{u_{i+1} - u_i}{\Delta x} \tag{10.7}$$

Similarly, in the mid-point of the interval (x_{i-1}, x_i), the approximation is

$$\left(\frac{du}{dx}\right)_{i-1/2} \approx \frac{u_i - u_{i-1}}{\Delta x} \tag{10.8}$$

Using (10.7) and (10.8), (10.1) can be approximated by

$$\frac{\lambda_{i+1/2}(u_{i+1} - u_i) - \lambda_{i-1/2}(u_i - u_{i-1})}{\Delta x^2} = f_i \tag{10.9}$$

For constant λ, this simplifies to

$$\lambda\frac{u_{i+1} - 2u_i + u_{i-1}}{\Delta x^2} = f_i \tag{10.10}$$

The Dirichlet boundary condition (10.2) is simply

$$u_0 = u_0 \tag{10.11}$$

The Neumann boundary condition can be introduced by the *image point method*. In this method, a point outside the domain $(N + 1)$ is defined which afterwards is eliminated. The discretization of the differential equation (1) in the end point of the domain is given by (10.9) for $i = N$.

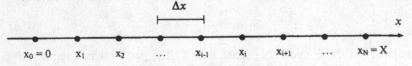

Fig. 10.1 Construction of a finite difference grid over the interval $0 \le x \le X$

The discretization of the Neumann boundary condition (10.3) is

$$\frac{1}{2}\frac{\lambda_{N+\frac{1}{2}}(u_{N+1}-u_N)}{\Delta x}+\frac{1}{2}\frac{\lambda_{N-\frac{1}{2}}(u_N-u_{N-1})}{\Delta x}=q$$

Combination with the discretized differential equation gives

$$q-\lambda_{N-\frac{1}{2}}\frac{(u_N-u_{N-1})}{\Delta x}=\frac{1}{2}f_N\Delta x \tag{10.12}$$

The resulting discretization is of second order. By taking the Taylor expansion of (10.10), this is obvious (for constant λ) for points inside the domain. At the Neumann boundary, the Taylor expansion up to second order (for constant λ) gives

$$u_{N-1}\approx u_N-\Delta x\left(\frac{du}{dx}\right)_N+\frac{1}{2}\Delta x^2\left(\frac{d^2u}{dx^2}\right)_N$$

Using the Neumann boundary condition

$$\lambda\left(\frac{du}{dx}\right)_N=q$$

and the differential equation in node N

$$\lambda\left(\frac{d^2u}{dx^2}\right)_N=f_N$$

this becomes

$$u_{N-1}\approx u_N-\frac{\Delta x}{\lambda}q+\frac{1}{2}\frac{\Delta x^2}{\lambda}f_N$$

For constant λ, this equation is identical to (10.12).

The originally continuous boundary value problem is now replaced by a *discrete problem*, consisting of the solution of the *set of algebraic equations*

$$K\,U=F \tag{10.13}$$

where U is the vector consisting of the elements $(u_1, u_2, \ldots, u_N)$, K is a matrix given by (in the case λ is a constant)

$$K=\begin{bmatrix} 2 & -1 & & & \\ -1 & 2 & -1 & & \\ & & \cdots & & \\ & & -1 & 2 & -1 \\ & & & -1 & 1 \end{bmatrix}$$

and F is the right hand side, given by

$$
F = \begin{bmatrix}
u_0 - \dfrac{\Delta x^2}{\lambda} f_1 \\[2mm]
-\dfrac{\Delta x^2}{\lambda} f_2 \\[2mm]
\vdots \\[2mm]
-\dfrac{\Delta x^2}{\lambda} f_{N-1} \\[2mm]
\dfrac{\Delta x}{\lambda} q - \dfrac{\Delta x^2}{2\lambda} f_N
\end{bmatrix}
$$

The most typical feature of the finite difference method is that it only gives information about the function values at the grid points, but no information on the function values between these points.

10.2.2 Weighted Residual Formulation

The first basic ingredient of the finite element method is that an approximate solution is sought which belongs to some *finite dimension function space*. This function space is to be specified more in detail later. For the time being, we look for an approximate solution of the boundary value problem (10.1, 10.2 and 10.3) which has the form

$$
\hat{u} = \psi + \sum_{k=1}^{N} \phi_k \, u_k \tag{10.14}
$$

where ψ is a function which satisfies the boundary conditions (10.2) and (10.3). For the given problem, the construction of ψ is obvious. The functions ϕ_k are called *basis functions* or *shape functions*. Since the dimension of the function space $\Phi = \{\phi_k; \ k = 1, 2, \ldots, N\}$ is finite, in general, an expression of type (10.14) cannot satisfy the differential equation (10.1) in each point of the domain. This means that the approximate solution $\hat{u}$ cannot be identical with the exact solution u. Of course, the shape functions should be chosen so that by enriching the function space Φ, i.e. letting N grow, the approximation obtained by (10.14) becomes better. This means that the approximate solution *converges* to the exact solution. This is called the *completeness requirement* of the function space.

Since a function $\hat{u}$ given by (10.14) cannot satisfy the differential equation (10.1), upon substitution of (10.14) into (10.1), a *residual* is left:

$$
r_\Omega = a(\hat{u}) - f \quad \text{in } \Omega \tag{10.15}
$$

An approximate solution of the boundary value problem now is obtained by finding a way to make this residual small in some sense. In the finite element method this is done by requiring that an appropriate number of *weighted integrals* of the residual over Ω be zero:

$$\int_{\Omega} w_i r_{\Omega} d\Omega = 0; \quad i = 1,2,\ldots,N \tag{10.16}$$

where $W = \{w_i; \; i = 1, 2,\ldots, N\}$ is a set of *weighting functions*.

Obviously, the convergence requirement now also implies a requirement of completeness of the space of weighting functions, i.e. (10.16) should imply $r_{\Omega} \rightarrow 0$ for $N \rightarrow \infty$.

Clearly, with satisfaction of the completeness, for $N \rightarrow \infty$, the *weighted residual formulation* (10.16) for a function of form (10.14) is completely equivalent to the strong formulation of the problem (10.1, 10.2 and 10.3). An approximate solution then is obtained for N being finite.

10.2.3 Galerkin Formulation

Among the possible choices for the set of weighting functions, the following ones are the most obvious.

The weighting functions can be chosen to be Dirac-delta functions in N points. This choice means making the residual equal to zero in a number of chosen points. The method is called the *point collocation method*. Obviously, it has much in common with the finite difference methodology.

A second possible choice of weighting functions is given by

$$w_i = 1 \quad \text{for } x_i \leq x \leq x_{i+1}$$
$$= 0 \quad \text{for } x < x_i \text{ or } x > x_{i+1}$$

The weighted residual statements (10.16) now require the integral of the residual to be zero on N subdomains. This method is called the *subdomain collocation method*. The finite volume method, in which not the differential form of the equation but the integral form of the equation is discretized, is a special form of this method.

The most popular choice for the weighting functions in the finite element method is the shape functions themselves:

$$w_i = \phi_i$$

This method is called the *Galerkin method*. Its meaning is that the residual is made to be orthogonal to the space of the shape functions.

To illustrate the Galerkin method, consider the boundary value problem (10.1–10.3) with constant λ. Then:

$$\psi = u_0 + \frac{q}{\lambda}x$$

Consider further as an example of (10.14) a Fourier-sine expansion of u:

$$\hat{u} = \psi + \sum_{k=1}^{N} u_k \sin \frac{\pi k' x}{X}, \quad \text{with } k' = k - \frac{1}{2}$$

Then:

$$r_\Omega = -\lambda \sum_{k=1}^{N} u_k \left(\frac{\pi k'}{X}\right)^2 \sin\frac{\pi k' x}{X} - f$$

The Galerkin method then gives

$$\lambda \sum_{k=1}^{N} u_k \left(\frac{\pi k'}{X}\right)^2 \int_0^X \sin\frac{\pi k' x}{X} \sin\frac{\pi i' x}{X} dx = -\int_0^X \sin\frac{\pi i' x}{X} f \, dx$$

Then noting that

$$\int_0^X \sin\frac{\pi k' x}{X} \sin\frac{\pi i' x}{X} dx = \frac{X}{2} \quad \text{for } k' = i'$$

$$= 0 \quad \text{for } k' \neq i'$$

we find

$$u_i = -\frac{2X}{\lambda \pi^2 i'^2} \int_0^X f \sin\frac{\pi i' x}{X} dx$$

The foregoing method used to determine an approximate solution of the boundary value problem (10.1, 10.2 and 10.3) is not a finite element method, but a *spectral method*. The finite element method however has the same starting point.

Before going on with the study of the building blocks of the finite element method, we should remark that a fourth weighted residual statement exists on which finite element methods can be based. The *least squares formulation* is based on the minimization of the integral

$$\int_\Omega r_\Omega^2 d\Omega$$

10.2.4 Weak Formulation

In many problems, it is not practical to construct a function which satisfies the boundary conditions in order to arrive at an expression for the approximate solution, as is done in (14). More generally, an approximate solution can be expressed as

$$\hat{u} = \sum_{k=1}^{N} \phi_k u_k \tag{10.17}$$

Now the approximate solution not only has a residual with respect to the field equation (10.4), but also with respect to the boundary equations (10.5) and (10.6):

$$r_0 = b_0(\hat{u}) - g_0 \qquad (10.18)$$

and

$$r_1 = b_1(\hat{u}) - g_1 \qquad (10.19)$$

A weighted residual statement is now to be of the form

$$\int_\Omega w_i r d\Omega + \int_{\Gamma_0} w_i^0 r_0 d\Gamma + \int_{\Gamma_1} w_i^1 r_1 d\Gamma = 0 \qquad (10.20)$$

This complicates the formulation since now additional weighting functions on the boundaries are to be chosen. Since the number of degrees of freedom of the approximate solution (10.17) is N, an equal number of independent weighting functions w_i can be chosen, while w_i^0 and w_i^1 are to depend on w_i. There is however a natural way to choose the dependent weighting functions on the boundary.

For the problem (10.1, 10.2 and 10.3), (10.20) becomes

$$\int_0^X w_i \left[\frac{d}{dx}\left(\lambda \frac{d\hat{u}}{dx}\right) - f \right] dx + w_i^0 [\hat{u}(0) - u_0] + w_i^1 \left[\lambda \frac{d\hat{u}}{dx}(X) - q \right] = 0 \qquad (10.21)$$

where the weighting functions at the boundary reduce to weighting factors w_i^0 and w_i^1.

By one integration by parts on the first term, (10.21) becomes

$$w_i \lambda \frac{d\hat{u}}{dx}\bigg|_0^X - \int_0^X \lambda \frac{dw_i}{dx}\frac{d\hat{u}}{dx} dx - \int_0^X w_i f \, dx + w_i^0 [\hat{u}(0) - u_0] + w_i^1 \left[\lambda \frac{d\hat{u}}{dx}(X) - q \right] = 0$$

This weighted residual statement is simplified by choosing the weighting factors on the Neumann boundary by

$$w_i^1 = -w_i(X)$$

The weighted residual statement then becomes

$$-\int_0^X \lambda \frac{dw_i}{dx}\frac{d\hat{u}}{dx} dx - w_i(0)\lambda \frac{d\hat{u}}{dx}(0) - \int_0^X w_i f \, dx + w_i^0 [\hat{u}(0) - u_0] + w_i(X)q = 0$$

Furthermore, if the Dirichlet boundary condition can be imposed on the approximate solution, the weighting functions and the weighting factors can be chosen to be zero at the Dirichlet boundary, so that the weighted residual statement further simplifies to

$$-\int_0^X \lambda \frac{dw_i}{dx}\frac{d\hat{u}}{dx} dx - \int_0^X w_i f \, dx + w_i(X)q = 0 \qquad (10.22)$$

subject to the Dirichlet boundary conditions

$$\hat{u}(0) = u_0 \quad w_i(0) = 0 \tag{10.23}$$

The weighted residual statement in form (10.22) is called the *weak formulation*.

The weak formulation (10.22 and 10.23) is not completely equivalent to the strong formulation (10.1, 10.2, 10.3), even not for $N \rightarrow \infty$. By the construction of the weak formulation, any solution of the strong formulation satisfies the weak formulation. The reverse, however, is not true. The weak formulation allows solutions which have a lower degree of regularity than required for the strong solution. This is the origin of the term *weak* and *strong*. For instance for the problem (10.1, 10.2, 10.3), the solution must have continuous first derivatives. We express this by saying that the *degree of regularity* is to be C^1. The corresponding weak formulation (10.22 and 10.23) only requires continuity of the function value itself. The necessary degree of regularity is here C^0. This means that functions with discontinuous first derivatives are allowed by (10.22). *We remark that this is precisely, certainly in fluid mechanics, what we want!* Indeed, in fluid mechanics, the governing equations are obtained from integral statements, i.e. conservation laws, requiring a lower degree of regularity than the partial differential equations which are obtained from these statements.

To conclude, we remark that the weak formulation (10.22), in case of sufficient regularity, through reverse integration by parts leads to a simplification of (10.21):

$$\int_0^X w_i \left[\frac{d}{dx} \left(\lambda \frac{d\hat{u}}{dx} \right) - f \right] dx - w_i(X) \left[\lambda \frac{d\hat{u}}{dx}(X) - q \right] = 0 \tag{10.24}$$

For an infinite number of degrees of freedom ($N \rightarrow \infty$), this implies exact satisfaction of the differential equation and the Neumann boundary condition.

In the weak formulation (10.22 and 10.23), the Neumann boundary condition need not be imposed in an explicit way to the solution. Boundary conditions of this type enter through the integration by parts in a natural way into the formulation. Therefore these boundary conditions are called *natural boundary conditions*. The boundary conditions which have to be imposed explicitly in the weak formulation are called *essential boundary conditions*.

10.2.5 *Variational formulation*

For elliptic self-adjoint boundary value problems, the weak formulation is equivalent to the minimization of the functional associated to the boundary value problem. Historically, this minimization formulation, or variational formulation, has played a big role in the development of the finite element method. Variational methods still have an important role in, for instance, structural mechanics. Also the variational formulation plays an important role in the mathematical theory of finite element methods, for instance, with respect to questions on solvability and uniqueness. In this introductory text we do not enter these aspects of the finite element method and refer the reader to Refs. [5,6].

10.2.6 Conclusion

The first basic ingredient of a finite element method generally is the *weak formulation of the boundary value problem*. Although, as discussed, other formulations are possible (more general weighted residual formulations and least squares formulations), a standard finite element method is based on a weak formulation. In this introductory text, we shall restrict ourselves to this formulation. If possible, the Galerkin approach is chosen with weighting functions equal to shape functions. We shall however see that sometimes modifications of this standard choice are necessary. The standard choice is denoted by the term Bubnov-Galerkin method. When modified weighting functions are used, the method is denoted by the term Petrov-Galerkin method (see later).

10.3 Piecewise Defined Shape Functions

10.3.1 The Finite Element Interpolation

A second basic ingredient of the finite element method is the *piecewise manner* in which the shape and weighting functions are constructed. The domain Ω is subdivided into *non-overlapping subdomains*, Ω_e, called *elements*, of simple geometrical form. For example, for the one dimensional domain shown in Fig. 10.1, an obvious choice for an element Ω_e is the interval $x_{e-1} \leq x \leq x_e$.

The integrals in the weak formulation (10.22) can be split into a sum of integrals over elements:

$$\int_{\Omega} (\,)\,d\Omega = \sum_e \int_{\Omega_e} (\,)\,d\Omega$$

Then, obviously, in the piecewise contributions to the integrals, it is computationally advantageous to have as many zero contributions as possible. This is achieved when the shape functions and weighting functions associated to some subscript are only non-zero in as few as possible elements associated to this subscript. Shape and weighting functions which are only non-zero in a small set of elements are said to have *compact support*.

In the finite element method, shape and weighting functions with compact support are constructed from an *interpolation problem* over the domain. For instance, a function û which is obtained through linear interpolation between function values u_k defined in the grid points of the grid of Fig. 10.1 can be written as

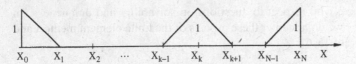

Fig. 10.2 Piecewise linear shape functions for a one-dimensional domain

$$\hat{u} = \sum_{k=0}^{N} \phi_k u_k \tag{10.25}$$

The shape functions ϕ_k in this expression have the hat-like form, shown in Fig. 10.2.

For a function representation based on an interpolation, the values u_k in the expression (10.25) have the meaning of function values in grid points. Obviously, other interpolation schemes are possible. For instance, the function u could be obtained by piecewise constant interpolation, as shown in Fig. 10.3. The values u_k are now to be seen as function values in mid-points of the elements.

Similarly, the interpolation could be piecewise quadratic as shown in Fig. 10.4.

In all these cases, the values u_k represent function values in some points associated to the elements. In the finite element technique, these points are called *nodes*. Interpolation formulas can be built in which the values u_k do not necessarily represent function values (or values of derivatives) in nodes. These are then called *nodeless variables*. For simplicity, in this introductory text, we shall only consider interpolation formulas with *nodal variables*.

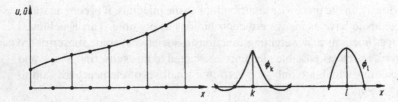

Fig. 10.3 Piecewise constant shape functions for a one-dimensional domain

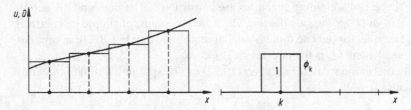

Fig. 10.4 Piecewise quadratic interpolation in one dimension

As a simple example, we first consider the problem (10.1, 10.2 and 10.3) with piecewise linear shape functions and a standard Galerkin weak formulation. The approximate solution is then represented by

$$\hat{u} = \sum_{k=0}^{N} \phi_k\, u_k = \sum_e \sum_{j=1}^{2} \phi_j^e u_j. \tag{10.26}$$

In (10.26) the sum over the nodes is rearranged as a double sum, first over the elements and then over the nodes belonging to the element. The shape functions ϕ_k, associated with the nodes are called *global shape functions*. On the element level, the shape functions ϕ_j^e are called *local shape functions* or *element shape functions*.

Figure 10.5 shows the shape functions on an element basis. The basis functions can be written as

$$\phi_1^e = \frac{x_e - x}{\Delta x_e} \quad , \qquad \phi_2^e = \frac{x - x_{e-1}}{\Delta x_e}$$

Hence:

$$\frac{d\phi_1^e}{dx} = -\frac{1}{\Delta x_e} \quad , \qquad \frac{d\phi_2^e}{dx} = \frac{1}{\Delta x_e}$$

The Galerkin weak formulation (10.22) is

$$\int_0^X \lambda \frac{dw_i}{dx} \frac{d\hat{u}}{dx} dx = -\int_0^X w_i f\, dx + w_i(X)q \tag{10.27}$$

For $i = c \neq N$, the integral in the left hand side is (for constant λ)

$$I_1^e = \lambda \left\{ \int_{\Omega_e} \frac{d\phi_2^e}{dx} \left[\frac{d\phi_1^e}{dx} u_{e-1} + \frac{d\phi_2^e}{dx} u_e \right] dx + \int_{\Omega_{e+1}} \frac{d\phi_1^{e+1}}{dx} \left[\frac{d\phi_1^{e+1}}{dx} u_e + \frac{d\phi_2^{e+1}}{dx} u_{e+1} \right] dx \right\}$$

$$= \lambda \left\{ \int_{\Omega_e} \frac{1}{\Delta x_e} \left[-\frac{1}{\Delta x_e} u_{e-1} + \frac{1}{\Delta x_e} u_e \right] dx + \int_{\Omega_{e+1}} \frac{1}{\Delta x_{e+1}} \left[-\frac{1}{\Delta x_{e+1}} u_e + \frac{1}{\Delta x_{e+1}} u_{e+1} \right] dx \right\}$$

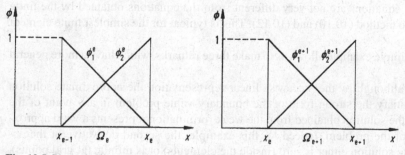

Fig. 10.5 Piecewise linear element shape functions

$$= \lambda \left\{ -\frac{1}{\Delta x_e} u_{e-1} + \frac{1}{\Delta x_e} u_e + \frac{1}{\Delta x_{e+1}} u_e - \frac{1}{\Delta x_{e+1}} u_{e+1} \right\}$$

For i = N, this integral is

$$I_1^N = \lambda \left[-\frac{1}{\Delta x_N} u_{N-1} + \frac{1}{\Delta x_N} u_N \right]$$

We remark that due to the essential boundary condition in node 0, $w_0 = 0$. Hence $I_1^0 = 0$.

Interpolating f in the same way as û, the integral on the right hand side becomes, for i = e ≠ N:

$$I_2^e = \int_{\Omega_e} \phi_2^e [\phi_1^e f_{e-1} + \phi_2^e f_e] dx + \int_{\Omega_{e+1}} \phi_1^{e+1} [\phi_1^{e+1} f_e + \phi_2^{e+1} f_{e+1}] dx$$

$$= \left[\frac{\Delta x_e}{6} f_{e-1} + \frac{\Delta x_e}{3} f_e \right] + \left[\frac{\Delta x_{e+1}}{3} f_e + \frac{\Delta x_{e+1}}{6} f_{e+1} \right]$$

For i = N, this integral is

$$I_2^N = \frac{\Delta x_N}{6} f_{N-1} + \frac{\Delta x_N}{3} f_N$$

For i = 1, ..., N − 1, the weak formulation (10.27) becomes, in the case of constant interval length

$$\frac{\lambda}{\Delta x} [-u_{i-1} + 2u_i - u_{i+1}] = -\Delta x [1/6 f_{i-1} + 2/3 f_i + 1/6 f_{i+1}] \qquad (10.28)$$

The equation associated to the last node is

$$\frac{\lambda}{\Delta x} [-u_{N-1} + u_N] = -\Delta x [1/6 f_{N-1} + 1/3 f_N] + q \qquad (10.29)$$

These equations are not very different from the equations obtained by the finite difference method (10.10) and (10.12). This is typical for the simplest finite element methods.

This simple example allows us to make three remarks, which have a more general validity.

First, although by the piecewise linear representation the approximate solution cannot satisfy the strong form of the boundary value problem in any point of the domain, the solution obtained from the weak formulation represents a valid approximation of the problem. Indeed for this example, the second derivative of the approximate solution either is zero (inside the elements) or is infinite (at grid points). So there is no way to satisfy the differential equation by such a function.

Fig. 10.6 Piecewise constant interpolation of the non-derivative term in an interweaving finite element grid, leading to lumping

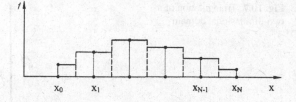

Second, by comparison of the finite element expressions (10.28, 10.29) with the corresponding finite difference expressions (10.10, 10.12) we see that the accuracy is not penalized by adding the contributions for undifferenced terms, such as f, from non-central nodes to the central node. This process is called *lumping*. It is often used to simplify finite element expressions. The result of this lumping, for this example replacing $^1/_6 f_{i-1} + ^2/_3 f_i + ^1/_6 f_{i+1}$ by f_i and replacing $^1/_6 f_{N-1} + ^1/_3 f_N$ by $^1/_2 f_N$, could automatically have been obtained if f would have been presented by a piecewise constant function in an interweaving finite element grid, as shown in Fig. 10.6.

This remark is also essential in the sense that it shows that *variables appearing with different order of derivatives can be approximated in different ways*, i.e. with different interpolation structures or even in different grids. This means that the finite element method is not a rigid method but allows many variants.

Finally, we can remark that the finite element method can be interpreted as a *systematic way to generate difference approximations*. For the simple example treated here, for constant mesh spacing, constant field parameter λ, and using lumping, the finite element method with piecewise linear shape functions reproduces the second order finite difference approximation. Therefore it is clear that in a more general application (non-constant mesh spacing, non-constant field parameter), the finite element method with piecewise linear shape functions still generates a second order difference approximation. This systematic result of the finite element method is further illustrated in the following sections.

10.3.2 *Finite Elements with* C_0 *Continuity in Two-Dimensions*

10.3.2.1 Triangular Elements

Figure 10.7 shows a domain subdivided into non-overlapping elements of rectilinear triangular form. In each element a local interpolation is defined. We consider now interpolation formulas which guarantee the continuity of the interpolated functions.

Figure 10.8 shows a triangular element with nodes at the corners of the triangle. A function can be interpolated in the triangle in a linear way based on the nodal values of the function. In the local coordinate system (ξ, η), an interpolated function can be written as

$$u = \sum_{j=1}^{3} \phi_j^e(\xi, \eta) u_j \tag{10.30}$$

Fig. 10.7 Triangulation of a two-dimensional domain

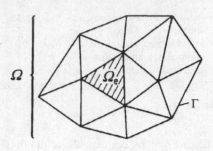

where ϕ_j^e are local interpolation functions.

In this case, the ϕ_j^e have to satisfy

$$\phi_j^e = a + b_1\xi + b_2\eta$$

i.e. ϕ_j^e is a linear function of ξ and η, with

$$\phi_j^e(\xi_i, \eta_i) = 1 \quad \text{for} \quad j = i$$
$$= 0 \quad \text{for} \quad j \neq i$$

It is easy to verify that for the element in Fig. 10.8, the local interpolation functions are

$$\phi_1^e(\xi, \eta) = 1 - \xi - \eta$$

$$\phi_2^e(\xi, \eta) = \xi$$
$$\phi_3^e(\xi, \eta) = \eta$$

These are shown in Fig. 10.9.

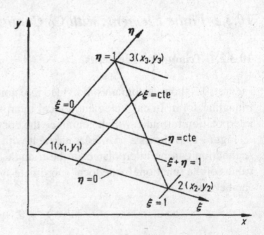

Fig. 10.8 The linear triangle

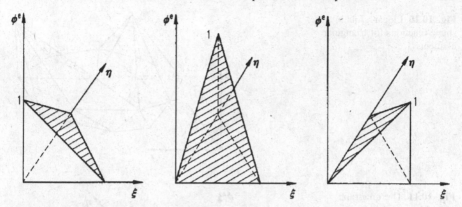

Fig. 10.9 Linear element shape functions for the triangle

The local interpolation functions also can be expressed as a function of the global coordinates by a coordinate transformation between the systems (ξ, η) and (x,y):

$$x = \sum_{j=1}^{3} \phi_j^e(\xi, \eta)x_j \quad , \qquad y = \sum_{j=1}^{3} \phi_j^e(\xi, \eta)y_j \tag{10.31}$$

In the coordinate transformation formulas (10.31), the same local interpolation functions are used as in the interpolation of a function value (10.30).

It is clear that with the above interpolation structure, if applied in each element, C_0 continuity (i.e. continuity of the function value) is reached in the whole domain. The interpolation itself is piecewise linear.

By summing the interpolation over all elements, we obtain

$$u = \sum_{e} \sum_{j,e} \phi_j^e(\xi, \eta)u_j \tag{10.32}$$

where $\sum_{e}$ denotes the sum over all elements Ω_e of the domain Ω and where $\sum_{j,e}$ denotes the sum over all nodes of the element Ω_e.

In (10.32), the summations can be reversed to write

$$u = \sum_{k} \sum_{e,k} \phi_k^e u_k = \sum_{k} \phi_k u_k \tag{10.33}$$

where $\sum_{k}$ denotes the sum over all nodes of the domain Ω and where $\sum_{e,k}$ denotes the sum over all elements adjacent to node k. In (10.33) the ϕ_k denote global interpolation functions or shape functions. Figure 10.10 shows some examples.

The order of interpolation within each triangle can be increased by adding nodes. In order to represent quadratic functions, six nodes are needed. For cubic functions

Fig. 10.10 Linear global shape functions for triangular elements

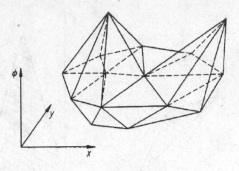

Fig. 10.11 The quadratic triangle

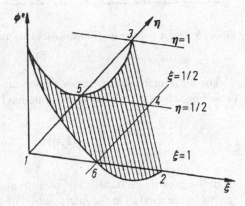

ten nodes are needed, etc. Figure 10.11 shows a quadratic element with six nodes: three nodes at the corners and three nodes at the mid-sides.

The local interpolation functions have to satisfy

$$\phi_j^e = a + b_1\xi + b_2\eta + c_{11}\xi^2 + c_{12}\xi\eta + c_{22}\eta^2$$

$$\phi_j^e(\xi_i, \eta_i) = 1 \quad \text{for} \quad j = i$$

$$= 0 \quad \text{for} \quad j \neq i$$

For example:

$$\phi_1^e = (1 - \xi - \eta)(1 - 2\xi - 2\eta)$$

Obviously, although a higher order interpolation is used within the elements, the interelement continuity remains C_0, as in the linear case. Also, the coordinate transformation still can be given with the first order basis functions (10.31). However, also the quadratic basis functions can be used, as discussed in the section on isoparametric elements.

Figure 10.12 shows the Pascal triangle and the associated C_0 triangular elements. These elements form the so-called *Lagrange family* of triangular elements. The

Fig. 10.12 The Lagrange
family of triangular elements

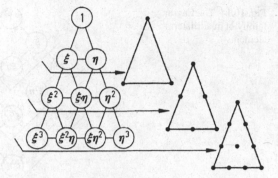

restriction of these elements to one dimension and the extension to three dimensions
(on tetrahedra) is clear.

The advantage of using higher order elements lies in their ability to represent
more accurately arbitrarily varying functions, for a given element size. Of course, a
better representation of a function can also be reached by low order elements with
a smaller element size. In practice, element size and order of the elements are to be
chosen to optimize accuracy with respect to computational work. Usually it is found
that quadratic elements are to be preferred.

10.3.2.2 Quadrilateral Lagrange Elements

Figure 10.13 shows a rectilinear quadrilateral element with a local affine coordinate
system, using four nodes. Local interpolation functions can be defined which are
bilinear:

$$\phi_j^e = a + b_1 \xi + b_2 \eta + c_{12} \xi \eta$$

$$\phi_1^e = (1 - \xi)(1 - \eta), \qquad \phi_2^e = \xi(1 - \eta), \qquad \phi_3^e = \xi \eta, \quad \phi_4^e = (1 - \xi)\eta$$

By adding nodes, biquadratic, bicubic, etc., elements can be constructed. For
example, a biquadratic element contains nine nodes. Figure 10.14 shows the Pascal

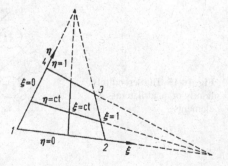

Fig. 10.13 The bilinear
quadrilateral element

Fig. 10.14 The Lagrange
family of quadrilateral
elements

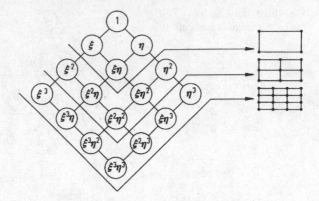

triangle and the associated quadrilateral elements. These quadrilaterals also form a
so-called *Lagrange family*.

10.3.2.3 Quadrilateral Serendipity Elements

Inspection of the Pascal triangle for quadrilateral Lagrange elements reveals that,
for instance, the quadratic element has third and fourth order terms in the expression of the local shape functions. These higher order terms do not form a complete
polynomial and as a consequence do not contribute to an enlargement of the order
of interpolation.

On the other hand, the high order terms in the local basis functions can generate
undesirable oscillations in interpolated data. Therefore, it is generally preferred to
eliminate the highest order terms by leaving out the internal nodes in the elements.
This generates the so-called *serendipity family* for which the Pascal triangle and
some examples are shown on Fig. 10.15. In particular, the quadratic serendipity
element is an attractive element and is widely used.

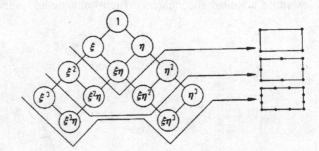

Fig. 10.15 The serendipity
family of quadrilateral
elements

10.3.2.4 Isoparametric Elements

It is, of course, more accurate to represent bodies with curved boundaries using elements with curved sides. Curved elements can be generated by applying a mapping between the rectilinear affine coordinate system used in the previous sections and a curvilinear coordinate system. In order not to complicate this mapping too much, usually the coordinate transformation formulas, in analogy with equations (10.31), are chosen based on the local shape functions:

$$x = \sum_j \phi_j^e(\xi, \eta)x_j \quad , \qquad y = \sum_j \phi_j^e(\xi, \eta)y_j \qquad (10.34)$$

Figure 10.16 shows how a coordinate transformation of type (10.34) can deform a quadratic serendipity element, in comparison with a linear element.

It is clear that with quadratic elements complicated boundaries can be represented.

Elements in which the coordinate transformation formulas are identical to the interpolation formulas are called *isoparametric elements*. As shown in Fig. 10.16, these elements can be considered as being mapped from a square element in a rectangular ξ, η coordinate system. The basic element in undistorted local coordinates is called the *parent element*.

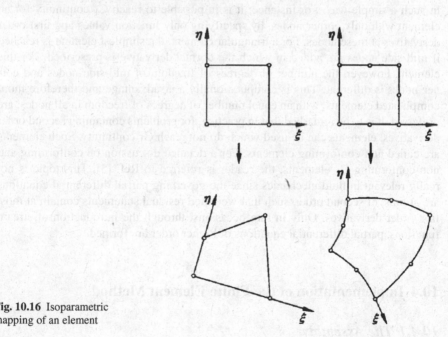

Fig. 10.16 Isoparametric mapping of an element

10.3.3 Finite Elements with C_1 Continuity

When the weighted residual statements contain second order derivatives, strictly, continuity up to first order derivatives is necessary. In one dimension, this can be reached by including the nodal values of the slopes of the function.

For an element with two corner nodes, the interpolation is written as

$$u = \psi_1(\xi)u_1 + \psi_2(\xi)u_2 + \psi_3(\xi)u'_1 + \psi_4(\xi)u'_2$$

with

$$u_1 = u(\xi = 0) \qquad u_2 = u(\xi = 1)$$
$$u'_1 = u'(\xi = 0) \qquad u'_2 = u'(\xi = 1)$$

Using a cubic polynomial, one finds

$$\psi_1 = (1 + 2\xi)(1 - \xi)^2, \qquad \psi_2 = (3 - 2\xi)\xi^2$$
$$\psi_3 = \xi(1 - \xi)^2, \qquad\qquad \psi_4 = (\xi - 1)\xi^2$$

An element of this type is called a *Hermite element*.

In more dimensions, continuity up to first order derivatives cannot be reached in such a simple way. For instance, it is impossible to reach C_1 continuity for an element with only corner nodes, by specifying only function values and first order derivatives at these nodes. For a triangular element, the simplest element is reached if mid-side nodes are added in which the normal derivative is prescribed. For this element, however, the number of degrees of freedom of mid-side nodes and corner nodes is different. This is computationally a disadvantage and therefore more complicated elements, with an equal number of degrees of freedom in all nodes, are preferred. It is to be remarked that in practice, for problems containing second order derivatives, elements can be used which do not reach C_1 continuity. Such elements are called non-conforming elements. For a detailed discussion on conforming and non-conforming C_1 elements, the reader is referred to Ref. [5]. This topic is not really relevant in fluid mechanics since the governing partial differential equations are at most of second order, such that weighted residual statements contain at most first order derivatives. Only in rare occasions, through the introduction of stream functions, partial differential equations of higher order are formed.

10.4 Implementation of the Finite Element Method

10.4.1 The Assembly

The third and final basic ingredient of the finite element method is the way in which the nodal equations are constructed.

As seen in the example of the previous section, the integration involved in the weak formulation leads to a sum of contributions on elements adjacent to the node which is treated. Instead of performing the integration on the set of elements adjacent to a node, as done in the previous section, we could of course first perform the integration on all elements separately and then afterwards construct the nodal equations by adding contributions from adjacent elements. This process is called *assembly*. It has the advantage that the integration on all elements can be done in the same way and, as a consequence, with the same routine.

For example, for (10.27), the integral in the left hand side, on an element Ω_e is (dropping the subscript i)

$$\lambda \int_{\Omega_e} \frac{dw}{dx} \frac{d\hat{u}}{dx} dx = \lambda \int_{\Omega_e} \left[\frac{d\phi_1^e}{dx} w_1^e + \frac{d\phi_2^e}{dx} w_2^e \right] \left[\frac{d\phi_1^e}{dx} u_1^e + \frac{d\phi_2^e}{dx} u_2^e \right] dx$$

An element matrix K^e can be defined by

$$K_{ij}^e = \lambda \int_{\Omega_e} \frac{d\phi_i^e}{dx} \frac{d\phi_j^e}{dx} dx \qquad (10.35)$$

Obviously, this matrix is

$$K^e = \frac{\lambda}{\Delta x_e} \begin{bmatrix} 1 & -1 \\ -1 & 1 \end{bmatrix}$$

The components of the system matrix in the global discrete system

$$K U = F \qquad (10.36)$$

can now be found by adding components from the element matrices.

For example:

$$K_{i,i-1} = K_{21}^i$$
$$K_{i,i} = K_{22}^i + K_{11}^{i+1}$$
$$K_{i,i+1} = K_{12}^{i+1}$$

The system matrix K in (10.36) generally is called the *stiffness matrix*. This term has its origin in structural mechanics.

Figure 10.17 shows a schematic representation of the assembly process. A similar assembly can be defined for the element vectors corresponding to the right hand side in (10.36).

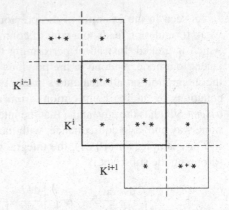

Fig. 10.17 Schematic representation of the assembly of the stiffness matrix for a one-dimensional problem

10.4.2 Numerical Integration

In general problems, the computation of the element stiffness matrix and the element right hand side cannot be performed analytically due to the complexity of the integrand and due to the complexity of the element.

Using curved elements, in two dimensions, the coordinate transformation leads to

$$\begin{bmatrix} \partial/\partial\xi \\ \partial/\partial\eta \end{bmatrix} = \begin{bmatrix} \partial x/\partial\xi & \partial y/\partial\xi \\ \partial x/\partial\eta & \partial y/\partial\eta \end{bmatrix} \begin{bmatrix} \partial/\partial x \\ \partial/\partial y \end{bmatrix} = J \begin{bmatrix} \partial/\partial x \\ \partial/\partial y \end{bmatrix}$$

or

$$\begin{bmatrix} \partial/\partial x \\ \partial/\partial y \end{bmatrix} = J^{-1} \begin{bmatrix} \partial/\partial\xi \\ \partial/\partial\eta \end{bmatrix} \tag{10.37}$$

where J is the Jacobian.

Using isoparametric mapping (10.34), this Jacobian is easily evaluated as a function of ξ and η.

An infinitesimal area $d\Omega_e$ is then given by

$$d\Omega_e = \text{Det}(J)d\xi\,d\eta \tag{10.38}$$

By (10.37) and (10.38) all element integrals are reduced to integrals on the parent element.

When the local shape functions are of order p, i.e. when they include as the highest complete polynomial a polynomial of order p, function values are represented up to order p. The integrand in (10.35), for example, is then represented up to order $2(p-1)$. In general for m-th order derivatives, this would be an integrand of order $2(p-m)$. It can be shown [5] that the order of convergence (i.e. the accuracy of the solution reached when the element size goes to zero) is h^{p-m+1} for an exact integration of the terms in the weighted residual principle and that there is no loss of convergence, even if curved elements are used, if the numerical quadrature shows an error of $0(h^{2(p-m)+1})$ or less, where h denotes a typical dimension of the element. Therefore for C_0 problems, the integration formulas should be:

linear elements   : O(h)
quadratic elements   : $O(h^3)$
cubic elements   : $O(h^5)$.

This means that, respectively, Gauss formulas with 1 point, 2×2 points and 3×3 points are sufficient in two dimensions.

10.4.3 Solution Procedure

The procedure outlined here, implicitly assumes that the discrete set of equations (10.36) actually is constructed and is solved with a direct solver. This is the usual procedure for linear problems. In fluid mechanics, due to the non-linearity of the equations, this is generally inefficient, since it at least implies a global iteration with, in each iteration step, the solution of a linearized system of form (10.36). It is then usually much more efficient to use directly an iterative technique on the non-linear form of (10.36). In a relaxation technique, for instance, the nodal equation in some node is only constructed when it is needed in the relaxation procedure and the global set of equations (10.36) is never formed.

10.5 Practical Construction of a Weak Formulation

For many applications, the weak formulation can be constructed in a shorter way than discussed up to now. To illustrate the short formulation, we consider again the example equation (10.1). We multiply with a weighting function w defined in the whole field and integrate over the whole field:

$$\int_0^X w \left[\frac{d}{dx} \left(\lambda \frac{du}{dx} \right) - f \right] dx = 0 \tag{10.39}$$

We perform the integration by parts:

$$-\int_0^X \lambda \frac{dw}{dx} \frac{du}{dx} dx + w(X) \lambda \frac{du}{dx}(X) - w(0) \lambda \frac{du}{dx}(0) - \int_0^X wf \, dx = 0 \tag{10.40}$$

By the integration by parts, the admissible boundary conditions show up. A Dirichlet boundary condition (10.2) at the left boundary is admissible. It is implemented by imposing it on the function value u(0) and by setting the value of the weighting function w(0) at the boundary equal to zero. A Neumann boundary condition (10.3) at the right boundary is admissible. It is implemented by filling in the value of the flux. The resulting formulation becomes

$$-\int_0^X \lambda \frac{dw}{dx}\frac{du}{dx}dx + w(X)q - \int_0^X w\,f\,dx = 0 \qquad (10.41)$$

The formula (10.41) is identical to the formula (10.22), but where the functions u and w are still general functions in the field. As a next step, we restrict these functions to piecewise interpolated functions. In a Bubnov-Galerkin formulation, the interpolation structure is the same for the weighting function as for the approximate solution. The nodal values of the interpolated weighting functions are arbitrary. This gives N equations of form (10.22):

$$-\int_0^X \lambda \frac{dw_i}{dx}\frac{d\hat{u}}{dx}dx - \int_0^X w_i f\,dx + w_i(X)q = 0$$

where $\hat{u}$ is the interpolated approximate solution and w_i is the local weighting function (shape function) associated to node i.

10.6 Examples

In this section some aspects of the application of the finite element method in fluid mechanics are discussed. The aim of the section is to provide some insight into the way a finite element method can be constructed. This section also demonstrates that a universal finite element methodology which can be applied as a recipe to all problems does not exist. Like the finite difference method, the finite element method is a framework in which, for each problem, an adapted formulation is necessary.

10.6.1 Steady Incompressible Potential Flow

For steady irrotational flow of an incompressible fluid, the equations of motion are

$$\nabla.\vec{V} = 0 \qquad (10.42)$$

$$\nabla \times \vec{V} = 0 \qquad (10.43)$$

where $\vec{V}$ is the velocity.
    Equation (10.43) allows the introduction of a velocity potential

$$\vec{V} = \nabla\phi \qquad (10.44)$$

Inserting (10.44) into (10.42) gives

$$\nabla^2\phi = 0 \qquad (10.45)$$

Fig. 10.18 Internal flow problem

Possible Dirichlet and Neumann boundary conditions are

$$\phi = g_0, \qquad \frac{\partial \phi}{\partial n} = g_1 \qquad\qquad (10.46)$$

For a problem of internal flow, as shown in Fig. 10.18, the boundary conditions can be, where u and v denote the Cartesian velocity components:

- at the inflow boundary AB: parallelism of the flow:

$$v = 0 \rightarrow \frac{\partial \phi}{\partial y} = 0 \rightarrow \phi = \text{constant} = \phi_i$$

- at the outflow boundary CD: parallelism of the flow: $v = 0 \rightarrow \phi = \phi_o$
- at solid boundaries AC and BD: impermeability: $\vec{V}.\vec{n} = 0 \rightarrow \frac{\partial \phi}{\partial n} = 0$

For these boundary conditions, the potential difference $\phi_i - \phi_o$ drives the flow. If desired, either at inlet or at outlet, a possible boundary condition is also the prescription of $u = \partial\phi/\partial x$.

The weighted residual statement is

$$\int_\Omega w\nabla^2\phi d\Omega = -\int_\Omega \nabla w.\nabla\phi d\Omega + \int_{\Gamma_0} w\frac{\partial \phi}{\partial n}d\Gamma + \int_{\Gamma_1} w\frac{\partial \phi}{\partial n}d\Gamma = 0$$

By introduction of the boundary conditions, all integrals over the boundary become zero for the chosen boundary conditions: essential boundary conditions and zero natural boundary conditions.

Next, the piecewise polynomial interpolation is introduced, resulting in

$$\phi = \sum \hat{\phi}_k \phi_k \quad \text{and} \quad w = \hat{\phi}_k$$

where $\hat{\phi}_k$ denotes the shape function associated to node k.

In order to gain insight into the coefficient structure of the stiffness matrix, we calculate the element stiffness matrix for bilinear elements on a square grid (Fig. 10.19):

$$K_{ij}^e = \int_{\Omega_e} \nabla\hat{\phi}_i \nabla\hat{\phi}_j dxdy$$

Fig. 10.19 Element and
assembly pattern for bilinear
square elements

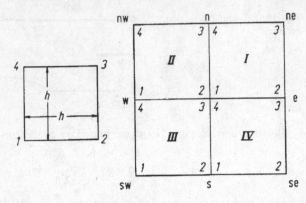

The result is

$$K^e = \begin{bmatrix} 2/3 & -1/6 & -1/3 & -1/6 \\ -1/6 & 2/3 & -1/6 & -1/3 \\ -1/3 & -1/6 & 2/3 & -1/6 \\ -1/6 & -1/3 & -1/6 & 2/3 \end{bmatrix}$$

According to the assembly pattern, some coefficients are

$$K_{k,k} = K_{1,1}^I + K_{2,2}^{II} + K_{3,3}^{III} + K_{4,4}^{IV} = 8/3$$
$$K_{k,n} = K_{1,4}^I + K_{2,3}^{II} = -1/3$$
$$K_{k,ne} = K_{1,3}^I = -1/3$$

where n and ne denote the nodes north and north-east of k.

The coefficient-molecule is

$$\begin{matrix} (-1/3) \cdots (-1/3) \cdots (-1/3) \\ \vdots \qquad \vdots \qquad \vdots \\ (-1/3) \cdots (8/3) \cdots (-1/3) \\ \vdots \qquad \vdots \qquad \vdots \\ (-1/3) \cdots (-1/3) \cdots (-1/3) \end{matrix} \qquad (10.47)$$

The stiffness matrix is of positive type, i.e. the diagonal coefficient is positive, the non-diagonal coefficients are negative and there is a weak diagonal dominance. As a consequence, the resulting system can be solved by a direct solver, but also by an iterative method. When deformed elements are used, the coefficient molecule deforms, but the system remains of positive type.

Further, it is instructive to note that the coefficient molecule by linear finite elements (10.47) represents an order h^2 approximation to the Laplace equation and not an order h^4 approximation, which is possible on a 9-point molecule on a square grid.

10.6.2 Incompressible Navier-Stokes Equations in ω-ψ Formulation

The equations of motion, in two dimensions, are

$$\frac{\partial u}{\partial t} + u\frac{\partial u}{\partial x} + v\frac{\partial u}{\partial y} + \frac{\partial p}{\partial x} = \nu\left(\frac{\partial^2 u}{\partial x^2} + \frac{\partial^2 u}{\partial y^2}\right) \tag{10.48}$$

$$\frac{\partial v}{\partial t} + u\frac{\partial v}{\partial x} + v\frac{\partial v}{\partial y} + \frac{\partial p}{\partial y} = \nu\left(\frac{\partial^2 v}{\partial x^2} + \frac{\partial^2 v}{\partial y^2}\right) \tag{10.49}$$

$$\frac{\partial u}{\partial x} + \frac{\partial v}{\partial y} = 0 \tag{10.50}$$

where p is the so-called kinematic pressure (pressure divided by density). The continuity equation (10.50) can be satisfied by the introduction of a stream function ψ by

$$u = \frac{\partial \psi}{\partial y} \quad , \qquad v = -\frac{\partial \psi}{\partial x} \tag{10.51}$$

By introduction of the vorticity ω by

$$\omega = \frac{\partial v}{\partial x} - \frac{\partial u}{\partial y} \tag{10.52}$$

and by taking the y-derivative of equation (10.48) and subtracting the x-derivative of equation (10.49), pressure can be eliminated to give

$$\frac{\partial \omega}{\partial t} + u\frac{\partial}{\partial x}\omega + v\frac{\partial}{\partial y}\omega = \nu\left(\frac{\partial^2 \omega}{\partial x^2} + \frac{\partial^2 \omega}{\partial y^2}\right) \tag{10.53}$$

Combination of (10.51) and (10.52) further gives

$$\frac{\partial^2 \psi}{\partial x^2} + \frac{\partial^2 \psi}{\partial y^2} = -\omega \tag{10.54}$$

For the problem shown in Fig. 10.20, possible boundary conditions are:

- at inflow (Γ_a): u can be prescribed (parabolic profile) and derivatives of velocity components in the x-direction can be assumed to be zero (fully developed flow). This allows the calculation of ψ by (10.51) and ω by (10.52):

$$\psi = \psi_a(y) \quad , \qquad \omega = \omega_a(y)$$

- at outflow (Γ_b): v can be assumed to be zero and derivatives of velocity components in the x-direction can be assumed to be zero (fully developed flow). This gives

$$\frac{\partial \psi}{\partial n} = 0 \quad , \qquad \frac{\partial \omega}{\partial n} = 0$$

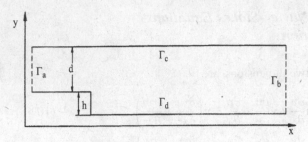

Fig. 10.20 A backward-facing step problem

- at solid boundaries (Γ_c and Γ_d):

$$\psi = \psi_c \quad \text{or} \quad \psi = \psi_d \quad \text{and} \quad \frac{\partial\psi}{\partial n} = 0$$

We remark that the boundary conditions for the Poisson problem (10.54) are over-specified, at solid boundaries, while in the problem (10.53) at these boundaries, conditions in ω are lacking.

The problem (10.54) can be treated with Γ_a as essential boundary and Γ_{b+c+d} as natural boundary. The weak formulation for (10.54) is then

$$\int_{\Omega} -\nabla w.\nabla\psi d\Omega = -\int_{\Omega} w\omega d\Omega \tag{10.55}$$

There is no contribution from the natural boundary since $\partial\psi/\partial n = 0$ there.

Using bilinear elements, on the grid shown in Fig. 10.21, for given ω, ψ can be determined in the interior of the domain and on Γ_b. We remark that by lumping in the right hand side of (10.55), knowledge of ω on Γ_c and Γ_d is not necessary to determine ψ. The formulation comes then very close to a finite difference formulation. Moreover, using the given value of ψ on Γ_c and Γ_d, the nodal equations obtained from (10.55) determine ω on these boundaries. With lumping, the expression for ω is even explicit.

After having determined ω on Γ_{c+d}, the problem (10.53) becomes well-posed. The weak formulation of (10.53) then becomes

$$\int_{\Omega} w\frac{\partial\omega}{\partial t}d\Omega + \int_{\Omega} w\vec{V}.\nabla\omega d\Omega = \int_{\Omega} vw\nabla^2\omega d\Omega$$

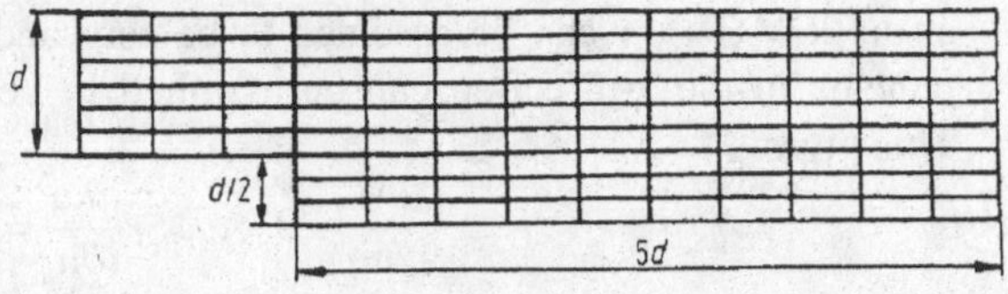

Fig. 10.21 Discretization
of the backward-facing step
problem

After integration by parts for the viscous term, this becomes

$$\int_\Omega w\frac{\partial\omega}{\partial t}d\Omega + \int_\Omega w\vec{V}.\nabla\omega d\Omega + \nu\int_\Omega \nabla w.\nabla\omega d\Omega = 0 \qquad (10.56)$$

There is no contribution from the natural boundary since $\partial\omega/\partial n = 0$ there.

The structure of the nodal equations obtained from (10.56) is very similar to the structure of a central finite difference method, if lumping is used on the time derivative term.

The discrete equation corresponding to (10.56), using lumping, has the form:

$$M_i\frac{\partial\omega_i}{\partial t} + \sum_j A_{ij}\omega_j = 0 \qquad (10.57)$$

where the subscript j describes nodes in the vicinity of the node i. The equation (10.57) can for instance, like a finite-difference equation, be integrated with a two-stage time stepping scheme (for second order accuracy in time):

$$M_i[\omega_i(t + {}^1\!/\!_2\Delta t) - \omega_i(t)] = -\frac{\Delta t}{2}\sum_j A_{ij}(t)\omega_j(t)$$

$$M_i[\omega_i(t + \Delta t) - \omega_i(t)] = -\Delta t\sum_j A_{ij}(t + {}^1\!/\!_2\Delta t)\omega_j(t + {}^1\!/\!_2\Delta t)$$

After each time step, ψ is to be evaluated from (10.55), new values of ω on the solid boundary are to be determined and a new velocity field is to be calculated from (10.51).

This can be done in a finite element way by

$$\int_\Omega wud\Omega = \int_\Omega w\frac{\partial\psi}{\partial y}d\Omega \quad , \qquad \int_\Omega wvd\Omega = -\int_\Omega w\frac{\partial\psi}{\partial x}d\Omega$$

Of course lumping is necessary on the left hand side to make the expressions explicit.

The foregoing procedure can be continued until a steady state is reached.

The procedure followed for this example is deliberately kept as close as possible to a finite difference method in order to show that a lot of known techniques from the finite difference method can be adapted to the finite element method.

By this it is also clear that for small values of ν, i.e. for *convection dominated flow*, equation (10.56) allows oscillating solutions, so-called "wiggles".

This easily can be seen on the one-dimensional analogue of (10.56), for steady state:

$$u\frac{d\omega}{dx} = \nu\frac{d^2\omega}{dx^2} \qquad (10.58)$$

Discretization of (10.58) with central differences, with constant mesh spacing Δx, leads to

$$\frac{u}{2\Delta x}(\omega_{i+1} - \omega_{i-1}) = \frac{v}{\Delta x^2}(\omega_{i+1} - 2\omega_i + \omega_{i-1})$$

or

$$(1 - \text{Re}_c/2)\omega_{i+1} - 2\omega_i + (1 + \text{Re}_c/2)\omega_{i-1} = 0 \qquad (10.59)$$

where the cell-Reynolds number is

$$\text{Re}_c = u\Delta x/v$$

It is obvious that for $|\text{Re}_c| > 2$, (10.59) allows oscillating solutions.

It is known from finite-difference techniques (see previous chapters) that oscillation-free solutions can be obtained by a partial upwind discretization of (10.58), which is for $u > 0$:

$$(1 - \alpha)\frac{u}{2\Delta x}(\omega_{i+1} - \omega_{i-1}) + \alpha\frac{u}{\Delta x}(\omega_i - \omega_{i-1}) = \frac{v}{\Delta x^2}(\omega_{i+1} - 2\omega_i + \omega_{i-1}) \qquad (10.60)$$

By the choice

$$\alpha = \coth(\text{Re}_c/2) - 2/\text{Re}_c$$

the scheme (10.60) produces the exact nodal values and the solution is oscillation-free for

$$\alpha \geq 1 - 2/\text{Re}_c \qquad \text{for} \quad \text{Re}_c \geq 2 \qquad (10.61)$$

It was shown by Christie et al. [7] that this difference scheme can be reproduced by a Petrov-Galerkin formulation with weighting functions that have a quadratic modification with respect to the basis functions, as shown in Fig. 10.22:

$$w_1 = \phi_1 - \alpha F(\xi), \quad w_2 = \phi_2 + \alpha F(\xi), \quad F(\xi) = 3\xi(1 - \xi) \qquad (10.62)$$

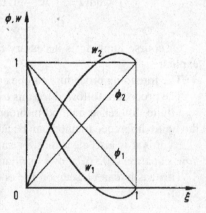

Fig. 10.22 Weighting functions and shape functions in the Petrov-Galerkin formulation

Fig. 10.23 Streamline pattern obtained with the Petrov-Galerkin finite element method for ω-ψ formulation ($Re_d = 250$)

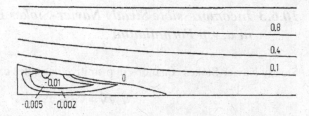

In two-dimensions, this Petrov-Galerkin formulation can be used on quadrilateral elements with weighting functions that are products of the one-dimensional weighting functions (10.62).

Figure 10.23 shows the solution obtained using these weighting functions, with the upwind factors equal to their minimum value according to (10.61), on the grid of Fig. 10.21, for $Re_d = Ud/v = 250$, where U is the mean velocity upstream of the step and d is the channel height at the inlet. As can be seen, even for such a coarse grid, a reasonable solution can be obtained [8].

The foregoing example is considered by the author as an excellent example to start with the FEM. A few remarks have to be added to this example. Nowadays, it has become common practice for convection-diffusion problems like (10.53) to introduce the upwinding in the streamline direction and not to treat both velocity directions independently as in the foregoing example. The method is then called a *streamline-upwind/Petrov-Galerkin method (SUPG)*. With the upwinding in the streamline direction, numerically caused artificial diffusion in the crosswise direction is avoided. Also nowadays, streamline upwinding is in most methods not introduced anymore through the explicit modification of the weighting functions. A more efficient approach is to use a convected form of the Galerkin weighting function, i.e.

$$w + \tau \vec{V}.\nabla w$$

instead of w, where τ is a time-wise parameter. With respect to the convective part in (53), this weighting function can be seen as a combination of a Galerkin weighting function and a least-squares weighting function. Therefore, also the name *Galerkin/Least squares method (GLS)* is used. For a basic discussion on SUPG and GLS, the reader is referred to Ref. [4]. The author considers such methods as already too advanced for an introductory level and therefore recommends to first implement upwinding in the more primitive form as described above.

An example of a finite element method for the incompressible Navier-Stokes equations in ω-ψ formulation using a SUPG-method is given in Ref. [9]. The time integration is a mixed implicit-explicit scheme. The necessary amount of upwinding depends on the time integration scheme. An example of a Bubnov-Galerkin method stable to very high Reynolds numbers with a fully implicit time integration scheme is given in Ref. [10].

10.6.3 Incompressible Steady Navier-Stokes Equations in u, v, p Formulation

Steady Navier-Stokes equations in primitive variables can be written as

$$\vec{V}.\nabla\vec{V} + \nabla p = \nu\nabla^2\vec{V} \tag{10.63}$$

$$\nabla.\vec{V} = 0 \tag{10.64}$$

By the introduction of a weighting function $\vec{W}$ associated to (10.63) and a weighting function q associated to (10.64), the weighted residual formulation is

$$\int\limits_{\Omega}\vec{W}.(\vec{V}.\nabla\vec{V})d\Omega + \int\limits_{\Omega}\vec{W}.\nabla p d\Omega - \int\limits_{\Omega}q\nabla.\vec{V}d\Omega = \int\limits_{\Omega}\nu\vec{W}.\nabla^2\vec{V}d\Omega$$

With integration by parts, this becomes

$$\int\limits_{\Omega}\vec{W}.(\vec{V}.\nabla\vec{V})d\Omega + \nu\int\limits_{\Omega}\nabla\vec{W}:\nabla\vec{V}d\Omega - \int\limits_{\Omega}(p\nabla.\vec{W} + q\nabla.\vec{V})d\Omega$$

$$= \int\limits_{\Gamma}\nu\vec{W}.(\nabla\vec{V}.\vec{n})d\Gamma - \int\limits_{\Gamma}p\vec{W}.\vec{n}d\Gamma \tag{10.65}$$

The formulation (10.65) involves a bilinear form

$$a(\vec{V},\vec{W}) = \int\limits_{\Omega}\nabla\vec{W}:\nabla\vec{V}d\Omega = \sum_{i,j}\int\limits_{\Omega}\frac{\partial W_i}{\partial x_j}\frac{\partial V_i}{\partial x_j}d\Omega$$

and a trilinear form

$$b(\vec{U},\vec{V},\vec{W}) = \int\limits_{\Omega}\vec{W}.(\vec{U}.\nabla\vec{V})d\Omega = \sum_{i,j}\int\limits_{\Omega}W_j U_i\frac{\partial V_j}{\partial x_i}d\Omega$$

Equation (10.65) can be written as

$$\nu a(\vec{V},\vec{W}) + b(\vec{V},\vec{V},\vec{W}) - \int\limits_{\Omega}(p\nabla.\vec{W} + q\nabla.\vec{V})d\Omega = c \tag{10.66}$$

The right hand side in (10.66) is (in two dimensions)

$$c = \int\limits_{\Gamma}[\nu w_x(\nabla u.\vec{n}) + \nu w_y(\nabla v.\vec{n}) - p(w_x n_x + w_y n_y)]d\Gamma$$

For a fluid domain as shown on Fig. 10.20, possible boundary conditions are

- at the solid boundaries: no-slip condition: u = 0, v = 0

This set of boundary conditions is a set of essential boundary conditions.
- at the inflow boundary: prescribed velocity profile: $u = u_0$, $v = 0$
This set is also a set of essential boundary conditions.
- at the outflow boundary ($n_x = 1$, $n_y = 0$) the right hand side of (10.66) becomes:

$$c = \int_\Gamma [-pw_x + v(w_x \frac{\partial u}{\partial x} + w_y \frac{\partial v}{\partial x})]d\Gamma$$

Thus, natural boundary conditions can be prescribed values of

$$f_n = -p + v\frac{\partial u}{\partial x}, \quad f_t = v\frac{\partial v}{\partial x} \tag{10.67}$$

Since pressure is only to be determined up to an additive constant, possible boundary conditions at the outflow boundary are the natural boundary conditions $f_n = 0$ and $f_t = 0$. With this choice, for the problem of Fig. 10.20, the right hand side in (10.66) vanishes since the boundary conditions on the weighting functions, associated to essential boundary conditions in u and v are: $w_x = w_y = 0$.

Clearly, both in the field equations (10.66) and in the boundary conditions (10.67), pressure enters in non-derivative form while velocity components enter in first-derivative form. This justifies the choice of different basis functions for velocity components and pressure:

$$u = \sum u_k \phi_k$$
$$v = \sum v_k \phi_k \tag{10.68}$$
$$p = \sum p_N \tilde{\phi}_N$$

Inserting the interpolation formulas (10.68) into the weak formulation results in the discrete set of equations

$$K(U)\,U + C\,P = f$$
$$C^T\,U = g \tag{10.69}$$

where U is the global vector of nodal velocities (u and v) and P is the global vector of nodal pressures. The vectors f and g contain the boundary conditions. K(U) is a positive definite symmetric matrix. C is an asymmetric and indefinite rectangular matrix.

It is obvious that due to this indefinite matrix, (10.69) can allow irregular solutions. For instance for $u = 0$, $v = 0$, $f = 0$, $g = 0$, $C\,P = 0$ can allow non-trivial solutions for P.

For instance, if bilinear interpolation is used both for velocity components and pressure, the integrals on the element level, contributing to the system $C\,P = 0$ are:

$$\int_{\Omega_e} p\nabla.Wd\Omega = \int_{-1}^{+1}\int_{-1}^{+1} p[C_1(\alpha + \beta\eta) + C_2(\gamma + \delta\xi)]d\xi d\eta$$

where α, β, γ and δ are coefficients associated to the mapping of the element to the parent element ($-1 \leq \xi \leq 1$, $-1 \leq \eta \leq 1$). Hence for $p \sim \xi\eta$ these contributions can be zero for non-zero p. These solutions are called *spurious pressure modes*.

In order to avoid these spurious modes, it is necessary to use a *mixed interpolation*, i.e. the interpolation structure of the velocity components is to be different from the interpolation structure of the pressure. Since the regularity conditions on pressure are lower than on velocity, pressure is to be approximated by interpolation polynomials with a lower degree than the polynomials for the velocity components. Additional requirements come from the observation that in (10.69), the continuity equation should not completely specify the velocity field.

For this introductory text it would be too long to enter into a discussion on the choice of the elements. The elements have to satisfy a so-called LBB-condition (after O.A. Ladyshenskaya, F. Brezzi and I. Babuska). The reader is referred to Refs. [3–5, 11, 12] for a discussion on this topic.

Two elements usually are preferred. The first one is the triangle with 7 degrees of freedom for velocity components: the values at the vertices and the mid-sides, defining a quadratic polynomial, enriched with a function of degree 3 which is zero at the boundary of the element (the so-called bubble function), and three degrees of freedom for pressure: the value at the centre and two derivatives, defining a piecewise linear discontinuous polynomial. The second one is the quadrilateral with 9 degrees of freedom for velocity components, obtained from a biquadratic interpolation, and three degrees of freedom for pressure: the value at the centre and two derivatives, defining a piecewise linear discontinuous polynomial.

It is also to be remarked that the solution of the system (10.69) is not easy since it contains zeros on the diagonal. For solution techniques, we again refer to Refs. [3–5].

Another remark concerns the form of the momentum equation (10.63). Many practitioners of the FEM prefer not to write this equation in its simplified form (10.63), but use the more primitive form

$$\vec{V}.\nabla\vec{V} + \nabla p = \nabla.\overset{\leftrightarrow}{\tau}$$

where $\overset{\leftrightarrow}{\tau}$ stands for the stress tensor (divided by ρ).

With this formulation, the natural boundary conditions become the normal components of the stress, i.e. the tractions. For problems with a free outlet, often the tractions can be set to zero (traction-free boundary conditions). For the outlet of a channel (Fig. 10.23), the tangential traction is not zero and cannot be prescribed. Instead, the boundary conditions (10.67) can be used. These terms usually are called pseudo-tractions.

10.6.4 Compressible Euler and Navier-Stokes Equations

Euler equations are a system of form

$$\frac{\partial U}{\partial t} + \frac{\partial f}{\partial x} + \frac{\partial g}{\partial y} = 0 \qquad (10.70)$$

where

$$U = \begin{bmatrix} \rho \\ \rho u \\ \rho v \\ \rho E \end{bmatrix}, \qquad f = \begin{bmatrix} \rho u \\ \rho uu + p \\ \rho uv \\ \rho Hu \end{bmatrix}, \qquad g = \begin{bmatrix} \rho v \\ \rho uv \\ \rho vv + p \\ \rho Hv \end{bmatrix}$$

where E is total energy and H is total enthalpy.

The system (10.70) is hyperbolic in time with Jacobians

$$A = \frac{\partial f}{\partial U}, \qquad B = \frac{\partial g}{\partial U}$$

Using Taylor expansion, we have, up to second order

$$U(t + \Delta t) \approx U(t) + \Delta t \frac{\partial U}{\partial t} + \frac{\Delta t^2}{2} \frac{\partial^2 U}{\partial t^2}$$

With

$$\frac{\partial^2 U}{\partial t^2} = \frac{\partial}{\partial t}\left(\frac{\partial U}{\partial t}\right) = -\frac{\partial}{\partial t}\left(\frac{\partial f}{\partial x} + \frac{\partial g}{\partial y}\right) = -\frac{\partial}{\partial x}\left(A\frac{\partial U}{\partial t}\right) - \frac{\partial}{\partial y}\left(B\frac{\partial U}{\partial t}\right)$$

$$= \frac{\partial}{\partial x}\left[A\left(\frac{\partial f}{\partial x} + \frac{\partial g}{\partial y}\right)\right] + \frac{\partial}{\partial y}\left[\left(B\left(\frac{\partial f}{\partial x} + \frac{\partial g}{\partial y}\right)\right)\right]$$

this becomes

$$U(t + \Delta t) \approx U(t) - \Delta t\left(\frac{\partial f}{\partial x} + \frac{\partial g}{\partial y}\right) + \frac{\Delta t^2}{2}\left\{\frac{\partial}{\partial x}\left[A\left(\frac{\partial f}{\partial x} + \frac{\partial g}{\partial y}\right)\right] + \frac{\partial}{\partial y}\left[\left(B\left(\frac{\partial f}{\partial x} + \frac{\partial g}{\partial y}\right)\right)\right]\right\}$$
$$\tag{10.71}$$

A central type finite difference discretization on (10.71) is called a one step Lax-Wendroff method. The analogue in the finite element technique is a Galerkin formulation on (10.71):

$$\int_\Omega W^T[U(t + \Delta t) - U(t)]d\Omega = -\Delta t \int_\Omega W^T\left(\frac{\partial f}{\partial x} + \frac{\partial g}{\partial y}\right)d\Omega$$

$$- \frac{\Delta t^2}{2} \int_\Omega \left(\frac{\partial W^T}{\partial x}A + \frac{\partial W^T}{\partial x}B\right)\left(\frac{\partial f}{\partial x} + \frac{\partial g}{\partial y}\right)d\Omega \quad (10.72)$$

$$+ \frac{\Delta t^2}{2} \int_\Gamma W^T\left(\frac{\partial f}{\partial x} + \frac{\partial g}{\partial y}\right)(A\vec{I}_x + B\vec{I}_y).n d\Gamma$$

where W is a vector of weighting functions.

In (10.72), the left hand side can be lumped. The vectors f and g can be represented by piecewise linear functions and the matrices A and B by piecewise constant functions.

A finite element method based on this procedure usually is called a *Taylor-Galerkin* method.

Although, in principle, there is nothing special in using a Lax-Wendroff time-stepping method as time-integration for a semi-discrete set of equations obtained by the FEM, these formulations mostly are indicated explicitly by the term *Lax-Wendroff/Taylor-Galerkin methods*. Obviously, other time-stepping methods like for instance Runge-Kutta methods also can be used. An example with two-stage Runge-Kutta stepping is given in Ref. [13].

Furthermore, by the example given in this section, it is demonstrated that many methods that have been developed in the FDM or in the FVM, with central discretizations can be adapted to the FEM. Also, the problems remain the same. For instance, Lax-Wendroff time-stepping to a Bubnov-Galerkin formulation on (10.70) as described above is linearly stable but necessitates the introduction of artificial viscosity to stabilize shocks. As for the FDM and FVM, Runge-Kutta stepping methods already need artificial viscosity for linear stability. Obviously, by adding artificial viscosity, certainly when it is done for linear stability, as in Ref. [13], much of the rigour of the FEM is lost and the formulations come very close to FVM-formulations.

10.7 Current Evolutions

As in other branches of computational fluid dynamics, there is almost no work done nowadays on potential flow. There is still some activity in the field of Euler equations, but the major part of the work is on Navier-Stokes equations. Algorithms typically are first tested on laminar flows and structured grids. Real life applications however are mostly in turbulent flow and in geometries that necessitate unstructured grids. In this introductory text we cannot discuss the coupling of the Navier-Stokes equations and the turbulence equations. The aspects of turbulence modelling and the coupling with the flow equations are similar to those in the FDM and FVM. Further, the formulation of a finite element method on unstructured grids is more or less straightforward. Of course, techniques for generating these grids are necessary but, again, the grid generation aspects are not particular for the FEM.

On the algorithmic side there are two current tendencies. The first is that upwind techniques gain popularity. This tendency is similar to what is observed in the FDM and the FVM. As already discussed for the scalar advection-diffusion equation in Sect. 10.6.2, upwinding can be introduced by SUPG and GLS. These techniques are equivalent for a scalar equation, but their extension to systems like incompressible Navier-Stokes equations or compressible Euler and Navier-Stokes equations is different. For a fundamental discussion of the SUPG and GLS methods, the reader is referred to Refs. [14, 15]. The main advantage of Galerkin/Least Squares or Least Squares methods for incompressible flows is that equal order interpolation of all variables is possible. These methods do not suffer from the instability encountered with Galerkin methods (LBB stability condition). Therefore they are indicated with the name *stabilized finite element methods*. For compressible flows, the advantage is, like in the FDM and FVM, the good representation of shocks thanks to the upwinding. A recent review of stabilized finite element methods is Ref. [16]. For a discussion of the use of least squares principles in compressible and incompressible flows, the reader is referred to Refs. [17, 18]. As is shown in these papers, the Galerkin part in the formulation is even not always necessary.

A second way to introduce upwinding is by a Galerkin formulation with discontinuous weighting functions. The *discontinuous Galerkin method* comes very close to finite volume upwind methods of flux-difference splitting type. For examples, the reader is referred to Refs. [19–21]. The discontinuous Galerkin method can be applied to compressible flow Euler and Navier-Stokes equations and to convection-diffusion equations like the vorticity equation in the ω-ψ formulation of incompressible flow Navier-Stokes equations.

A second tendency observed in algorithms for incompressible flows is the wish to avoid mixed interpolation. One way to make equal order interpolation possible is, as just discussed, the use of Galerkin/Least Squares methods. Another way, only applicable to incompressible Navier-Stokes equations, is the so-called *projection method*. The technique is the FEM variant of what usually is called a *pressure correction method* or a fractional step method in the FDM. The momentum equations are advanced in time with a stepping technique for advection-diffusion equations, under frozen pressure. The newly obtained velocity field is in general not divergence free. It is then projected to a divergence free space resulting in a velocity correction and an associated pressure correction. There are different techniques to calculate the correction. Also, many stepping techniques can be used. An example, using a two stage Runge-Kutta Taylor/Galerkin stepping is given in Ref. [22]. A very simple example, but restricted to low Reynolds numbers is given in Ref. [23].

References

1. Zienkiewicz O.C. and Morgan K., Finite elements and approximation. John Wiley, 1983.
2. Reddy J.N., An introduction to the finite element method. McGraw-Hill, 1984.

References [1] and [2] are introductory texts on the finite element method with the same level as used in this chapter. The texts are more elaborate and treat more examples but do not introduce much more basic notions. This chapter is, in some sense, a short version of Refs. [1, 2]. References [3] and [4] discuss finite element methods for flow problems on a level which is somewhat higher than the level used in this chapter. After obtaining a good understanding of this chapter, the reader should be able to study Refs. [3, 4]. Reference [3] treats only incompressible fluids with the mixed finite element approach (interpolation polynomials for velocity components with higher order than for pressure). Reference [4] discusses stabilized finite element approaches for incompressible fluids (allowing the same interpolation polynomials for all variables), discontinuous Galerkin approaches and methods for compressible fluids. Reference [5] is the fifth edition of a standard text on finite element methods, covering all basic aspects and applications to solid and fluid mechanics. For the fluid mechanics examples, the characteristic-based split method (not explained in this chapter) is almost exclusively used.

3. Reddy J.N. and Gartling D.K., The finite element method in heat transfer and fluid dynamics. CRC Press, 1994.
4. Donea J. and Huerta A., Finite element methods for flow problems. John Wiley, 2003.
5. Zienckiewicz O.C. and Taylor R.L., The finite element method. Volume 1: The basics, Volume 2: Solid mechanics, Volume 3: Fluid dynamics. Butterworth-Heineman, 2000.
6. Oden J.T. and Reddy J.N., An introduction to the mathematical theory of finite elements. John Wiley, 1976.
7. Christie I., Griffiths D.F., Mitchell A.R. and Zienkiewicz O.C., Finite element methods for second order differential equations with significant first derivatives. Int. J. Num. Meth. Engng., Vol. 10, pp. 1389–1396, 1976.
8. Dick E., Steady laminar flow over a downwind-facing step as a critical test case for the upwind Petrov-Galerkin finite element method. Applied Scientific Research, Vol. 39, pp. 321–328, 1982.
9. Tezduyar T.E., Liou J., Ganjoo D.K. and Behr M., Solution techniques for the vorticity-streamfunction formulation of two-dimensional unsteady incompressible flows. Int. J. Num. Meth. Fluids, Vol. 11, pp. 515–539, 1990.
10. Comini G., Manzan M. and Nonino C., Finite element solution of the stream function-vorticity equations for incompressible two-dimensional flows. Int. J. Num. Meth. Fluids, Vol. 19, pp. 513–525, 1994.
11. Sani R.L., Gresho P.M., Lee R.L., Griffiths D.F. and Engelman M., The cause and cure of the spurious pressures generated by certain FEM solutions of the incompressible Navier-Stokes equations. Int. J. Num. Meth. Fluids, Vol. 1, pp. 17–43 & pp. 171–204, 1981.
12. Fortin M., Old and new finite elements for incompressible flows. Int. J. Num. Meth. in Fluids, Vol. 1, pp. 347–364, 1981.
13. Löhner R., Morgan K., Peraire J. and Vahdati M., Finite element flux-corrected transport (FEM-FCT) for the Euler and Navier-Stokes equations. Int. J. Num. Meth. Fluids, Vol. 7, pp. 1093–1109, 1987.
14. Hughes T.J.R., Recent progress in the development and understanding of SUPG methods with special reference to the compressible Euler and Navier-Stokes equations. Int. J. Num. Meth. Fluids, Vol. 7, pp. 1261–1275, 1987.
15. Hughes T.J.R., Franca L.P. and Hulbert G.M., A new finite element formulation for computational fluid dynamics: VIII. The Galerkin/least squares method for advective-diffusive equations. Comput. Meth. Appl. Mech. Engng., Vol. 73, pp. 173–189, 1989.
16. Tezduyar T.E., Finite elements in fluids: Stabilized formulations and moving boundaries and interfaces. Computers and Fluids, Vol. 36, pp. 191–206, 2007.
17. Lefebre D., Peraire J. and Morgan K., Least squares finite element solutions of compressible and incompressible flows. Int. J. Num. Meth. Heat and Fluid Flow, Vol. 2, pp. 99–113, 1992.
18. Pontaza J.P., A least-squares finite element formulation for unsteady incompressible flows with improved velocity-pressure coupling. J. Comput. Phys., Vol. 217, pp. 199–224, 2006.
19. Bassi F. and Rebay S., A high-order accurate discontinuous finite element method for the numerical solution of the compressible Navier-Stokes equations. J. Comput. Phys., Vol. 131, pp. 267–279, 1997.
20. Cockburn B. and Shu C.W., The Runge-Kutta discontinuous Galerkin finite element method for conservation laws V: Multidimensional systems. J. Comput. Phys., Vol. 141, pp. 199–224, 1998.
21. Bassi F., Crivellini A., Rebay S., Savini M., Discontinuous Galerkin solution of the Reynolds-averaged Navier-Stokes and k-ω turbulence model equations. Computers and Fluids, Vol. 34, pp. 507–540, 2005.
22. Ren G. and Utnes T., A finite element solution of the time dependent incompressible Navier-Stokes equations using a modified velocity correction method. Int. J. Num. Meth. Fluids, Vol. 17, pp. 349–364, 1993.
23. Nonino C., Comini G. and Groce G., Three-dimensional flows over backward facing steps. Int. J. Num. Meth. Heat and Fluid Flow, Vol. 9, pp. 224–239, 1999.

Chapter 11
Introduction to Finite Volume Methods in Computational Fluid Dynamics

E. Dick

11.1 Introduction

The basic laws of fluid dynamics are conservation laws. They are statements that express the conservation of mass, momentum and energy in a volume closed by a surface. Only with the supplementary requirement of sufficient regularity of the solution can these laws be converted into partial differential equations. Sufficient regularity cannot always be guaranteed. Shocks form the most typical example of a discontinuous flow field. In case discontinuities occur, the solution of the partial differential equations is to be interpreted in a weak form, i.e. as a solution of the integral form of the equations. For example, the laws governing the flow through a shock, i.e. the Hugoniot-Rankine laws, are combinations of the conservation laws in integral form. For a correct representation of shocks, also in a numerical method, these laws have to be respected.

There are additional situations where an accurate representation of the conservation laws is important in a numerical method. A second example is the slip line which occurs behind an airfoil or a blade if the entropy production is different on streamlines on both sides of the profile. In this case, a tangential discontinuity occurs. Another example is incompressible flow where the imposition of incompressibility, as a conservation law for mass, determines the pressure field.

In the cases cited above, it is important that the conservation laws in their integral form are represented accurately. The most natural method to accomplish this is to *discretize the integral form of the equations* and not the differential form. This is the basis of a *finite volume method*. Further, in cases where strong conservation in integral form is not absolutely necessary, it is still physically appealing to use the basic laws in their most primitive form.

The flow field or *domain* is subdivided, as in the finite element method, into a set of *non-overlapping cells* that *cover the whole domain*. In the finite volume method (FVM) the term *cell* is used instead of the term *element* used in the finite element method (FEM). The conservation laws are applied to determine the flow

E. Dick
Department of Flow, Heat and Combustion Mechanics, Ghent University, Sint-Pietersnieuwstraat 41, 9000 Gent, Belgium, e-mail: Erik.Dick@ugent.be

Fig. 11.1 Typical choice of
grids in the FVM; (**a**): structured quadrilateral grid; (**b**):
structured triangular grid; (**c**):
unstructured triangular grid

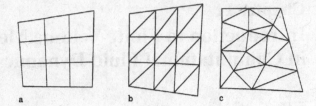

a   b   c

variables in some discrete points of the cells, called *nodes*. As in the FEM, these
nodes are at typical locations of the cells, such as cell-centres, cell-vertices or mid-
sides. Obviously, there is considerable freedom in the choice of the cells and the
nodes. Cells can be *triangular, quadrilateral*, etc. They can form a *structured grid*
or an *unstructured grid*. The whole geometrical freedom of the FEM can be used in
the FVM. Figure 11.1 shows some typical grids.

The choice of the nodes can be governed by the wish to represent the solution
by an interpolation structure, as in the FEM. A typical choice is then *cell-centres*
for representation as piecewise constant functions or *cell-vertices* for representation
as piecewise linear (or bilinear) functions. However, in the FVM, a function space
for the solution need not be defined and nodes can be chosen in a way that does
not imply an interpolation structure. Figure 11.2 shows some typical examples of
choices of nodes with the associated definition of variables.

The first two choices imply an interpolation structure, the last two do not. In
the last example, function values are not defined in all nodes. The grid of nodes
on which pressure and density are defined is different from the grid of nodes on
which velocity-x components and velocity-y components are defined. This approach
commonly is called the *staggered grid approach.*

The third basic ingredient of the method is the choice of the volumes on which
the conservation laws are applied. In Fig. 11.2 some possible choices of control

Fig. 11.2 Typical choice
of nodes in the FVM. The
marked nodes are used in the
flux balance of the control volume. (**a**): piecewise constant
interpolation structure; (**b**):
piecewise linear interpolation
structure; (**c**): no interpolation
structure with all variables
defined in each node; (**d**): no
interpolation structure with
not all variables defined in
each node; (Cartesian grid), o:
ρ and p,: u, Δ: v

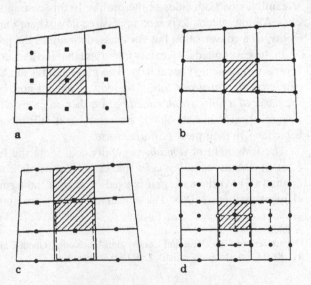

a   b

c   d

volumes are shown (shaded). In the first two examples, control volumes coincide with cells. The third example in Fig. 11.2 shows that the *volumes* on which the conservation laws are applied *need not coincide with the cells* of the grid. Volumes even can be overlapping. Figure 11.3 shows some typical examples of volumes not coinciding with cells, for overlapping and non-overlapping cases. The term volume denotes the control volume to which the conservation laws are applied (i.e. connected to function value determination), while the term cell denotes a mesh of the grid (i.e. connected to geometry discretization). A consistency requirement for the cells is that they are non-overlapping and that they span the whole domain. The consistency requirement for the volumes is weaker. They can be overlapping so that families of volumes are formed. Each family should consist of non-overlapping volumes which span the whole domain. The consistency requirement is that a flux leaving a volume should enter another one.

Obviously, by the decoupling of volumes and cells, the freedom in the determination of the function representation of the flow field in the finite volume method becomes much larger than in both the finite element and finite difference method. It is in particular the combination of the formulation of a flow problem on control volumes which is the most *physical* way to obtain a discretization, with the *geometric* flexibility in the choice of the grid and the flexibility in defining the

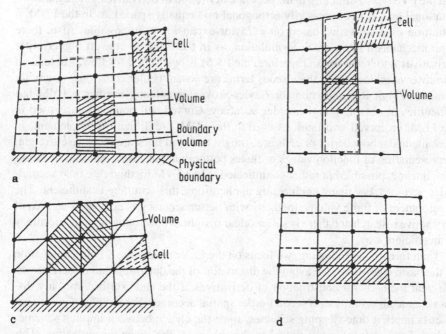

Fig. 11.3 Choice of volumes not coinciding with cells, overlapping and non-overlapping cases. (**a**): volumes staggered with respect to cells, non-overlapping case; (**b**): volumes non-staggered with respect to cells, overlapping case; (**c**): volumes non-staggered with respect to cells, overlapping case; (**d**): volumes staggered with respect to cells, overlapping case

discrete *flow variables* which makes the finite volume method attractive for engineering applications.

The finite volume method (FVM) tries to combine the best from the *finite element method* (FEM), i.e. the *geometric flexibility*, with the best of the *finite difference method* (FDM), i.e. the flexibility in defining the *discrete flow field* (discrete values of dependent variables and their associated fluxes). Some formulations are near to finite element formulations and can be interpreted as subdomain collocation finite element methods (e.g. Fig. 11.2a). Other formulations are near to finite difference formulations and can be interpreted as conservative finite difference methods (e.g. Fig. 11.3a). Other formulations are in between these limits.

The mixture of FEM-like and FDM-like approaches sometimes leads to confusion in terminology. Some authors with an FEM-background use the term *element* for *cell* and then often use the term (control) *cell* for (control) *volume*. Strictly speaking, the notion element is different from the notion cell. A grid is subdivided into meshes. A mesh has the significance of a cell if it only implies a subdivision of the geometry. If it also implies, in the FEM-sense, a definition of a function space, it is an element.

From the foregoing, it could be concluded that the FVM only has advantages over the FEM and the FDM and thus one could ask why all of computational fluid dynamics (CFD) is not based on the FVM. From the foregoing, it is already clear that the FVM has a difficulty in the accurate definition of derivatives. Since the computational grid is not necessarily orthogonal and equally spaced, as in the FDM, a definition of a derivative based on a Taylor-expansion is impossible. Also, there is no mechanism like a weak formulation, as in the FEM, to convert higher order derivatives into lower ones. Therefore, the FVM is best suited for flow problems in primitive variables, where the viscous terms are absent (Euler equations) or are not dominant (high Reynolds number Navier-Stokes equations). Further, a FVM has difficulties in obtaining higher order accuracy. Curved cell boundaries, as used in the FEM, or curved grid lines, as used in the FDM, are difficult to implement. In most methods, boundaries of cells are straight and grid lines are piecewise straight. Representation of function values or fluxes better than piecewise constant or piecewise linear is possible but rather complicated. Most FVM methods are only second-order accurate. For many engineering applications, this accuracy is sufficient. The development of finite volume methods with better accuracy is nowadays an area of very active research and there is still no clear insight in how to reach higher accuracy in an efficient way.

Therefore, in the following, we focus on the Euler equations. So, for explanation of the basic algorithms, we avoid the discussion of the determination of derivatives. We treat methods for construction of derivatives at the end. Further, we only discuss classic algorithms with second-order spatial accuracy. For simplicity we do not discuss implicit time stepping schemes, since the choice between implicit schemes and explicit schemes is not linked to the choice of the space discretization. This introductory text also does not aim to give a complete overview of the FVM. It only aims to illustrate some of the basic properties on examples of methods that are widely used.

11.2 Fem-Like Finite Volume Methods

FEM-like finite volume methods use cells to which an interpolation structure is associated. So, the cells form elements in the FEM-sense. Two interpolation structures can be used: piecewise constant interpolation and piecewise linear (or bilinear) interpolation. Figure 11.4 shows some possibilities on (structured) quadrilateral and triangular grids. The piecewise constant interpolation is denoted by the *cell-centred* method, while the piecewise linear interpolation is denoted by the *cell-vertex* method. In both methods, the cells and a group of cells around a node are used as volumes. In the first method, data are at cell centres. In the second method, data are al cell vertices.

We illustrate here some formulations for the Euler equations. The set of Euler equations can be written in two dimensions as

$$\frac{\partial U}{\partial t} + \frac{\partial f}{\partial x} + \frac{\partial g}{\partial y} = 0 \tag{11.1}$$

with

$$U = \begin{bmatrix} \rho \\ \rho u \\ \rho v \\ \rho E \end{bmatrix}, \qquad f = \begin{bmatrix} \rho u \\ \rho uu + p \\ \rho uv \\ \rho Hu \end{bmatrix}, \qquad g = \begin{bmatrix} \rho v \\ \rho uv \\ \rho vv + p \\ \rho Hv \end{bmatrix}$$

where ρ is density, u and v are Cartesian components of velocity, p is pressure, E is total energy and H is total enthalpy (γ is the adiabatic constant).

$$E = \frac{1}{\gamma - 1} \frac{p}{\rho} + \tfrac{1}{2} u^2 + \tfrac{1}{2} v^2$$

$$H = E + \frac{p}{\rho}$$

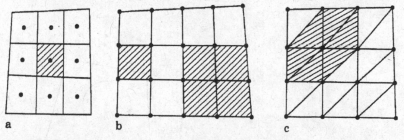

a   **b**   **c**

Fig. 11.4 FEM-like finite volume methods. (**a**): cell-centred; (**b**): cell-vertex with non-overlapping and overlapping volumes on quadrilateral cells; (**c**): cell-vertex on triangular cells

11.2.1 Cell-Centred Formulation

For a cell as shown in Fig. 11.5, the values of the dependent variables are stored in
the centre of the cell. These values do not necessarily have to be seen as nodal values, but can also be seen as mean values over the cell. Therefore, in the cell-centred
method, for visualization purposes, often, after completion of the calculations, values are attributed to the vertices of the grid by taking a weighted mean of the values
in adjacent cells. Further, the interpretation as mean values allows higher order formulation, as we discuss in Sect. 11.6. First, we discuss the typical second-order
accurate formulations.

Using the control volume of Fig. 11.5, a semi-discretization of (11.1) is obtained by

$$\Omega_{i,j}\frac{\partial U}{\partial t} + \int_{abcd} \vec{F}.\vec{n}\,dS = 0 \tag{11.2}$$

where $\Omega_{i,j}$ denotes the volume (area) of the control volume. $\vec{F}$ is the flux vector:
$\vec{F} = f\,\vec{1}_x + g\,\vec{1}_y$, dS is a surface element and $\vec{n}$ is the outward normal. By taking
the positive sense as indicated in the figure, we have

$$\vec{n}\,dS = dy\,\vec{1}_x - dx\,\vec{1}_y \tag{11.3}$$

Inserting (11.3) into (11.2) gives

$$\Omega_{i,j}\frac{\partial U}{\partial t} + \int_{abcd} (f\,dy - g\,dx) = 0 \tag{11.4}$$

Further, f and g have to be defined on the boundary of the volume. A mean value
between adjacent nodes looks to be the simplest choice, for example:

$$f_{ab} = \tfrac{1}{2}(f_{i,j} + f_{i,j-1}), \qquad g_{ab} = \tfrac{1}{2}(g_{i,j} + g_{i,j-1}) \tag{11.5}$$

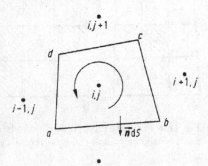

Fig. 11.5 Cell-centred
formulation

Since the flux functions are non-linear functions of the dependent variables, an alternative for (11.5) is

$$f_{ab} = f[\tfrac{1}{2}(U_{i,j} + U_{i,j-1})], \qquad g_{ab} = g[\tfrac{1}{2}(U_{i,j} + U_{i,j-1})] \tag{11.6}$$

With (11.6) is meant that the dependent variables are first averaged and that afterwards flux vectors are calculated. This is not a popular choice, since it implies about twice as many flux evaluations as (11.5). Indeed, when in a structured quadrilateral grid, there are n_x subdivisions in longitudinal direction and n_y subdivisions in transversal direction, then there are $n_x n_y$ cells, but $n_x(n_y + 1) + n_y(n_x + 1)$ cell faces. This does not imply that the work involved in (11.6) is twice as much as the work involved in (11.5). A lot of computational effort can be gained by remarking that a momentum flux is a mass flux multiplied by an average velocity, etc. Nevertheless, the definition (11.5) is the cheapest. Therefore, (11.5) is the only *central flux definition* used in the following (one-sided flux definitions are also possible, as discussed later).

With the definition of the discrete fluxes f and g, the semi-discretization (11.4) is completed. It is now to be integrated in time.

11.2.1.1 Lax-Wendroff Time Stepping

Since Lax-Wendroff time-stepping is a very classic explicit time integration method in the finite difference method, explained in previous chapters, we begin by discussing how this time-stepping can be applied to a finite volume formulation. We first recall the principles of a Lax-Wendroff method with the use of the one-dimensional scalar model equation

$$\frac{\partial u}{\partial t} + \frac{\partial f(u)}{\partial x} = 0 \tag{11.7}$$

A Taylor series expansion to second order gives

$$u^{n+1} \approx u^n + \Delta t \left(\frac{\partial u}{\partial t}\right)^n + \frac{\Delta t^2}{2}\left(\frac{\partial^2 u}{\partial t^2}\right)^n \tag{11.8}$$

and

$$\frac{\partial^2 u}{\partial t^2} = \frac{\partial}{\partial t}\left(\frac{\partial u}{\partial t}\right) = -\frac{\partial}{\partial t}\left(\frac{\partial f}{\partial x}\right) = -\frac{\partial}{\partial x}\left(\frac{\partial f}{\partial t}\right)$$

or

$$\frac{\partial^2 u}{\partial t^2} = -\frac{\partial}{\partial x}\left(\frac{\partial f}{\partial u}\frac{\partial u}{\partial t}\right) = \frac{\partial}{\partial x}\left(a\frac{\partial f}{\partial x}\right) \tag{11.9}$$

with

$$a = \frac{\partial f}{\partial u}$$

Combination of (11.8) and (11.9) gives

$$u^{n+1} \approx u^n + \Delta t \left(-\frac{\partial f^n}{\partial x} \right) + \frac{\Delta t^2}{2} \frac{\partial}{\partial x} \left(a \frac{\partial f^n}{\partial x} \right) \tag{11.10}$$

The two-dimensional analogue of (11.10) on the Euler equations (11.1) is

$$U^{n+1} \approx U^n + \Delta t \left(-\frac{\partial f^n}{\partial x} - \frac{\partial g^n}{\partial y} \right) + \frac{\Delta t^2}{2} \left\{ \frac{\partial}{\partial x} \left[A^n \left(\frac{\partial f^n}{\partial x} + \frac{\partial g^n}{\partial y} \right) \right] + \frac{\partial}{\partial y} \left[B^n \left(\frac{\partial f^n}{\partial x} + \frac{\partial g^n}{\partial y} \right) \right] \right\} \tag{11.11}$$

where A and B are the Jacobian matrices of the flux vectors:

$$A = \frac{\partial f}{\partial U} \quad , \quad B = \frac{\partial g}{\partial U}$$

In the finite-difference method, a discretization of (11.10) or (11.11) is called a one-step Lax-Wendroff method. As explained in previous chapters, a possible procedure is to expand the second-order derivatives in space in (11.10) or (11.11) and to replace these derivatives by central difference approximations. In principle, a finite volume formulation on (11.10) or (11.11) is possible since these equations take the form of a flux-balance. The fluxes contain however derivatives. Since the definition of derivatives is not simple in the finite volume method, one-step methods are never used. The most popular two-step formulations, such as the Richtmyer variant and the MacCormack variant, can however be used without problems in the FVM. Further, in the one-step method the primitive flux balances are lost while these are visible in the two-step formulations. Since the MacCormack variant was explained in previous chapters, we illustrate here how this variant can be formulated in finite volume form.

In the MacCormack variant of the Lax-Wendroff method, (11.8) is written as

$$u^{n+1} = {}^1\!/_2 u^n + {}^1\!/_2 \Delta t \left(\frac{\partial u}{\partial t} \right)^n + {}^1\!/_2 u^n + {}^1\!/_2 \Delta t \left[\frac{\partial}{\partial t} \left(u + \Delta t \frac{\partial u}{\partial t} \right) \right]^n \tag{11.12}$$

With (predictor)

$$\overline{u^{n+1}} = u^n + \Delta t \left(\frac{\partial u}{\partial t} \right)^n \tag{11.13}$$

(11.12) can be written as (corrector)

$$u^{n+1} = {}^1\!/_2 \left[u^n + \overline{u^{n+1}} + \Delta t \frac{\partial}{\partial t} \overline{u^{n+1}} \right] \tag{11.14}$$

The discretization by MacCormack of (11.13) and (11.14) is

$$u_i^{\overline{n+1}} = u_i{}^n - \Delta t \left(\frac{f_{i+1}^n - f_i^n}{\Delta x} \right) \tag{11.15}$$

$$u_i^{n+1} = {}^1\!/_2 \left[u_i{}^n + u_i^{\overline{n+1}} - \Delta t \left(\frac{f_i^{\overline{n+1}} - f_{i-1}^{\overline{n+1}}}{\Delta x} \right) \right] \tag{11.16}$$

Equations (11.15) and (11.16) form the forward-backward variant. Obviously the forward and backward discretizations can be interchanged. In the terminology of ordinary differential equations, the MacCormack method is a *predictor-corrector method*.

The implementation of the MacCormack variant of the Lax-Wendroff method is rather straightforward. In the forward-backward formulation, in the predictor step on Fig. 11.5, the fluxes at the sides ab, bc, cd and da are evaluated with function values in the nodes (i,j), $(i+1,j)$, $(i,j+1)$ and (i,j), respectively. In the corrector step this is $(i,j-1)$, (i,j), (i,j) and $(i-1,j)$.

At inflow and outflow boundaries, the FVM can be used as the FDM. This means that, in general, extrapolation formulas are used to define values in nodes outside the domain. For instance, for a subsonic inflow, it is common practice to extrapolate the Mach number from the flow field and to impose stagnation properties and flow direction. At a subsonic outflow, the reverse can be done, i.e. extrapolation of stagnation properties and flow direction and fixing of a Mach number. Very often, pressure is imposed at outflow.

At solid boundaries, the convective flux can be set to zero. This means that in the flux through a cell surface on a solid boundary, only the pressure comes in:

$$f\,dy - g\,dx = p \begin{bmatrix} 0 \\ dy \\ -dx \\ 0 \end{bmatrix}$$

The pressure at the boundary can be taken to be the pressure in the cell. Sometimes, as in the FDM, an extrapolation of pressure is used. It is however not always easy to define extrapolation formulas on distorted or unstructured grids.

Obviously, four geometrical variants in the choice of the biasing of the fluxes are possible. Figure 11.6 shows schematically the possibilities for the predictor step. In

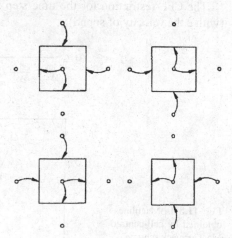

Fig. 11.6 Possible variants of the biasing for flux functions in the predictor step of a MacCormack method

Fig. 11.7 GAMM-channel
test problem

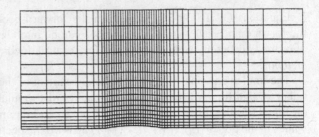

the corrector step, the biasing is inverted. In practice, the four possibilities are used alternatively.

We illustrate now the cell-centred MacCormack scheme on the well-known GAMM-channel test problem for transonic flows [1]. This problem is shown in Fig. 11.7, discretized with a 49×17 grid. The result shown in Fig. 11.8 is however obtained on a once refined grid, i.e. a 97×33 grid. The channel of Fig. 11.7 is almost straight except for a small circular perturbation on the lower boundary with height 4.2% of the chord. The result of Fig. 11.8 is obtained with the MacCormack method described above. Pressure is imposed at the outlet, corresponding to an isentropic Mach number of 0.85.

As in the finite-difference method, to obtain this result, some artificial viscosity is needed to stabilize the solution in the shock region (see discussion in previous chapters). This is done here in a rather primitive way by adding to each step a smoothing of form

$$\mu \left[U^n_{i+1,j} + U^n_{i-1,j} + U^n_{i,j+1} + U^n_{i,j-1} - 4U^n_{i,j} \right],$$

where μ is a very small coefficient. For the result in Fig. 11.8: $\mu = 0.001$. This is enough to stabilize the shock. Of course, by increasing μ, the observed wiggles can be eliminated completely, but this increases the smearing of the shock. Therefore it is preferred to keep some of the wiggles in the solution.

The CFL-restriction for the time step in the MacCormack scheme is given by (with c the velocity of sound):

$$\Delta t \leq \frac{1}{\frac{|u|}{\Delta x} + \frac{|v|}{\Delta y} + c \sqrt{\frac{1}{(\Delta x)^2} + \frac{1}{(\Delta y)^2}}}$$

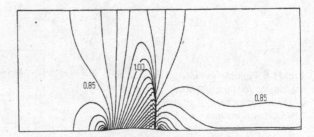

Fig. 11.8 IsoMachlines
obtained by cell-centred
MacCormack scheme

where

$$\Delta x = \frac{x_{i+1,j} - x_{i-1,j}}{2} \quad , \qquad \Delta y = \frac{y_{i,j+1} - y_{i,j-1}}{2}$$

11.2.1.2 Runge-Kutta Time Stepping – Multi-Stage Time Stepping

Runge-Kutta time stepping schemes for ordinary differential equations are unstable when applied to the semi-discretization (11.4) with the central flux (11.5):

$$
\begin{aligned}
\Omega_{i,j}\frac{\partial U}{\partial t} &+ \tfrac{1}{2}(\Delta y_{ab}\ f_{i,j-1} - \Delta x_{ab}\ g_{i,j-1}) \\
&+ \tfrac{1}{2}(\Delta y_{bc}\ f_{i+1,j} - \Delta x_{bc}\ g_{i+1,j}) \\
&+ \tfrac{1}{2}(\Delta y_{cd}\ f_{i,j+1} - \Delta x_{cd}\ g_{i,j+1}) \\
&+ \tfrac{1}{2}(\Delta y_{da}\ f_{i-1,j} - \Delta x_{da}\ g_{i-1,j}) = 0
\end{aligned}
\tag{11.17}
$$

There is no contribution of the central node in the flux balance in (11.17), since the flux balance for a constant flux on a closed surface is zero. As a consequence, (11.17) is an exact analogue of a central type finite difference discretization.

The instability of Runge-Kutta time stepping can be seen by considering a Fourier analysis on a central space discretization of the model equation (11.7) for the case of constant $a = \partial f / \partial u$:

$$\frac{\partial u_i}{\partial t} = -a\frac{u_{i+1} - u_{i-1}}{2\Delta x} \tag{11.18}$$

Inserting

$$u = Z\, e^{j\omega x}$$

where ω is the wave-number and j now stands for $\sqrt{-1}$, gives

$$Z' = -Z\, a\frac{e^{j\theta} - e^{-j\theta}}{2\Delta x} = -Z\, ja\frac{\sin\theta}{\Delta x} \tag{11.19}$$

where $\theta = \omega\Delta x$.

Equation (11.19) has the form

$$Z' = \lambda Z$$

with

$$\lambda = -j\, a\frac{\sin\theta}{\Delta x} \tag{11.20}$$

Figure 11.9 shows the stability domain for $\lambda\Delta t$ for the Runge-Kutta second, third and fourth-order, time-integration methods, according to [2].

Since λ according to (11.20) is on the imaginary axis, the second-order Runge-Kutta method is unstable. Higher order Runge-Kutta methods are marginally stable. Higher order Runge-Kutta methods can be stabilized by introducing a small amount of artificial viscosity. For example, equation (11.18) can be modified to

Fig. 11.9 Stability regions in the complex plane for classic explicit Runge-Kutta methods

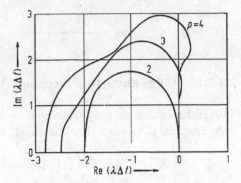

$$\frac{\partial u}{\partial t} = -a\frac{u_{i+1} - u_{i-1}}{2\Delta x} + \varepsilon\frac{u_{i+1} - 2u_i + u_{i-1}}{\Delta x^2}$$

The value of λ according to the previous analysis now becomes

$$\lambda = -j\,a\frac{\sin\theta}{\Delta x} - \frac{2\varepsilon}{\Delta x^2}(1 - \cos\theta)$$

Since there is now a small negative real part in λ, higher order Runge-Kutta time stepping now becomes stable, according to Fig. 11.9, when subject to a CFL-condition which restricts the time step. Note that a modification of equation (11.18) by adding a fourth-order derivative term instead of a second-order derivative term leads to a similar stabilization effect.

Runge-Kutta time stepping was introduced in the finite volume method by Jameson et al. in 1981 [3] and is nowadays a very popular method.

The fourth-order method, with simplifications, is mostly used since it gives the best ratio of allowable time step to computational work per time step. A simplified fourth-order scheme can be written as

$$U_{i,j}^0 = U_{i,j}^n$$

$$U_{i,j}^1 = U_{i,j}^0 - \alpha_1\frac{\Delta t}{\Omega_{i,j}}R^0$$

$$U_{i,j}^2 = U_{i,j}^0 - \alpha_2\frac{\Delta t}{\Omega_{i,j}}R^1$$

$$U_{i,j}^3 = U_{i,j}^0 - \alpha_3\frac{\Delta t}{\Omega_{i,j}}R^2$$

$$U_{i,j}^4 = U_{i,j}^0 - \alpha_4\frac{\Delta t}{\Omega_{i,j}}R^3$$

$$U_{i,j}^{n+1} = U_{i,j}^4 \tag{11.21}$$

with $\alpha_1 = {}^1/_4$, $\alpha_2 = {}^1/_3$, $\alpha_3 = {}^1/_2$, $\alpha_4 = 1$
and where the residual R is given by

$$R = \int (f\,dy - g\,dx)$$

and where the superscript denotes the (intermediate) time level.

Obviously (11.21) is not a classic fourth-order Runge-Kutta scheme. In a Runge-Kutta scheme, the fourth step is

$$U_{i,j}^4 = U_{i,j}^0 - \alpha_4 \frac{\Delta t}{\Omega_{i,j}} \left(\frac{R^0 + 2R^1 + 2R^2 + R^3}{6} \right)$$

with the choice of coefficients

$$\alpha_1 = {}^1/_2, \quad \alpha_2 = {}^1/_2, \quad \alpha_3 = 1, \quad \alpha_a = 1$$

The accuracy of the fourth-order Runge-Kutta scheme is fourth order in time. This is unnecessarily high since the space accuracy of the discretization is only second order. The simplification (11.21) has second-order accuracy in time for a non-linear equation, which is sufficient. *The simplified multi-stage time-stepping* (11.21) requires less storage than a classic Runge-Kutta time-stepping. Originally, Jameson used the classic Runge-Kutta method. The *low storage modification*, later introduced by Jameson, is nowadays universally used. For a discussion of it the reader is referred to [4].

The scheme (11.21) can be constructed by considering a Taylor expansion up to fourth order

$$U^{n+1} \approx U^n + \Delta t \frac{\partial U}{\partial t} + {}^1/_2 \Delta t^2 \frac{\partial^2 U}{\partial t^2} + {}^1/_6 \Delta t^2 \frac{\partial^3 U}{\partial t^3} + {}^1/_{24} \Delta t^4 \frac{\partial^4 U}{\partial t^4}$$

The following grouping defines (11.21):

$$U^{n+1} \approx U^n + \Delta t \frac{\partial}{\partial t} \left[U^n + {}^1/_2 \Delta t \frac{\partial}{\partial t} \left\{ U^n + {}^1/_3 \Delta t \frac{\partial}{\partial t} \left(U^n + {}^1/_4 \Delta t \frac{\partial U}{\partial t} \right) \right\} \right]$$

The stability domain of the multi-stage time stepping is the same as that of the fourth-order Runge-Kutta scheme shown in Fig. 11.9.

The artificial viscosity introduced by Jameson is a blend of a second-order and a fourth-order term. It is used in all steps of (11.21).

In order to keep the calculation conservative, the added dissipative term is, for a structured quadrilateral grid:

$$d_{i+1/2,j} - d_{i-1/2,j} + d_{i,j+1/2} - d_{i,j-1/2} \tag{11.22}$$

where

$$d_{i+1/2,j} = \varepsilon_{i+1/2,j}^{(2)} (U_{i+1,j} - U_{i,j}) - \varepsilon_{i+1/2,j}^{(4)} (U_{i+2,j} - 3U_{i+1,j} + 3U_{i,j} - U_{i-1,j}) \tag{11.23}$$

with similar definitions of the other terms in (11.22).

The coefficients of the second-order term $\varepsilon^{(2)}$ and the fourth-order term $\varepsilon^{(4)}$ are chosen in a self-adaptive way.

As a detector of the smoothness of the flow field, for the definition of the coefficients in (11.23), Jameson uses

$$v^i_{i,j} = \frac{\left| p_{i+1,j} - 2p_{i,j} + p_{i-1,j} \right|}{p_{i+1,j} + 2p_{i,j} + p_{i-1,j}}$$

and then defines

$$\varepsilon^{(2)}_{i+1/2,j} = \kappa^{(2)} \max(v^i_{i+1,j}, v^i_{i,j})$$

$$\varepsilon^{(4)}_{i+1/2,j} = \max(0, \kappa^{(4)} - \varepsilon^{(2)}_{i+1/2,j})$$

with $\kappa^{(2)} = {}^1/_4$, $\kappa^{(4)} = {}^1/_{256}$.

By this definition, the second-order term is only significant in shock regions. In smooth regions of the flow, the second-order term has a very small coefficient and the fourth-order term dominates. The fourth-order term constitutes the so-called *background dissipation*. For equal stabilization effect, it diffuses the solution less than a second-order term. Therefore it is used in smooth regions of the flow. In shock regions, the fourth-order dissipation has to be switched off since it causes wiggles and the second-order dissipation is to be used to eliminate wiggles. Therefore the second-order dissipation is called the *shock dissipation*.

At solid boundaries, the dissipative terms in (11.22) in the direction normal to the boundary are to be set equal to zero. In the foregoing definition of the dissipative terms (11.22, 11.23) the so-called second-order and fourth-order terms only correspond to second-order derivatives and fourth-order derivatives on a smooth grid. However, the expressions (11.22, 11.23) do not have to be changed on an irregular grid. First, they are not meant to simulate a physical viscosity. Second, they are also meant to eliminate *spurious modes*, i.e. the *non-physical solutions of the discretization*. Figure 11.10 shows the perturbation patterns in fluxes, and as a consequence also in dependent variables, not detected by the central type flux balance for quadrilateral and triangular grids.

Authors using Jameson's Runge-Kutta scheme often have their own variant of the dissipative term. Also very often, the dissipative correction in the second to fourth step is taken to be the same as in the first step.

A formulation of the artificial viscosity applicable to unstructured grids, which is a slight extension of the formulation given by Jameson and Mavriplis [5], is given hereafter.

The time-step limit is calculated from (for CFL = 1)

$$\Delta t = \frac{\Omega_i}{\sum\limits_e (|V_n| + c)\Delta s} \tag{11.24}$$

Fig. 11.10 Spurious modes for cell-centred central discretization

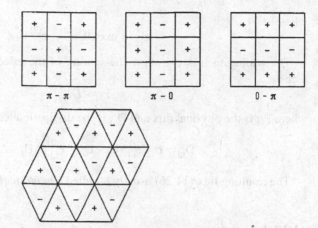

where the subscript i denotes the node, V_n is the normal velocity on an edge, obtained by averaging, c is the velocity of sound obtained in a similar way, and Δs is the length of the edge. Ω_i is the volume and the summation is taken over all edges.

The second-order smoothing operator is then, similar to (11.23), obtained by a sum of terms:

$$\varepsilon_{i,j}^{(2)} \, \sigma_{i,j}(U_j - U_i) \tag{11.25}$$

where the subscript j denotes the surrounding nodes. The weight function $\varepsilon_{i,j}$ is obtained from

$$\varepsilon_{i,j}^{(2)} = \kappa^{(2)} \max(v_i, v_j)$$

where v_i and v_j are pressure switches. The pressure switch v_i is defined by

$$v_i = \frac{\text{abs}\{\sum\limits_j (p_j - p_i)\}}{\sum\limits_j (p_j + p_i)}$$

$\sigma_{i,j}$ is a scaling factor given by

$$\sigma_{i,j} = \max\left(\frac{\Omega_i}{\Delta t}, \frac{\Omega_j}{\Delta t}\right)$$

with Δt the time step obtained from (11.24) for CFL = 1.

To define the fourth-order smoothing, first un-weighted pseudo-Laplacians are constructed by

$$\Delta U_i = \sum_j (U_j - U_i)$$

The fourth-order term is then given by a sum of terms:

$$\varepsilon_{i,j}^{(4)} \sigma_{i,j}(\Delta U_j - \Delta U_i)$$

where

$$\varepsilon_{i,j}^{(4)} = \max(0, \kappa^{(4)} - \varepsilon_{i,j}^{(2)})$$

The scaling factors $\sigma_{i,j}$ allow the writing of the effective flux through a cell-face as

$$F_{i,j} - D_{i,j} \qquad (11.26)$$

where $F_{i,j}$ is the physical flux and $D_{i,j}$ is the dissipation term, given by

$$D_{i,j} = \sigma_{i,j}[\varepsilon_{i,j}^{(2)}(U_j - U_i) - \varepsilon_{i,j}^{(4)}(\Delta U_j - \Delta U_i)] \qquad (11.27)$$

The resulting flux (11.26) usually is called a *numerical flux*.

11.2.1.3 Accuracy

The stencils obtained by the finite volume cell-centred formulation are very similar to the stencils obtained by the analogous finite difference methods. This means that if the grid is sufficiently smooth, such that the cell-centres are themselves on a grid which is sufficiently smooth, i.e. a grid which can be obtained by a continuous mapping from a square grid, the methods discussed in the previous sections are second-order accurate in space in a finite difference sense. This can easily be seen by comparison of the result in Fig. 11.8 with the result obtained by second-order finite difference methods [1]. Since, however, the representation of the solution is done in a piecewise constant way, on an irregular grid the accuracy is formally of first order. In practice, the order is between one and two.

11.2.2 Cell-Vertex Formulation

In the cell-vertex formulation, the variables are stored at the vertices of the grid. The control volumes either coincide with cells (non-overlapping case) or consist of a group of cells around a node (overlapping case). Figure 11.11 shows some of the possibilities. In all cases, a linear interpolation of the fluxes is now possible. Therefore, cell-vertex formulations have the possibility to be second-order accurate in space, irrespective of the irregularity of the grid.

11.2.2.1 Multi-Stage Time Stepping – Overlapping Control Volumes

For the overlapping cases, the methods discussed in the previous sections can be adapted directly. Very popular nowadays is the formulation of the multi-stage time stepping scheme. For the overlapping control volumes of Fig. 11.11, the semi-discretization is very similar to (11.17), now involving, however, six or eight surrounding nodes. At solid boundaries, half volumes are formed. The impermeability

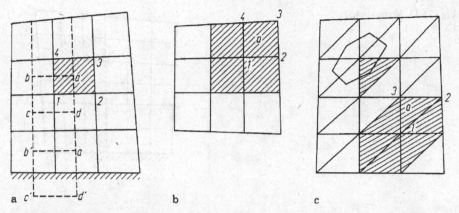

Fig. 11.11 Cell-vertex formulation. (**a**): quadrilateral cells, non-overlapping volumes (with inter-weaving grid); (**b**): quadrilateral cells, overlapping volumes; (**c**): triangular cells, overlapping and non-overlapping volumes

can be expressed by setting the convective fluxes to zero. Another approach is to treat the control volume as permeable and to impose tangency. This means that, between steps, the normal component of velocity is set equal to zero.

Again, in order to stabilize the scheme, some form of artificial viscosity is necessary. The artificial viscosity is also necessary to eliminate the spurious modes in the solution. Figure 11.12 shows the spurious modes that are possible for the quadrilateral and triangular grids.

As in the basic method of Jameson, a blend of a second-order smoothing and a fourth-order smoothing can be used. Often, the dissipative operator of the cell-centred method is used. This operator is then a sum of terms of form (11.23) for a quadrilateral grid. The method loses then its pure cell-vertex character. The resulting flux balance of inviscid and dissipative terms is then a balance over a control volume centred around a vertex as shown on Fig. 11.13. Such a control volume is called a

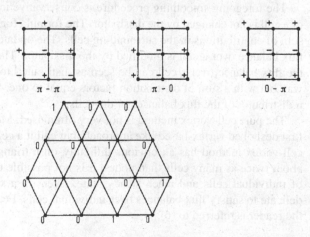

Fig. 11.12 Spurious modes
for cell-vertex central
discretization

Fig. 11.13 Vertex-based
FVM

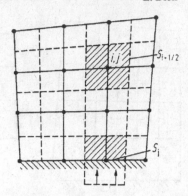

dual control volume. The inviscid flux balance over the dual control volume can be defined as one fourth of the flux balance over the volume formed by the four surrounding cells. Strictly, the method then becomes a vertex-centred or vertex-based method according to the terminology introduced in Sect. 11.3.

A pure cell-vertex method can be obtained by changing the construction of the dissipator. The same methodology as for the cell-centred method is used, but summations now run over cells surrounding a node rather than over surrounding nodes. This means that differences of values used in the expression (11.27) have to be modified.

For instance $U_j - U_i$ is to be replaced by

$$\tfrac{1}{2}(U_{j1} + U_{j2}) - U_i$$

or

$$\tfrac{1}{3}(U_{j1} + U_{j2} + U_{j3}) - U_i \quad \text{or} \quad \tfrac{1}{2}(U_{j1} + U_{j3}) - U_i$$

for triangular and quadrilateral cells respectively, where j1, j2 and j3 denote the nodes not coinciding with node i of the surrounding cells. Also the scaling factors $\sigma_{i,j}$ and the weight factors $\varepsilon_{i,j}^{(2)}$, $\varepsilon_{i,j}^{(4)}$ now involve maxima over all nodes of a cell.

The foregoing smoothing procedure is conservative in the sense that the content of a cell is not changed by the dissipator. The formula for the update of a node is the sum of contributions of the surrounding cells. The update coming from the inviscid flux balance over a cell is modified by the dissipator. The modification is such that the flux balance over a cell can be seen as distributed to its vertices in an unequal way but with a sum of distribution factors equal to one. So, the dissipator acts as a redistributor of the flux balances of the cells.

The pure cell-vertex method is not very often used. Most researchers employ the first described vertex-based like approach, but call it a cell-vertex method. The pure cell-vertex method has an obvious difficulty on a triangular grid. Since there are about twice as many cells than nodes, it is not possible to satisfy the flux balances of individual cells and reach steady state. Even on a structured grid, it is rather delicate to satisfy flux balances over individual cells. For a discussion on this topic the reader is referred to [6].

11.2.2.2 Lax-Wendroff Time-Stepping Non-Overlapping Control Volumes

For the non-overlapping case, a Lax-Wendroff variant exists due to Ni, developed in 1981 [7]. It requires the use of a second set of control volumes centred around the nodes, obtained in the way as shown in Fig. 11.11a. Ni's variant starts from the Lax-Wendroff formulation (11.8), (11.9). Without loss of accuracy in (11.9), $\partial f/\partial t$ can be replaced by a first-order accurate difference $\Delta f/\Delta t$. The result is

$$u^{n+1} \approx u^n - \Delta t \frac{\partial f^n}{\partial x} - \frac{\Delta t}{2}\frac{\partial}{\partial x}(\Delta f)$$

In two dimensions, on the Euler equations, this is

$$\Omega_{i,j}(U^{n+1} - U^n) = -\Delta t\left[\int (f^n \, dy - g^n \, dx) + \tfrac{1}{2}\int (\Delta f \, dy - \Delta g \, dx)\right] \quad (11.28)$$

On the quadrilateral grid of Fig. 11a, the method is then as follows.

Based on the cell 1-2-3-4, using an Euler step, i.e. a step forward in time, a first-order approximation of the increment of the flux vectors is obtained from

$$\Omega_a \Delta U_a = -\Delta t \int_{1234} (f \, dy - g \, dx)^n \quad (11.29)$$

and

$$\Delta f_a = A \, \Delta U_a, \qquad \Delta g_a = B \, \Delta U_a$$

where A and B are the Jacobians of the flux vectors f and g with respect to U. A and B are taken to be the mean values of the Jacobians evaluated at the nodes 1, 2, 3 and 4.

The area-weighted mean value of the first-order increments given by (11.29) over the four cells surrounding the node 1, gives a first-order increment for the dependent variables:

$$\Delta U_1^1$$

The discretization of (11.28) on the cell abcd is then:

$$\Omega_1(U_1^{n+1} - U_1^n) = \Omega_1 \, \Delta U_1^1 - \tfrac{1}{2}\,\Delta t \int_{abcd} (\Delta f \, dy - \Delta g \, dx) \quad (11.30)$$

The spatial integration is again taken to be piecewise linear.

The CFL-restriction for the time step, given by Ni is

$$\Delta t \le \min\left(\frac{\Delta x}{|u| + c}, \frac{\Delta y}{|u| + c}\right)$$

with

$$\Delta x = \frac{x_{i+1,j} - x_{i-1,j}}{2}, \qquad \Delta y = \frac{y_{i,j+1} - y_{i,j-1}}{2}$$

The boundary conditions at solid boundaries for the first step (11.29) can be implemented by setting convective fluxes equal to zero, as in the previous methods. In the second step (11.30), a half-volume is needed around a boundary node. This half-volume can be seen to be half the complete volume shown in Fig. 11.11a. Step (11.30) can be done by setting the first-order changes in the fictitious cells c' and d' equal to zero. So the boundary node only receives both first-order and second-order contributions from the inward cells a' and b'. As a consequence, for a boundary node, there is no implicit imposition of impermeability in step (11.30). Tangency is then imposed afterwards by setting the normal component of the velocity equal to zero.

It is to be remarked that, although an intermediate grid is used, the Ni-method is a true cell-vertex method. Indeed, if the flux balance of a cell is satisfied, there is no contribution to both first- and second-order terms and flow parameters are not changed. Therefore step (11.30) often is called the *distribution step* since its function can be seen to be the distribution of changes in the control volumes to the nodes.

As already mentioned, in a triangular grid, there are about twice as many cells as nodes. This means that in a cell-vertex formulation, flux-balances cannot be satisfied for all cells. The steady state result of a cell-vertex time stepping scheme then corresponds to some combinations of flux balances being zero. In a quadrilateral grid, all flux-balances can be satisfied at steady state. We also note that the distribution of the changes in the control volumes for triangular cells can be done with upwind methods. For a discussion on these much more complex methods we refer to [8].

11.3 FDM-Like Finite Volume Methods

In the finite difference method, the nodes are at the vertices of the grid. This is particularly attractive with respect to data on boundaries. For instance, pressure extrapolation at solid boundaries is then not necessary. A cell-centred FVM is therefore less attractive. A cell-vertex FVM does not have this drawback, but on the other hand the flux through a volume surface is continuous. This does not allow an upwind definition of a flux.

More freedom in the definition of a flux, combined with nodes at the vertices of the grid, can be obtained by using an interweaving grid, as shown in Fig. 11.13. The interweaving grid can be constructed by connecting the cell-centres. The cells of this interweaving grid can now be considered as control volumes for the nodes inside them. Fluxes at volume faces can, for instance, be defined as averages of fluxes calculated with function values in adjacent nodes. The semi-discretization is then very close to a finite difference semi-discretization and can be called a *conservative finite difference method*. We prefer here to call a finite volume method of this type a *vertex-based FVM* or a *vertex-centred FVM*. The method has gained much popularity in recent years. The central type discretization obtained with it is the same as the discretization by a Galerkin-FEM. So it is very easy to bring concepts from FEM into this type of FVM. Moreover, it is very easy to use upwinding in this type of FVM.

11.3.1 Central Type Discretizations

The adaptation of a Lax-Wendroff time-stepping or a multi-stage time-stepping, as discussed for the cell-centred FVM, to the vertex-based FVM is straightforward. The formulations obtained with both methods are very similar, except at solid boundaries.

11.3.2 Upwind Type Discretizations

As an example of an upwind discretization we treat here the flux-difference splitting technique introduced by Roe [9].

The flux through a surface $(i + {}^1/_2)$ of the control volume on Fig. 11.13 can be written as

$$F_{i+1/2} = \Delta y_{i+1/2} f_{i+1/2} - \Delta x_{i+1/2} g_{i+1/2} \tag{11.31}$$

where $f_{i+1/2}$ and $g_{i+1/2}$ have to be defined using the values of the flux vectors in the nodes (i,j) and $(i+1,j)$. We switch here to the classic finite difference notation using halves in the subscripts to denote intermediate points. Also, non-varying subscripts are not written.

We denote by F_i the value of $F_{i+1/2}$ using the function values in (i,j) and by F_{i+1}, the value using the function values in $(i+1,j)$. The flux (11.31) can be written as

$$F_{i+1/2} - \Delta s_{i+1/2}(n_x f_{i+1/2} + n_y g_{i+1/2}) \tag{11.32}$$

with

$$n_x = \Delta y_{i+1/2}/\Delta s_{i+1/2}, \qquad n_y = -\Delta x_{i+1/2}/\Delta s_{i+1/2}, \qquad \Delta s^2_{i+1/2} = \Delta x^2_{i+1/2} + \Delta y^2_{i+1/2}$$

In order to define an upwind flux, we consider the flux-difference

$$\Delta F_{i,i+1} = \Delta s_{i+1/2}(n_x \Delta f_{i,i+1} + n_y \Delta g_{i,i+1}) \tag{11.33}$$

where

$$\Delta f_{i,i+1} = f_{i+1,j} - f_{i,j}, \qquad \Delta g_{i,i+1} = g_{i+1,j} - g_{i,j}$$

For construction of the flux, it is essential that the linear combination of Δf and Δg in (11.33) can be written as

$$\Delta \phi = n_x \Delta f + n_y \Delta g = A \Delta U \tag{11.34}$$

where A is a discrete Jacobian matrix with similar properties as the analytical Jacobians of the flux vectors. This means that the eigenvalues of A are real and that the matrix has a complete set of eigenvectors. Of course, for consistency, the eigenvalues and eigenvectors should be approximations of the eigenvalues and eigenvectors of the linear combination of the analytical Jacobians. The construction of the discrete Jacobian is not unique and many formulations have been proposed after the first formulation by Roe [9]. For the numerical illustration later in this section, we use the formulation by the author [10]. The algebraic manipulations in the construction of the discrete Jacobian are not relevant for a principal discussion of the methodology and we do not describe these here.

The matrix A can be split into positive and negative parts by

$$A^+ = R\Lambda^+ L, \qquad A^- = R\Lambda^- L \tag{11.35}$$

where R and L denote the right and left eigenvector matrices in orthonormal form and where

$$\Lambda^+ = \text{diag}(\lambda_1^+, \lambda_2^+, \lambda_3^+, \lambda_4^+), \qquad \Lambda^- = \text{diag}(\lambda_1^-, \lambda_2^-, \lambda_3^-, \lambda_4^-)$$

with $\lambda_i^+ = \max(\lambda_i, 0)$, $\lambda_i^- = \min(\lambda_i, 0)$.

Positive and negative matrices denote matrices with, respectively, non-negative and non-positive eigenvalues.

This allows a splitting of the flux-difference (11.34) by

$$\Delta\phi = A^+ \Delta U + A^- \Delta U$$

As a consequence (11.33) can be written as

$$\Delta F_{i,i+1} = F_{i+1} - F_i = \Delta s_{i+1/2} A_{i,i+1} \Delta U_{i,i+1}$$

where the matrix $A_{i,i+1}$ can be split into positive and negative parts.

The absolute value of the flux-difference is defined by

$$\left| \Delta F_{i,i+1} \right| = \Delta s_{i+1/2} (A_{i,i+1}^+ - A_{i,i+1}^-) \Delta U_{i,i+1} \tag{11.36}$$

Based on (11.36) an upwind definition of the flux is

$$F_{i+1/2} = \frac{1}{2} \left[F_i + F_{i+1} - \left| \Delta F_{i,i+1} \right| \right] \tag{11.37}$$

That this represents an upwind flux can be verified by writing (11.37) in either of the two following ways, which are completely equivalent:

$$F_{i+1/2} = F_i + \frac{1}{2}\Delta F_{i,i+1} - \frac{1}{2}\left| \Delta F_{i,i+1} \right| = F_i + \Delta s_{i+1/2} A_{i,i+1}^- \Delta U_{i,i+1} \tag{11.38}$$

$$F_{i+1/2} = F_{i+1} - \frac{1}{2}\Delta F_{i,i+1} - \frac{1}{2}\left| \Delta F_{i,i+1} \right| = F_{i+1} - \Delta s_{i+1/2} A_{i,i+1}^+ \Delta U_{i,i+1} \tag{11.39}$$

Indeed, when $A_{i,i+1}$ has only positive eigenvalues, the flux $F_{i+1/2}$ is taken to be F_i and when $A_{i,i+1}$ has only negative eigenvalues, the flux $F_{i+1/2}$ is taken to be F_{i+1}.

The fluxes on the other surfaces of the control volume $S_{i-1/2}, S_{j+1/2}, S_{j-1/2}$ can be treated in a similar way as the flux on the surface $S_{i+1/2}$. With (11.38) and (11.39), the flux balance on the control volume of Fig. 11.13 can be brought into the form

$$\Delta s_{i+1/2} A_{i,i+1}^{-} [U_{i+1} - U_i] + \Delta s_{i-1/2} A_{i,i-1}^{+} [U_i - U_{i-1}]$$
$$+ \Delta s_{j+1/2} A_{j,j+1}^{-} [U_{j+1} - U_j] + \Delta s_{j-1/2} A_{j,j-1}^{+} [U_j - U_{j-1}] = 0 \tag{11.40}$$

or

$$CU_{i,j} = \Delta s_{i-1/2} A_{i,i-1}^{+} U_{i-1,j} + \Delta s_{i+1/2} (-A_{i,i+1}^{-}) U_{i+1,j}$$
$$+ \Delta s_{j-1/2} A_{j,j-1}^{+} U_{i,j-1} + \Delta s_{j+1/2} (-A_{j,j+1}^{-}) U_{i,j+1} \tag{11.41}$$

where C is the sum of the matrix-coefficients on the right-hand side.

The matrix coefficients in (11.41) have non-negative eigenvalues. The positivity of the coefficients on the right hand side of (11.41) and the (weak) dominance of the central coefficient guarantee that the solution can be obtained by a collective variant of any scalar relaxation method. By a collective variant is meant that in each node all components of the vector of dependent variables U are relaxed simultaneously.

In order to illustrate the boundary treatment, we consider now the half-volume on a solid boundary as shown in Fig. 11.13. This half-volume can be seen as the limit of a complete volume in which one of the sides tends to the boundary.

The flux on the side S_j of the control volume at the solid boundary can be expressed by

$$F_j - \Delta s_j A_{i,j}^{+} (U_j - U_{j-1}) \tag{11.42}$$

where the matrix $A_{i,j}$ is calculated with the function values in the node (i,j).

With the definition (11.42), the flux balance on the control volume takes the form (11.40) in which a node outside the domain comes in. This node, however, can be eliminated.

It is easily seen that on a solid boundary, three combinations of (11.42) exist, eliminating the outside node [10]. The combinations are the left eigenvectors corresponding to the zero eigenvalues in $A_{i,j}^{+}$. These equations are to be supplemented by the boundary condition of tangency.

As an illustration, Fig. 11.14 shows the solution obtained by the previous method for the test-case of Fig. 11.7 under the same conditions as for Fig. 11.8. Comparison of the upwind result with the central result shows the superiority of the upwind calculation with respect to sharpness of the shock.

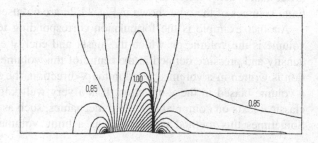

Fig. 11.14 IsoMachlines obtained by a vertex-based upwind FVM

In the above, the upwind discretization is used in first-order form. For more complex flows, of course, at least second-order accuracy is needed. In this introductory text we prefer not to enter the discussion of higher order upwinding. For second-order formulations on unstructured grids, the reader is referred to [11].

Examples of vertex-centred methods for Euler and Navier-Stokes equations can be found in [12]. Flux-difference splitting is used to define inviscid fluxes. The paper is in particular interesting for its discussion on treatment of viscous fluxes. An example of a vertex-centred method with central discretization of the inviscid fluxes and stabilization by artificial viscosity can be found in [13]. In this paper, viscous fluxes are treated by FEM. This becomes nowadays a widely accepted procedure and can be recommended. The vertex-centred FVM can be combined easily with a Galerkin-type FEM. References [11] and [13] use multigrid methods in order to obtain a steady solution in a fast way. The multigrid method is nowadays a standard method to accelerate the convergence to steady state.

For a general discussion on the choice between central and upwind finite volume methods, the reader is referred to [14]. In [15] a general discussion on the choice between cell-centred and vertex-centred methods and the choice between central and upwind methods is given. An interesting example of a cell-centred method using upwinding is given in [16]. Reference [17] discusses different time stepping algorithms for upwind methods both for vertex-centred and cell-centred formulations. Finally, the reader is referred to [18] for an overview of current finite volume methods. This reference dates from more than a decade ago, but there have not been major developments on basic algorithms in recent times.

11.4 Other Formulations

Finite volume methods that cannot be classified as FEM-like or FDM-like are methods which use nodes neither at cell-centres nor at cell-vertices. An example is given in Fig. 11.2c. Note that the volume also could be horizontal. For a control volume of this type, some fluxes are expressed with function values in nodes at the surfaces. Other fluxes require an averaging. Methods of this type are principally first-order accurate on an irregular grid. Since the volume contains two cells, the accuracy is lower than in cell-centred or vertex-based formulations. For this reason, overlapping finite volumes of the type shown in Fig. 11.2c are not used anymore.

Another example is the formulation corresponding to Fig. 11.2d. The shaded volume is the volume on which the mass- and energy equations are written with density and pressure defined in the centre of this volume. The x-momentum equation is written in a volume biased in the x-direction, the y-momentum equation on a volume biased in the y-direction. It is a very well-known method, described in classical texts on computational fluid mechanics, such as the book by Roache [19]. Sometimes this method is classified as a finite volume method because it uses control volumes. Since, however, the u- and v-components of velocity are not stored in the same points, principally, it only can be applied on a Cartesian grid. Using contra-variant velocity components, it can be extended to more general orthogonal grids. The method is popular in incompressible flow computations and convection-diffusion computations. This is mainly due to the work of Patankar [20]. The method allows the formulation of an incompressibility constraint. It also allows the introduction of upwinding.

11.5 Construction of Derivatives

When derivatives are needed for the definition of viscous terms, these commonly are calculated by the use of Gauss' theorem. For instance for the cell-centred formulation shown in Fig. 11.15, in order to define a derivative in the vertex a, an integration over the shaded volume gives

$$\left(\frac{\partial \phi}{\partial x}\right)_a \approx \frac{1}{\Omega_a} \int_{\Omega_a} \frac{\partial \phi}{\partial x} dxdy = \frac{1}{\Omega_a} \int_{S_a} \phi \, dy$$

Thus

$$\left(\frac{\partial \phi}{\partial x}\right)_a \approx \frac{1}{\Omega_a}\Big[\phi_{i+1,j+1}\frac{y_{i,j+1}-y_{i+1,j}}{2}+\phi_{i,j+1}\frac{y_{i,j}-y_{i+1,j+1}}{2}$$
$$+\phi_{i,j}\frac{y_{i+1,j}-y_{i,j+1}}{2}+\phi_{i+1,j}\frac{y_{i+1,j+1}-y_{i,j}}{2}\Big]$$

with

$$\Omega_a \approx \frac{y_{i+1,j+1}-y_{i,j}}{2}(x_{i+1,j}-x_{i,j+1})+\frac{y_{i,j+1}-y_{i+1,j}}{2}(x_{i+1,j+1}-x_{i,j})$$

A similar procedure can be used for the other vertices of the cell abcd. This allows a definition of the viscous terms on the boundary of the cell.

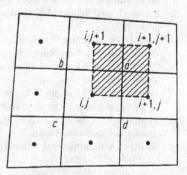

Fig. 11.15 Definition of a derivative

11.6 Higher Order Formulations

In the past decade, a lot of research has been done towards the formulation of finite volume methods of higher order accuracy in space and in time. In particular, higher order accuracy in space is difficult to obtain with a finite volume method. Work has been done, more or less through three kinds of approaches. A first class requires a regular Cartesian grid. Fluxes on the faces are obtained by higher order interpolation in co-ordinate directions. These methods aim at obtaining better accuracy for the node values. In principle, they are conservative formulations of finite difference methods and cannot easily be extended to irregular grids. A relevant example is found in [21]. A particular method, where higher accuracy is obtained by Richardson extrapolation, is presented in [22]. In a second class of methods, which is the most common, the function values in the nodes are seen as averages over the control volumes. Data necessary to calculate the fluxes are obtained by reconstruction. This means that higher order surfaces are constructed that satisfy the volume averages. In these methods, it is crucial that oscillations in the higher order surfaces are avoided. They are typically called ENO methods (essentially non oscillatory) or WENO methods (weighted essentially non oscillatory). A relevant example is [23]. A rather new class of methods are the spectral volume methods [24]. In such a method, as in a finite element method, the state is represented by higher order polynomials within the cell. The cell is subdivided into as many subcells as there are degrees of freedom in the polynomials. Averaged states are calculated in the subcells with the conservation laws. The polynomial is obtained by a reconstruction algorithm, such that the averaged states are satisfied. At cell boundaries, discontinuities are allowed as in the discontinuous Galerkin FEM.

References

1. Rizzi A. and Viviand H. (eds.), Numerical methods for the computation of inviscid transonic flows with shock waves. Notes on numerical fluid mechanics, Vol. 3, Vieweg, 1981.
2. Lapidus L. and Seinfeld J.H., Numerical solution of ordinary differential equations. Academic press, 1971.
3. Jameson A., Schmidt W. and Turkel E., Numerical solution of the Euler equations by finite volume methods using Runge-Kutta time stepping schemes. AIAA paper 81–1259, 1981.
4. Pike J. and Roe P.L., Accelerated convergence of Jameson's finite volume Euler scheme using Van Der Houwen integrators. Computers and Fluids, Vol. 13, pp. 223–236, 1985.
5. Jameson A. and Mavriplis D.J., Finite volume solution of the two-dimensional Euler equations on a regular triangular mesh. AIAA-Journal, Vol. 24, pp. 611–618, 1986.
6. Morton K.W., Crumpton P.I. and Mackenzie J.A., Cell vertex methods for inviscid and viscous flows. Computers and Fluids, Vol. 22, pp. 91–102, 1993.
7. Ni R.H., A multiple grid scheme for solving the Euler equations. AIAA-Journal, Vol. 20, pp. 1565–1571, 1982.
8. Deconinck H., Roe P.L. and Struijs R., A multidimensional generalisation of Roe's flux difference splitter for the Euler equations. Computers and Fluids, Vol. 22, pp. 215–222, 1993.
9. Roe P.L., Approximate Riemann solvers, parameter vectors and difference schemes. J. Comput. Phys., Vol. 43, pp. 357–372, 1981.

10. Dick E., A flux-difference splitting method for steady Euler equations. J. Comput. Phys., Vol. 76, pp. 19–32, 1988.
11. Riemslagh K. and Dick E., Multi-stage Jacobi relaxation in multigrid methods for the steady Euler equations. J. Comp. Fluid Dyn., Vol. 4, pp. 343–361, 1995.
12. Haselbacher A., McGuirk J.J. and Page G.J., Finite volume discretization aspects for viscous flows on mixed unstructured grids. AIAA-Journal, Vol. 37, pp. 177–184, 1999.
13. Mavriplis D.J. and Martinelli L., Multigrid solution of compressible turbulent flow on unstructured meshes using a two-equation model. Int. J. Num. Meth. Fluids, Vol. 18, pp. 887–914, 1994.
14. Swanson R.C. and Turkel E., On central-difference and upwind schemes. J. Comput. Phys., Vol. 101, pp. 292–306, 1992.
15. Mavriplis D.J., Unstructured mesh algorithms for aerodynamic calculations. Proceedings of the thirteenth international conference on numerical methods in fluid dynamics, Lectures Notes in Physics, Vol. 414, Springer, pp. 57–77, 1993.
16. Soltani S., Morgan K. and Peraire J., An upwind unstructured grid solution algorithm for compressible flow. Int. J. Num. Meth. Heat Fluid Flow, Vol. 3, pp. 283–304, 1993.
17. Slack D.C., Whitacker D.L. and Walters R.W., Time integration algorithms for the two-dimensional Euler equations on unstructured meshes. AIAA-Journal, Vol. 32, pp. 1158–1166, 1994.
18. Jameson A., Analysis and design of numerical schemes for gas dynamics, 1: artificial diffusion, upwind biasing, limiters and their effect on accuracy and multigrid convergence. J. Comp. Fluid Dyn., Vol. 4, pp. 171–218, 1995. Analysis and design of numerical schemes for gas dynamics, 2: artificial diffusion and discrete shock structure. J. Comp. Fluid Dyn., Vol. 5, pp. 1–38, 1995.
19. Roache P.J., Computational fluid dynamics. Hermosa Publishers, 1972.
20. Patankar S.V., Numerical heat transfer. Hemisphere, Washington D.C., 1980.
21. Lilek Z. and Peric M., A fourth-order finite volume method with colocated variable arrangement. Computers and Fluids, Vol. 24, pp. 239–252, 1995.
22. Verstappen R.W.C.P. and Veldman A.E.P., Direct numerical simulation of turbulence at lower costs. J. Eng. Math., Vol. 32, pp. 143–159, 1997.
23. Hu C. and Shu C.-W., Weighted essentially non-oscillatory schemes on triangular meshes. J. Comput. Phys., Vol. 150, pp. 97–127, 1999.
24. Zang Z. J., Spectral finite volume method for conservation laws on unstructured grids. J. Comput. Phys., Vol. 178, pp. 210–251, 2002.

Part III

Chapter 12
Aspects of CFD Computations
with Commercial Packages

J. Vierendeels and J. Degroote

12.1 Introduction

The purpose of this chapter is to give some insight into the steps that are needed to obtain a CFD solution of the flow field inside or around an object with the use of a commercial CFD software package. Note that it is not the intention to compare different commercial CFD software packages. The applications that are shown can be computed with most of the available software packages.

A CFD solution involves the following basic steps:

- Creation of the geometry (or import of the geometry from a CAD package)
- Grid generation
- Choice of the models
- Application of the boundary conditions
- Flow field computation
- Postprocessing

The first step is the creation of the geometry. Usually this is done with a separate CAD package. However, since the grid generator has some specific demands on the imported geometry, the imported geometry often has to be 'cleaned up'. Most CFD packages provide a CAD tool together with their grid generator. The geometry created with this embedded CAD tool is directly suitable for the grid generator. However, design engineers are using specific CAD packages for their needs and therefore the most common way to obtain the geometry in the grid generation package is the import from a CAD package. The 'cleaning up' phase is treated in Sect. 12.2.

The next phase is the grid generation process. A choice has to be made as to which kind of grid will be used: structured, block structured, unstructured, hybrid. For viscous calculations, a boundary layer mesh also has to be constructed. For turbulent flow calculations, the distance to the wall of the first cell in the boundary layer mesh depends on the near-wall treatment of the turbulence model. In cases where

J. Vierendeels
Ghent University, Ghent, Belgium, e-mail: jan.vierendeels@ugent.be

J.Degroote
Ghent University, Ghent, Belgium, e-mail: joris.degroote@ugent.be

J.F. Wendt (ed.), *Computational Fluid Dynamics*, 3rd ed.,
© Springer-Verlag Berlin Heidelberg 2009

the grid is not optimal for an accurate solution of the flow field, grid adaptation can be used in order to adapt the grid to the computed flow field features, such as shocks, slip lines, etc...These aspects will be discussed in Sect. 12.3.

The choice of the models depends on the kind of flow to be computed, and will have an impact on the grid generation process. The flow can be two- or three-dimensional, steady or unsteady, incompressible or compressible, laminar, turbulent or both and heat transfer can be important. These are the models used in the examples in this lesson. Other models that are often used, but which could not be dealt with in this introductory CFD course are mass transfer, chemical reactions, combustion, multiphase flows, discrete particle flows, flow in moving geometries, etc., and combinations of the above. Some modelling aspects on turbulent flows will be discussed in Sect. 12.4.

The next step is the application of the boundary conditions. Since the flow field is only computed in the region of interest, adequate boundary conditions have to be provided at the boundaries of the computed region. Frequently used boundary conditions are inlet, outlet and wall boundary conditions. Different implementations of these boundary conditions are considered in Sect. 12.5. More complex boundary conditions can be defined through user-written routines (Sect. 12.6).

The computation of the flow field with the solver becomes of less and less concern to standard users of a commercial CFD software package. So, the user can focus on the fluid dynamics without caring too much of the numerics behind it. However, the more experienced user who intends to write user routines that can be coupled with the software package needs to have a basic understanding of the underlying algorithms of the discretization and solution techniques, which is the subject of the other chapters in this text. Some solver aspects are discussed in Sect. 12.7.

Once the flow field is computed, it can be analyzed in the postprocessing phase. Many postprocessing means are available today. It is not the intention here to go into much detail on postprocessing features, but a short overview of possibilities is given in Sect. 12.8. If the user is not satisfied with the solution, a grid adaptation step can be performed as mentioned before.

More complex flow calculations e.g. with moving meshes and fluid-structure interactions can also be performed these days and will have an influence on the different steps outlined above, but are beyond the scope of this chapter.

12.2 Import of the Geometry from a CAD Package

There exist several CAD data exchange formats which are used to exchange data between different CAD packages. Most of these formats are also supported in a grid generator package. A non-exhaustive list is given below:

- IGES
- ACIS
- STEP
- Parasolid
- STL
- .....

Fig. 12.1 Edges with end
points that are not coincident

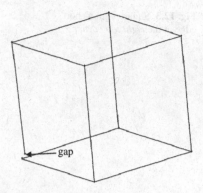

Sometimes direct input from a specific CAD package into the grid generator
package is also supported, e.g. CATIA, Pro/Engineer, Euclid, EDS/Unigraphics,
CADDS, I-DEAS, ...

When a CAD geometry is imported into a grid generator package an inconsistent
geometry can be the result. This can be due to e.g. tolerance differences between the
CAD package and the grid generator package. This imported geometry can then not
be meshed or not be meshed adequately. A geometry can be inconsistent because of
several reasons:

- a face can consist of edges with end points that are not coincident (Fig. 12.1)
- a volume can consist of faces with 'common' edges that are not coincident
  (Fig. 12.2)
- a geometry can consist of volumes with 'common' faces that are not coincident

Due to the inconsistencies, the geometry contains gaps between some of the en-
tities that make it unsuitable for creating a CFD mesh.

Very short edges (Fig. 12.3) and very small and/or sharp pointed faces (Figs. 12.4
and 12.5) may also be imported into the grid generator packages. This can lead to
very distorted face meshes on the related faces which can make it impossible to
generate a volume mesh starting from these distorted face meshes. And even if a
volume mesh could be created, its quality will be poor, i.e. distorted volume cells
will be present in the volume mesh. This can lead to less accurate solutions and even
to divergence in the solution process.

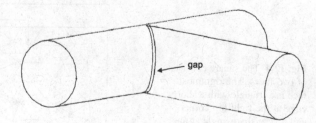

Fig. 12.2 Faces with edges
that are not coincident

Fig. 12.3 Square with a short edge at the *right top corner*

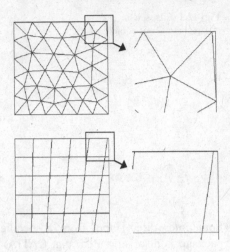

Because of the above mentioned reasons, an imported CAD geometry needs to be checked for 'clean up'. The most frequently encountered problems will be discussed.

Short edges can be eliminated by connecting its vertices or by merging them together with an adjacent edge. Faces with a sharp angle between its edges can also be merged together with an adjacent face (Fig. 12.6). Most of the gaps can be fixed either automatically during mesh import or subsequently manually by connecting coincident or almost coincident (dependent on a tolerance parameter) vertices, edges and faces (Figs. 12.7 and 12.8).

Coincident edges/faces that form an interface between two adjacent faces/volumes can either be connected or left unconnected. The first option will lead to a 'conformal'

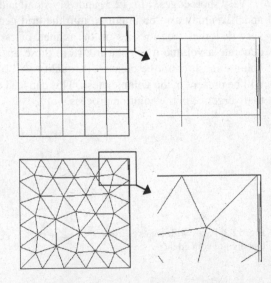

Fig. 12.4 The square consists of two faces. The rightmost face is a triangle with a short edge at the right top corner and a very sharp angle at the *right bottom corner*

Fig. 12.5 The square consists of two faces. The *right face* is a triangle with two sharp angles: one at the *right top corner* and one at the *right bottom corner*

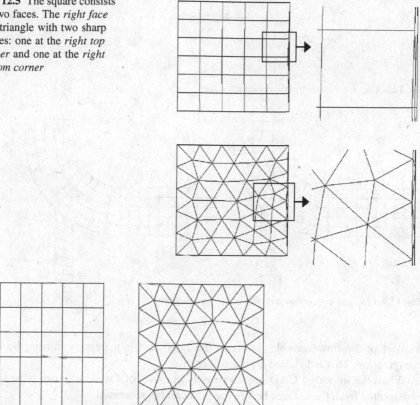

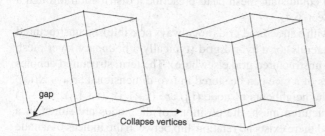

Fig. 12.6 Square consisting of one face. Edges and/or faces are merged together

mesh (Fig. 12.9), i.e. a unique mesh on this edge/face used by both adjacent faces/volumes. The second option will lead to a 'non-conformal' mesh (Fig. 12.9). Each interface edge/face of the adjacent faces/volumes can then be meshed independently. These edge/face meshes have then to be added to an interface list in order to allow the solver to know that flow is going through these interfaces. An adequate

Fig. 12.7 Gap is removed by collapsing the vertices

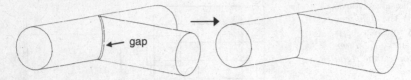

Fig. 12.8 Gap is removed by collapsing the edge

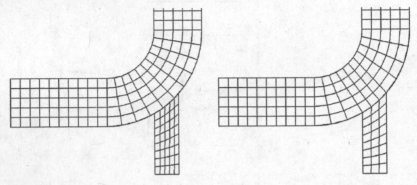

Fig. 12.9 *Left*: non-conformal grid. *Right*: conformal grid

transfer of the flow variables between the interfaces is necessary during the flow computation. This is handled by the solver.

When the imported CAD geometry only consists of faces, volumes have to be constructed from these faces before the grid can be generated.

12.3 Grid Generation

Once a valid geometry is obtained, the grid generation process can start. Normally the grid is constructed from lower topologies to higher topologies, i.e. first the edges are meshed, then the faces and finally the volumes.

The most common choices to mesh an edge are to prescribe the number of nodes or the interval size for an equidistant mesh or to prescribe a distribution through a stretching function (see Sect. 6.4).

Faces can be meshed with a structured grid (not always possible), an unstructured grid (triangles or quadrilaterals) or a hybrid grid (typically a boundary layer mesh near the boundary and an unstructured grid elsewhere). The term structured denotes that positions of the nodes of a face can be stored in two-dimensional arrays $X(i,j)$, $Y(i,j)$ and $Z(i,j)$ so that the neighbours of node (i,j) are $(i+1,j)$, $(i-1,j)$, $(i,j-1)$ and $(i,j+1)$. In an unstructured mesh, the positions of the nodes are stored in a one-dimensional array and there exists no relationship between the indices of a node and the indices of its neighbouring nodes. A separate list has then to be stored with

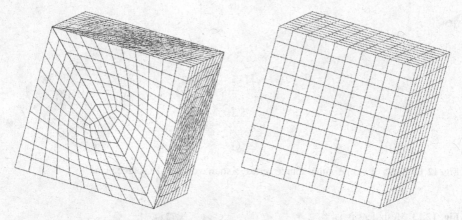

Fig. 12.10 *Left*: front square consisting of two faces. *Right*: faces are merged

edge information that describes the connectivity of the nodes. Most commercial CFD software packages use only the unstructured storage type, also for structured meshes. The face mesh is constructed starting from the edge meshes. Often it is interesting to merge some faces together to allow the use of different meshing schemes (Figs. 12.10 and 12.11). The density of the face mesh is computed from the density of the edge meshes. However the proximity of other faces can cause the need for a change in mesh density of the face mesh. This can be accounted for by introduction of sources. Point, line, face and volume sources can be used (Fig. 12.12). The mesh density can also be based on properties of the elements to be meshed, e.g. curvature (Fig. 12.13) or on the proximity of other objects (Figs. 12.14 and 12.15).

An important issue in grid generation is alignment of the grid edges to the flow direction. This leads to less numerical diffusion in the direction across the streamlines. In boundary layers and shear layers the behaviour of the flow is dominated by diffusion mechanisms (laminar or turbulent diffusion). The diffusion layer can be

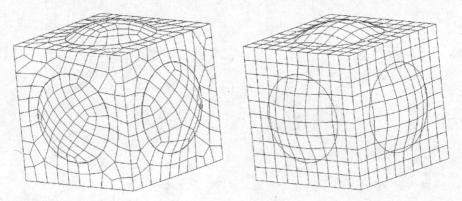

Fig. 12.11 Grid on bump not merged and merged with the surrounding planes

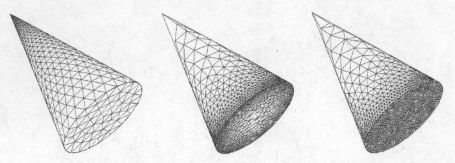

Fig. 12.12 Mesh of a cone with a source vertex, a source edge and a source plane

Fig. 12.13 Mesh density is
function of curvature

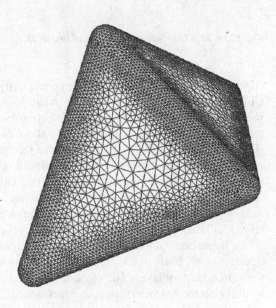

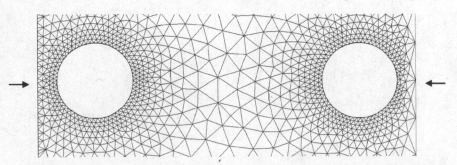

Fig. 12.14 Proximity of the hole is taken into account when meshing the *left edge*. Proximity was
not taken into account when meshing the *right edge*

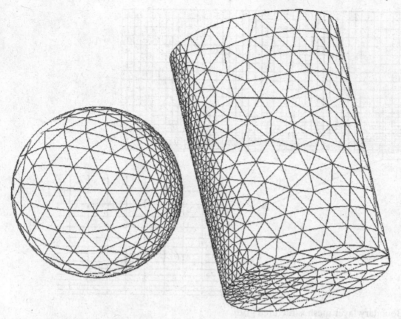

Fig. 12.15 Surface mesh taking into account the proximity of other surfaces

very thin for high Reynolds number flows. If numerical diffusion in the cross-wise direction is not suppressed, the boundary layer or shear layer diffuses too fast and a non-physical flow pattern is obtained. One way to suppress this numerical diffusion is the use of a huge amount of cells in these regions, another way is to use aligned cell layers, with moderate- to high-aspect-ratio cells, suppressing the numerical diffusion by the alignment and keeping the number of cells acceptable due to the larger aspect ratio. Nevertheless an adequate number of cells must be present across the diffusion layer. So, an indication of the thickness of the layer must be known in advance. For boundary layers, an estimation can often be obtained as a function of the Reynolds number. For shear layers this is more difficult for the main reason that the exact position of the shear layer is not known in advance. Therefore grid adaptation is used for shear layers and also for shocks. However, for flows where the boundary layers have to be resolved, a boundary layer grid is constructed in advance (Figs. 12.16, 12.17 and 12.18).

The error of a 2D first order upwind advection equation $(u, v \geq 0)$ is given by the right hand side of:

$$\frac{\partial \phi}{\partial t} + u\frac{\phi_{i,j} - \phi_{i-1,j}}{\Delta x} + v\frac{\phi_{i,j} - \phi_{i,j-1}}{\Delta y} = (u\Delta x)\frac{\phi_{i-1,j} - 2\phi_{i,j} + \phi_{i+i,j}}{\Delta x^2}$$
$$+ (v\Delta y)\frac{\phi_{i,j-1} - 2\phi_{i,j} + \phi_{i,j+i}}{\Delta y^2} + H.O.T.$$

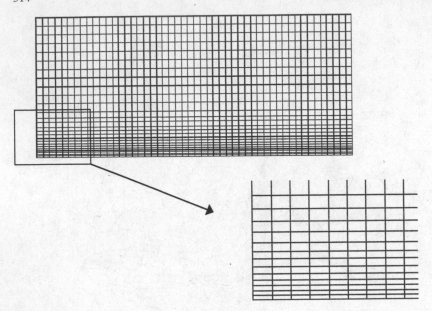

Fig. 12.16 Boundary layer mesh, structured grid

Consider a Cartesian grid. If the flow is aligned with the grid, either u or v is equal to zero and the dissipation across the streamlines disappears. In that case a contact discontinuity will be preserved. Otherwise the contact discontinuity is smeared out. A second order scheme will perform better in this case than a first order scheme (Figs. 12.19, 12.20, 12.21 and 12.22).

A volume grid is constructed starting from the face grids. Also here, sources can be used to influence the local mesh density. Volume grids with prismatic cells are often used in boundary layers and in geometries where one dimension is much larger than the other ones. Prismatic cells can be obtained by extruding a surface

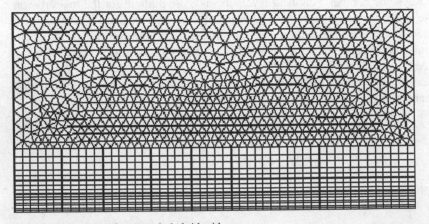

Fig. 12.17 Boundary layer mesh: hybrid grid

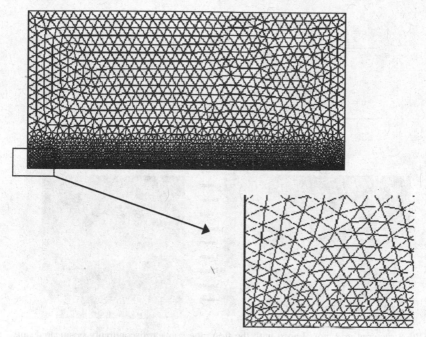

Fig. 12.18 Unstructured grid, stretched towards the boundary

grid in the direction away from the surface. For prismatic volumes, the surface grid of the top or the bottom of the prism can be projected along the side walls towards the opposite face (Fig. 12.23). For non-prismatic volumes, prismatic cells can be obtained near the walls, but in the inner part of the volume, tetrahedral cells will be used. If hexahedral cells are used in the boundary layer, then pyramidal cells or trimmed cells are necessary in the transition layer between the hexahedral cells and the tetrahedral cells (Figs. 12.24 and 12.25). When connecting cell centers and/or edge centers around the vertices of the grid, then polyhedral meshes are obtained (Fig. 12.26).

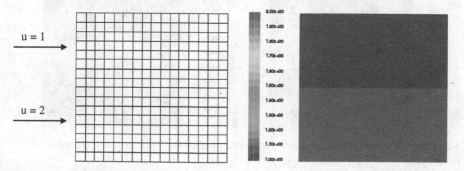

Fig. 12.19 Structured grid, aligned with the flow. The contact discontinuity is preserved

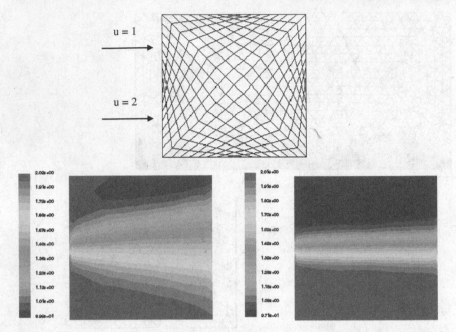

Fig. 12.20 Structured grid, not aligned with the flow, the contact discontinuity is smeared out. Left: first order discretization. Right: second order discretization

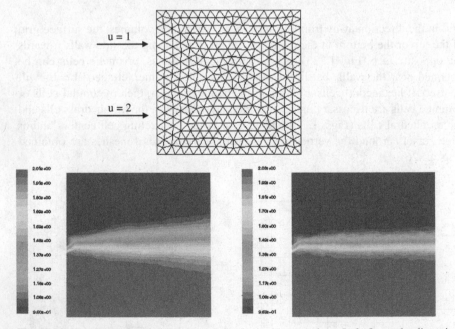

Fig. 12.21 Unstructured grid, the contact discontinuity is smeared out. Left: first order discretization. Right: second order discretization

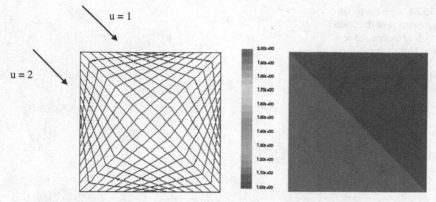

Fig. 12.22 Structured grid, aligned to the flow, the contact discontinuity is preserved

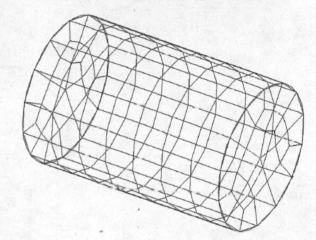

Fig. 12.23 Prismatic volume meshed with prismatic cells. The unstructured quadrilateral grid on the left face is projected onto the opposite face. This face mesh is repeated inside the volume leading to the formation of the prismatic cells

12.4 Choice of the Models

The next step in the process to obtain a flow field solution is the setting of the models in the solvers. Of course, the user should have already decided which models are to be used before the grid is generated, because some models will have their impact on the choice of the grid. It is obvious that the choice of a two- or three-dimensional, inviscid or viscous flow computation will have a direct impact on the grid generation process. In the majority of flow calculations, the flow will be turbulent. If the turbulent fluctuations are small, the mean flow can often be considered as steady. In order to take into account the turbulent interactions, a turbulence model is used. For large fluctuations in the flow field, the choice of a LES (Large Eddy Simulation) model can be more appropriate. This, however, implies a 3D unsteady computation. As already mentioned in the introduction, more complicated models can not be treated in this chapter. In the remaining part of this section, various turbulence models will

Fig. 12.24 *Top*: Cartesian mesh, trimmed to the boundary, with extrusion layer. *Middle*: Unstructured mesh with extrusion layer. *Bottom*: Hybrid mesh with extrusion layer

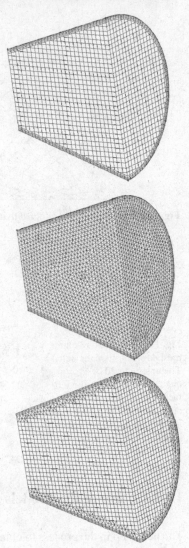

be discussed as well as the implications the models have on the construction of the boundary layer in the grid generation process.

When the flow is turbulent, the velocity in one point can vary as a function of time as in Fig. 12.27. Three cases are shown: steady mean flow with turbulence fluctuations superposed, unsteady mean flow with turbulence fluctuations, and transitional flow. For industrial applications, the details of the fluctuations are not important. Only the mean flow and the impact of the turbulence fluctuations on the mean flow are of importance.

Fig. 12.25 *Top*: regular cells used in unstructured mesh. *Bottom*: Trimmed hexahedral cells

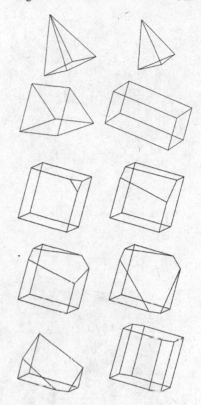

Mean values of the velocity are defined as follows:

$$\bar{v}_i(t_0) = \frac{1}{T} \int_{t_0-T/2}^{t_0+T/2} v_i dt$$

The time scale T has not to be defined precisely. It is chosen to be small with respect to the mean flow fluctuations and large with respect to the time scale of the turbulent fluctuations. The fluctuating part of the velocity is denoted by $v'_i = v_i - \bar{v}_i$ with $\overline{v'_i} = 0$. For incompressible flow, the Navier-Stokes equations (summation convention is used) are

$$\frac{\partial v_i}{\partial x_i} = 0$$

$$\frac{\partial v_i}{\partial t} + \frac{\partial}{\partial x_j} v_j v_i + \frac{1}{\rho} \frac{\partial p}{\partial x_i} = v \frac{\partial^2}{\partial x_j^2} v_i$$

The non-linear terms in the Navier-Stokes equations after averaging, result in

$$\overline{v_j v_i} = \overline{(\bar{v}_j + v'_j)(\bar{v}_i + v'_i)} = \bar{v}_j \bar{v}_i + \overline{v'_j v'_i}$$

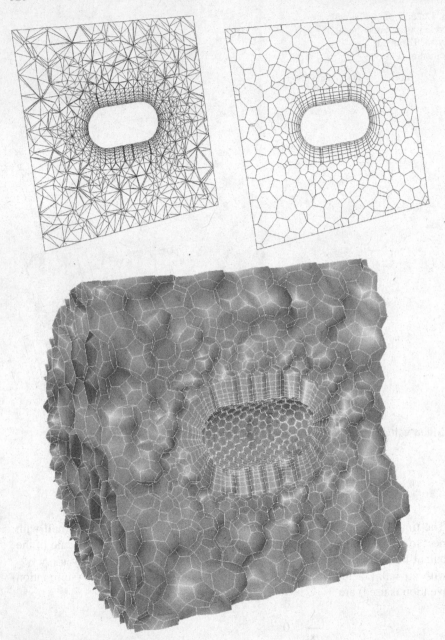

Fig. 12.26 Polyhedral mesh

Fig. 12.27 *Top*: steady
turbulent flow. Middle:
Unsteady turbulent flow.
Bottom: Transitional flow

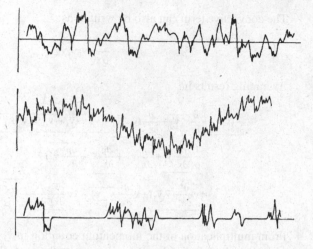

So, extra terms are present in the averaged equations.

$$\frac{\partial \overline{v}_i}{\partial x_i} = 0$$

$$\frac{\partial \overline{v}_i}{\partial t} + \frac{\partial}{\partial x_j} \overline{v}_j \overline{v}_i + \frac{1}{\rho} \frac{\partial \overline{p}}{\partial x_i} = \nu \frac{\partial^2}{\partial x_j^2} \overline{v}_i - \frac{\partial}{\partial x_j} \overline{v'_j v'_i}$$

These averaged equations are called the RANS (Reynolds-Averaged Navier-Stokes) equations. They are identical to the original Navier-Stokes equations except for the turbulent stress tensor:

$$\tau'_{ij} = -\rho \, \overline{v'_j v'_i}$$

The stress components are called the Reynolds stresses.

Equations for the Reynolds stresses can be derived from the Navier-Stokes equations. The tensor product of the momentum equation with the velocity vector results after averaging in an equation for the Reynolds stresses:

$$\frac{\partial v_i}{\partial t} + v_k \frac{\partial}{\partial x_k} v_i + \frac{1}{\rho} \frac{\partial p}{\partial x_i} = \nu \frac{\partial^2}{\partial x_k^2} v_i$$

$$\frac{\partial v_i}{\partial t} v_j + v_k \frac{\partial v_i}{\partial x_k} v_j + \frac{1}{\rho} \frac{\partial p}{\partial x_i} v_j = \nu \frac{\partial^2 v_i}{\partial x_k^2} v_j$$

After exchanging index i and j and summation of both equations, it follows that:

$$\frac{\partial}{\partial t}(v_i v_j) + v_k \frac{\partial}{\partial x_k}(v_i v_j) + \frac{1}{\rho} \left(\frac{\partial p}{\partial x_i} v_j + \frac{\partial p}{\partial x_j} v_i \right) = \nu \frac{\partial^2}{\partial x_k^2} v_i v_j - 2\nu \frac{\partial v_i}{\partial x_k} \frac{\partial v_j}{\partial x_k}$$

The convective term can also be written as:

$$\frac{\partial}{\partial x_k} v_i v_j v_k$$

Averaging results in:

$$\frac{\partial}{\partial t}(\bar{v}_i\bar{v}_j) + \frac{\partial}{\partial t}\overline{v'_i v'_j} + \frac{\partial}{\partial x_k}(\bar{v}_i\bar{v}_j\bar{v}_k + \overline{v'_i v'_j}\bar{v}_k + \overline{v'_i v'_k}\bar{v}_j + \overline{v'_j v'_k}\bar{v}_i + \overline{v'_i v'_j v'_k})$$

$$+ \frac{1}{\rho}(\frac{\overline{\partial \bar{p}}}{\partial x_i}\bar{v}_j + \frac{\overline{\partial p'}}{\partial x_i}v_j' + \frac{\overline{\partial \bar{p}}}{\partial x_j}\bar{v}_i + \frac{\overline{\partial p'}}{\partial x_j}v_i')$$

$$= v\frac{\partial^2}{\partial x_k^2}(\bar{v}_i\bar{v}_j) + v\frac{\partial^2}{\partial x_k^2}\overline{v'_i v'_j} - 2v\frac{\partial \bar{v}_i}{\partial x_k}\frac{\partial \bar{v}_j}{\partial x_k} - 2v\overline{\frac{\partial v'_i}{\partial x_k}\frac{\partial v'_j}{\partial x_k}}$$

From multiplication of the momentum equation in $\bar{v}_i$ with $\bar{v}_j$ and vice versa, we see that the full underlined terms and a part of the partially underlined terms cancel, so the result becomes:

$$\frac{\partial}{\partial t}\overline{v'_i v'_j} + \bar{v}_k\frac{\partial}{\partial x_k}\overline{v'_i v'_j} = -\overline{v'_i v'_j}\frac{\partial \bar{v}_j}{\partial x_k} - \overline{v'_j v'_k}\frac{\partial \bar{v}_i}{\partial x_k} + \frac{\partial}{\partial x_k}\overline{v'_i v'_j v'_k}$$

$$+ \frac{1}{\rho}\overline{\frac{\partial p'}{\partial x_i}v'_j} + \overline{\frac{\partial p'}{\partial x_j}v'_i} + v\frac{\partial^2}{\partial x_k^2}\overline{v'_i v'_j} - 2v\overline{\frac{\partial v'_i}{\partial x_k}\frac{\partial v'_j}{\partial x_k}}$$

The left hand side denotes the material derivative of the stress components. The first two terms in the right hand side are the interaction of the stress component with the mean velocity gradient. This represents the production of the Reynolds stresses. The next two terms represent the redistribution of turbulent fluctuations, which represents turbulent diffusion. The last term represents dissipation.

The equation can therefore be written as

$$\frac{D}{Dt}\overline{v'_i v'_j} = P_{ij} + d_{ij} - \varepsilon_{ij} \tag{12.1}$$

Diffusion and dissipation terms need further modelling. In Reynolds stress models (RSM), the dissipation terms are modelled through one convection-diffusion equation with source term, so seven scalar equations are to be solved in addition to the Reynolds-averaged Navier-Stokes equations.

The second kind of models used in RANS computations employ the Boussinesq hypothesis to relate the stresses to the mean velocity gradients:

$$-\rho\overline{v'_i v'_j} = 2\mu_t S_{ij} - \rho\frac{2k}{3}\delta_{ij}$$

where the turbulent kinetic energy k is defined as $^1/_2\overline{v'_k v'_k}$ and

$$S_{ij} = \frac{1}{2}\left(\frac{\partial v_i}{\partial x_j} + \frac{\partial v_j}{\partial x_i}\right) - \frac{1}{3}\frac{\partial v_k}{\partial x_k}\delta_{ij}.$$

This concept of a turbulent viscosity or eddy viscosity is based on the analogy between Brownian movement of molecules resulting in a viscosity term when the fluid is treated as a continuum and the chaotic movement of eddies in a turbulent flow. The eddy viscosity μ_t can be modelled algebraically or with a one-equation model such as Spalart-Allmaras or with two-equation models, such as the k-ε model and its variants and the k-ω model and its variants. For a one-equation model, the transport equation for k is derived from Eq. (12.1). For the two-equation models, the second equation describes the transport of the dissipation ε as in RSM or of the frequency ω, derived from k and ε. For the one-equation models μ_t is computed as a function of k and a length scale, modelled with an algebraic correlation. For the two-equation models, μ_t is computed as a function of the two transported quantities.

The use of a RANS turbulence model implies a constraint on the grid in the vicinity of a wall. Either the turbulent boundary layer is completely computed or the turbulent boundary layer is modelled. In the near-wall region, three layers can be observed. In the innermost layer, the viscous sublayer, the flow is almost laminar and momentum, heat and mass transfer are dominated by the (molecular) viscosity. In the outer layer, the fully-turbulent layer, turbulence plays a major role. In the region in between, both the effects of viscosity and turbulence are important. Figure 12.28 shows the different layers in the near-wall region. Here y^+ and u^+ are given by $^{yu_\tau}/_v$ and $^u/_{u_\tau}$ respectively, where $u_\tau = \sqrt{\tau_w/\rho}$.

There are two ways to model the near-wall flow. In the first approach, the flow in the viscous sublayer and buffer layer is not resolved, but 'wall functions' are used to correlate the viscous stress at the wall with flow data in the fully turbulent region (log-layer). This approach is called 'high-Reynolds' turbulence modelling. In a second approach, 'low-Reynolds' turbulence modelling, the flow through the inner

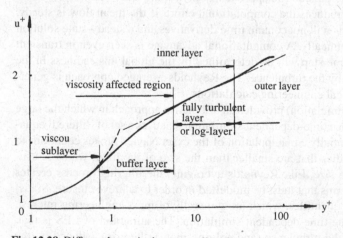

Fig. 12.28 Different layers in the near-wall region

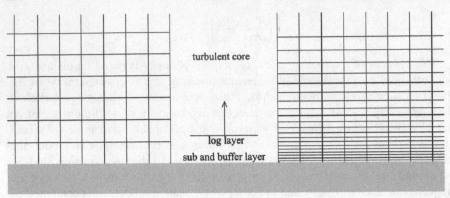

Fig. 12.29 *Left*: grid for high-Reynolds turbulence modelling. *Right*: grid for low-Reynolds turbulence modelling.

layer is resolved. The turbulence models have to be modified with near-wall models in order to account for the presence of the wall, since near the wall the flow is almost laminar. Damping functions have to be introduced in the turbulence equations to damp the turbulence adequately near the wall. Figure 12.29 shows different grids to be used for both approaches. The first approach results in lower cost computations and is often used for industrial applications, but is only valid for flows where the assumptions for the use of wall functions are valid. If not, the second approach can be used if the near-wall models are appropriate for the type of flow being considered. With the introduction of blending functions between the low-Reynolds approach and the high-Reynolds approach, the right choice for the distance from the wall for the first cell becomes more obsolete.

The above-mentioned Reynolds-averaged Navier-Stokes (RANS) equations represent transport equations for the mean flow quantities only, with all the scales of the turbulence being modelled. The approach of permitting a solution for the mean flow variables greatly reduces the computational effort. If the mean flow is steady, the governing equations will not contain time derivatives and a steady-state solution can be obtained economically. A computational advantage is seen even in transient situations, since the time step will be determined by the global unsteadiness in the mean flow rather than by the turbulence. The Reynolds-averaged approach is generally adopted for practical engineering calculations.

LES (Large Eddy Simulation) provides an alternative approach in which the large eddies are computed in a time-dependent simulation that uses a set of 'filtered' equations. Filtering is essentially a manipulation of the exact Navier-Stokes equations to remove only those eddies that are smaller than the size of the filter, which is usually taken as the mesh size. Like Reynolds averaging, the filtering process creates additional unknown terms that must be modelled in order to achieve closure. Statistics of the mean flow quantities, which are generally of most engineering interest, are gathered during the time-dependent simulation. The attraction of LES is that, by modelling less of the turbulence (and solving more), the error induced by the turbulence model will be reduced. One might also argue that it ought to be easier to find a 'universal' model for the small scales, which tend to be more isotropic and less affected by the macroscopic flow features than the large eddies. For LES computations, large computer resources are required since the mesh size and time step to be used are very small. This kind of computation can only be performed in 3D, due to the 3D nature of turbulent flow.

12.5 Boundary Conditions

Once the geometry is defined and the appropriate model is chosen, the boundary conditions have to be specified. Boundaries are typical inlets, outlet, walls, symmetry planes, periodic planes or an axis for axisymmetrical computations. For subsonic flow, $n - 1$ conditions have to be specified at the inlet (n is the number of degrees of freedom for each cell). For turbulent compressible flow, the velocity components or the total pressure and flow direction, together with the temperature and the turbulence variables are prescribed. For supersonic flow, all degrees of freedom need to be specified at the inlet. At a subsonic outlet, one condition has to be specified. The pressure or a combination of flow and pressure is then prescribed. If the outlet is supersonic, no boundary conditions have to be prescribed. If there is backflow at an outlet, convected quantities, such as temperature (or entropy) and turbulence variables have to be correctly prescribed. Sometimes it happens that backflow is detected only during the convergence process,. If the backflow conditions are not correctly prescribed, this can lead to divergence of the calculation.

More complex in- and outlet conditions such as fans or vents are often available. Then only the characteristic (pressure loss or gain as a function of the flow) has to be specified. If such boundary conditions are not available, these can be programmed in user subroutines that can be linked with the commercial package (see next section).

For incompressible flow with heat transfer or compressible flow, the temperature or heat flux is to be specified at the wall. For conjugate heat transfer problems no boundary conditions for the interface wall are to be specified. When a turbulence model is used, the turbulent quantities at the inlet need to be specified. This can e.g. be done in terms of turbulent intensity and hydraulic diameter, from which the variables of a two-equation model can be derived. The values at the inlet are generally not critical since the turbulence is strongly damped in uniform flow. Turbulence is mostly created in the vicinity of walls in boundary layers or in shear layers inside the computational domain.

12.6 User Written Routines

As mentioned before, user subroutines can be written to specify complex boundary conditions, such as space and time varying boundary conditions. These user routines can also be used to specify fluid properties or source terms in the equations.

Nowadays, user-developed turbulence models can be coupled with a commercial software package through user-written routines, since convection-diffusion equations with source terms for user-defined scalars can be linked to the package. If these scalars are chosen to be the variables of a turbulence model, then the model can be added on through user-defined diffusion and source terms and a user-defined turbulent viscosity as a function of these user-defined scalars.

12.7 Computation of the Flow Field

Knowledge of the solver becomes unnecessary and the user can increasingly focus on the fluid dynamics. However in case of non convergence, some parameters have to be tuned adequately. For explicit solvers the 'cfl' number, and for implicit solvers the underrelaxation factors can be changed. Since the flow equations are nonlinear, a good initial guess for the flow field is important. For turbulent flow calculations, it can be helpful to start with a low order scheme and without the turbulence models in the initial phase of the iterations. Afterwards the turbulence equations can be switched on and finally the order of the discretization method can be increased. If the solution is not satisfactory, grid adaptation can be used (Fig. 12.30).

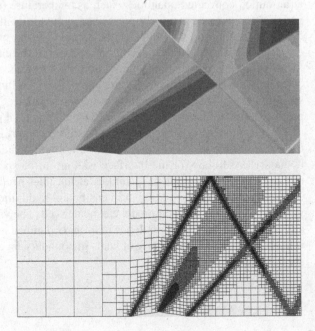

Fig. 12.30 Very sharp shock pattern obtained after several grid adaptations

12.8 Post-Processing

Once the flow field is computed, a discrete solution for the flow variables is available for the domain at each mesh element. This solution can be processed to obtain values of the flow variables at any location within the flow domain by standard interpolation techniques. It is common for CFD packages to provide powerful graphics capabilities for visually analyzing the solution, as well as to report values of various flow quantities. If the user is not satisfied with the solution the grid can be refined or modifications to the numerical or physical models can be made.

The most commonly employed postprocessing features are contour and vector plots, path lines and particle tracks (Figs. 12.31 and 12.32) and reports of fluxes, surface and volume integrals, or XY plots of extracted data. Animation sequences can be used for time dependent computations or for the analysis of 3D computations by a moving 2D cutting plane.

Fig. 12.31 Contour plot of velocity magnitude, cutting plane through a centrifugal compressor

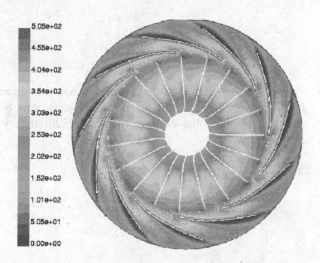

5.05e+02
4.55e+02
4.04e+02
3.54e+02
3.03e+02
2.53e+02
2.02e+02
1.52e+02
1.01e+02
5.05e+01
0.00e+00

12.9 Final Remarks

CFD has matured during the last two decades into a powerful tool to analyze fluid problems. However, one must never forget that the solution obtained can still be far from reality if the grid or the models used (e.g. turbulence models) are not adequate for the type of flow being studied.

A CFD result should always be verified with the flow results a fluid dynamicist can expect. Hand calculations are necessary to verify the computed results (e.g. in- and output fluxes). If the flow patterns are unexpected, an explanation should be sought. Often the explanation is an error somewhere in the input data.

Fig. 12.32 Path lines in a centrifugal pump

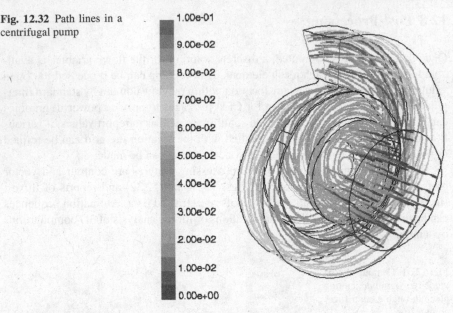

In conclusion, it may be stated that CFD has become an indispensable tool for the fluid dynamicist, but as with all valuable objects or tools, it should be handled with care. Keep in mind that also from incorrect solutions very impressive 3D animations can be shown.

Index